Nonlinear Optics: A student's perspective
– *With python problems and examples*

Mark G. Kuzyk
Department of Physics and Astronomy
Washington State University

August 9, 2017

This book was written using WinEdt, the best text editor on the planet, typeset with LaTeX and set in New Century Schoolbook

Copyright © 2017 by Mark G. Kuzyk
All rights reserved

Nonlinear Optics: A student's perspective
– With python problems and examples

Library of Congress Control Number: 2017900404
CreateSpace Independent Publishing Platform, North Charleston, SC
ISBN-13: 978-1523334636
ISBN-10: 1523334630

10 9 8 7 6 5 4 3 2 1

NLOSource.com

Dedication

We dedicate this book to everyone who is trying to learn nonlinear optics.

Acknowledgements

I thank my students who took nonlinear optics in 2010 for taking class notes that formed the core of the original book, from which this volume draws. The students were Benjamin R. Anderson, Nathan J. Dawson, Sheng-Ting Hung, Wei Lu, Shiva Ramini, Jennifer L. Schei, Shoresh Shafei, Julian H. Smith, Afsoon Soudi, Szymon Steplewski and Xianjun Ye. I also thank the students in my Nonlinear Optics class in the spring 2016 semester for their useful comments and would like to single out Thomas Ferron, Auberry Fortuner and Jerred Jesse for providing detailed edits. I also appreciate more recent feedback from Sean Mossman and Ethan Crowell as well as useful comments and edits from Joseph Lanska. I am especially indebted to Washington State University for providing me with an atmosphere that encourages me to think about things.

Preface

0.1 Preface to this book

As a new faculty member at WSU, I was trying to explain a concept to a confused student in an introductory physics class. I called upon analogies, attacked the ideas from various perspectives, and drew all sorts of diagrams, but that blank stare never went away.

On his way out, he ran into a graduate student who he recognized as a teaching assistant from a previous semester, and stepped into his office to chat. The source of his confusion came up. I eavesdropped, listening to the graduate student's bumbling and inaccurate explanation, fearing that the undergraduate's understanding would be irreversibly damaged. Instead, he excitedly exclaimed, "Now I understand. You're great at explaining physics."

Over the years, I have witnessed this phenomena many times and at all levels. I have concluded that the very best students learn from a clear and rigourous explanation; but, most mortals learn in a process that is akin to an image coming slowly into focus. Like a photographer adjusting her lens back and forth several times, I believe that students learn by overcoming stages of confusion. Compatriots with an understanding that is one step of fuzziness ahead are better able to mediate the next transition. This was my original motivation for publishing the book, "Lecture Notes in Nonlinear Optics: A student's perspective." What better way to explain complex topics than from the perspective of students who have just learned it, with recent memories of the obstacles that they had to overcome?

When I taught *Nonlinear Optics* in spring 2014, I read the book in parallel to preparing for class, which made me better able to judge the book's effectiveness. While the book is adequate for this purpose, I found three issues that compelled me to work on this volume. First, some of the nuances that make nonlinear optics interesting were not properly explained. Secondly, topics which are unique to the way I teach the course were absent. Finally, I felt it was time to add more original homework problems, the kinds that have helped me to understand the material better, and I hope will be useful to the students. Additionally, many of the typos and grammatical errors were fixed, poor quality figures were redrawn, and redundant materials were eliminated.

While textbooks should focus on the fundamentals, I felt that integrating some programming into the book would not only be instructive, but fun. New to this book are sections on using Python for modeling nonlinear-optical processes as well as some very simple problems that require only basic coding skills. This addition serves two purposes; it teaches students how to solve problems numerically, but most importantly, it allows a broader range of problems to be solved, giving the students a deeper appreciation of nonlinear optics.

My original intention was to produce a second edition of "Lecture Notes in Nonlinear Optics: A student's perspective." However, since the spirit of the book has changed, and the present volume is a far cry from lecture notes, I decided to start fresh with a new title and a new approach. Many of the students liked the raw feel of the original book while others were annoyed by it, including me. As such, I have decided to continue to make the older book available even after releasing this volume. Perhaps many of you will find this new volume objectionable, but I disagree. We'll see how it all plays out.

I believe that this book is amendable to self study. The student should read it cover to cover and solve all the problems along the way. Many of the problems are short and presented right after the relevant material, providing realtime practice while reading. I originally intended to produce a solution manual, but after decades of teaching, I am convinced that such a crutch is just too tempting. Students who peek prior to substantial effort will develop a false sense of understanding that will unravel when their knowledge is put to the test.

Some differences between this book and the older one are as follows:

Added to Chapter 1 is a section on the comparison of nonlinear optics between the bulk and microscopic perspectives. The correspondence between beams and photons; and, the correspondence between continuous media and molecules are discussed. An explanation of how energy is conserved at the quantum level when the macroscopic processes seem to be in violation leads to an introduction to stimulated processes. A section on using Python with simple code as an example has also been added, as is a homework problem using Python. Python problems now are found throughout the book.

Section 2.1.3 on the nonlinear spring has been significantly expanded and errors corrected. The extra level of detail included leads to a more nuanced appreciation for the origins of the nonlinear response. Section 2.3 on response functions has been both rewritten and expanded to make it more accurate and complete. The topic of monochromatic fields was omitted in the previous book, which is a critical part in understanding why the nonlinear susceptibility can be written in the standard way in terms of frequencies. This volume describes these concepts in semi-rigourous detail. Section 2.4 on cascading has been expanded to include a more detailed description of the interaction between molecules, two separate examples have been added and three techniques for solving problems described.

Section 2.5 has been added on the topic of magnetic dipole and electric quadruple contributions to nonlinear optics – two topics that are normally neglected in most text books, but are becoming important in modern research. This section is intended to introduce the topic in an intuitive way rather than to add the burden of more formalism.

Chapter 3 has an expanded explanation of how one solves the nonlinear wave equation. The steps are better organized and more detailed, and, the order of presentation is rearranged to make the explanations both more concise and clear. The first example, in Section 3.2, has been expanded and rewritten to more closely follow the general steps so that the student can more easily apply it to more complex cases.

The topic of sum frequency generation in materials with a spatially-varying second-order susceptibility profile has been added as Section 3.3. Covered is the general theory and several applications, including delta function profiles, sinusoidal profiles and quasi phase matching . Also added is a short description of phase matching in waveguides when the mode indices at the fundamental and second harmonic are equal. The problems in this section have been expanded and include python problems.

The description of birefringence phase matching in Section 3.7.3 has been overhauled with the goal of making this somewhat complex process easier to visualize and understand.

To Chapter 4 has been added a non-rigorous introduction to Dirac notation, which simplifies the derivations that follow and gives the student a general tool that strengthens the intuition. Section 4.2 on the quantized photon field has been overhauled. Section 4.3 on coherent states has been added, and applied to the quantized photon field to show how the classical field amplitude is associated with the coherent state's complex eigenvalues. Section 4.4 on molecules in a cavity with photons has been added, where Rabi oscillations are discussed. Section 4.13 on local fields has been overhauled, and the Lorentz-Lorenz local field model is introduced in an intuitive way. Nonlinear local fields are also treated, which can be viewed as being related to cascading. Finally, Section 4.14 describes surface plasmons, and how a surface plasmon resonance can enhance the nonlinear-optical response by several orders of magnitude.

Chapter 6 on applications has be modified substantially and new topics have been added. The topic of phase conjugation has been re-written and extended with new material, such as the description of the physical picture of the process, which has been added as Section 6.2.1. Also included is a microscopic view of the process at the level of photons interacting with a molecule. Section 6.3 on self focusing has been added, which describes the phenomenon in limiting cases that simplifies the discussion.

The interesting topic of negative temperature is added as Section 6.4, which analyzes how the polarizability depends on temperature, showing that absorption turns to lasing at the crossover between positive and negative temperature. The nonlinear case is relegated to a homework problem. Section 6.5 on multistability has been added, which starts with an introduction of bistability. Section 6.5.3 uses a ray-tracing approach to reflectors that straddle a nonlinear material to determine the output versus input characteristics to illustrate how multistability can result. This approach provides an alternate perspective. Finally, Section 6.5.5 shows how the nonlinear problem of determining the output versus input characteristics of multistability can be solved graphically.

While the nonlinear susceptibilities have upper bounds that originate from quantum principles, composite properties such as device figures of merit are not so constrained because they are sometimes quotients of susceptibilities, so can diverge when the denominator vanishes. Section 6.6.1 on the figure of merit and Section 6.6.2 on materials with permittivity near zero describe such interesting cases where practical applications can benefit from small denominators. The book ends with Section 6.7 on the topic of single-molecule nonlinear optics, which calculates signal levels that can be expected from molecules and nano-structures.

This volume does not cover every topic in nonlinear optics, and it certainly is not as comprehensive as many other books. I believe that the topics that are excluded can be more easily learned with a better fundamental grounding. As such, I have chosen to concentrate on topics that I believe develop a deeper understanding. My own fascination with the origin of the nonlinear response tilts the topics in this volume into the quantum realm, but macroscopic phenomena are not neglected.

My inability for noticing typos and errors convinces me that I suffer from some form of dyslexia, which I hope does not detract from the information that is conveyed in this book. My goal is to provide an inexpensive textbook whose price would significantly escalate if I employed professional proof readers. I would therefore appreciate hearing from readers by email at mgkuzyk@gmail.com if you find substantive errors or have suggestions for improvements.

Enjoy!

Mark G. Kuzyk
July, 2017
Pullman, WA

0.2 Preface to "Lecture Notes in Nonlinear Optics: A student's perspective"

This book grew out of lecture notes from the graduate course in Nonlinear Optics I taught in spring 2010 at Washington State University. Each student took notes for the equivalent of about two book sections, then prepared an electronic version using LaTeX. Each section was criticized by two students who acted as devil's advocates with particular attention to clarity of presentation and accuracy. After making edits, each student presented me with a draft, which I edited and in some cases rewrote. I contributed a chapter, merged all the sections together, and added bridge material when necessary. This book is the result of a process which started on Day 1 of class, and spilled over into the summer.

In addition to this book project, the students gave presentations on modern research that uses nonlinear optics. They were also assigned a class project on two original research topics that I felt were interesting. One project led directly to two publications in *Physics Review A* on cascading and the other one spawned several journal articles on quantum graphs, which are still a topic of research in my group. These additional topics are not included in the lecture notes.

In its present form, this book is a first cut that is still in need of editing – the existing material needs to be expanded, and new material needs to be added. The homework assignments and solutions are incomplete and will be added in a future edition. Nevertheless, the notes serve as a useful reference that contains the fundamentals from the perspective of students who made an effort to distill my lectures into a form that they believe best describes the ideas. In this, I believe they have succeeded. As a result, the pages that follow provide the student with the fundamentals needed to initiate new research in the field and to understand the more advanced topics.

I would like to thank Washington State University for providing a dynamic atmosphere where bright students, faculty, and visitors are encouraged to navigate uncharted waters in search of discovery, fueled solely by passion. I thank my fellow voyagers for sharing their love of learning, making lots of work seem effortless and downright fun.

A pdf file of this book is freely available at:

http://www.nlosource.com/LectureNotesBook.pdf

Mark G. Kuzyk
July, 2010
Pullman, WA

Contents

Dedication   i

Acknowledgements   iii

Preface   v
    0.1   Preface to this book v
    0.2   Preface to "Lecture Notes in Nonlinear Optics: A student's perspective" ix

1   Introduction to Nonlinear Optics   1
    1.1   History 1
        1.1.1   Kerr Effect 1
        1.1.2   Two-Photon Absorption 3
        1.1.3   Second Harmonic Generation 4
        1.1.4   Optical Kerr Effect 4
    1.2   Units 7
    1.3   Example: Second Order Susceptibility 9
    1.4   Maxwell's Equations 10
        1.4.1   Electric displacement 11
        1.4.2   The Polarization 12
    1.5   Interaction of Light with Matter 13
    1.6   Magnitudes of Nonlinear susceptibilities 19
    1.7   Nonlinear Optics at the Quantum Level 20
    1.8   Numerical Simulations using Python 23
    1.9   Goals 33

2   Models of the NLO Response   35
    2.1   Harmonic Oscillator 35
        2.1.1   Linear Harmonic Oscillator 35
        2.1.2   Nonlinear Harmonic Oscillator 37

		2.1.3 Non-Static Harmonic Oscillator 40
2.2	Macroscopic Propagation 49	
2.3	Response Functions 60	
	2.3.1 Time Invariance 60	
	2.3.2 Fourier Transforms of Response Functions and Electric Susceptibilities 61	
	2.3.3 Second-Order Polarization and Susceptibility 64	
	2.3.4 n^{th} Order Polarization and Susceptibility 67	
	2.3.5 Tensor Symmetries of Response Functions 68	
	2.3.6 Complex Susceptibilities 72	
	2.3.7 Kramers-Kronig 72	
	2.3.8 Permutation Symmetry 77	
	2.3.9 Symmetries 81	
2.4	Cascading 85	
	2.4.1 Two Interacting Molecules 87	
	2.4.2 Two Polarizable Molecules in a Static Electric Field 89	
2.5	Magnetic and Quadrupole Contributions to Nonlinear Optics . . 92	
	2.5.1 Multipole Moments 93	
	2.5.2 Magnetic Monopoles 96	
	2.5.3 Magnetic Dipoles 97	
	2.5.4 Electric Quadrupole 100	
	2.5.5 Mixed Moments 104	
2.6	Symmetry 107	
	2.6.1 Time Reversal Symmetry 108	

3 Nonlinear Wave Equation   111

3.1	General Technique 111
3.2	Sum Frequency Generation - Non-Depletion Regime 116
3.3	Spatially Varying Second-Order Susceptibility 123
	3.3.1 Applications 124
3.4	Sum Frequency Generation - Small Depletion Regime 131
	3.4.1 Physical Interpretation of the Manley-Rowe Equation . . 136
3.5	Sum Frequency Generation with Depletion of One Input Beam . 138
3.6	Difference Frequency Generation and Parametric Amplification 141
3.7	Second Harmonic Generation 146
	3.7.1 General Case 147
	3.7.2 No Depletion Approximation 149
	3.7.3 Phase Matching 150
3.8	Applications of Frequency Mixing 156
3.9	Practical Problems 157

4 Quantum Theory of Nonlinear Optics **161**

- 4.1 Dirac Notation 161
- 4.2 A Hand-Waving Introduction to Quantum Field Theory 170
  - 4.2.1 Continuous Theory 171
  - 4.2.2 Second Quantization 175
  - 4.2.3 Photon-Molecule Interactions 178
  - 4.2.4 Stimulated Emission 181
- 4.3 Coherent States 182
- 4.4 A Molecule in a Cavity with Photons 185
- 4.5 Nonlinear Susceptibilities and Dipole Moment Expectation Value 188
  - 4.5.1 Time-Dependent Perturbation Theory 189
  - 4.5.2 First-Order Susceptibility 192
  - 4.5.3 Nonlinear Susceptibilities and Permutation Symmetry . . 195
- 4.6 Using Feynman-like Diagrams 198
  - 4.6.1 Introduction 198
  - 4.6.2 Elements of Feynman Diagrams 200
  - 4.6.3 Rules of Feynman Diagram 201
  - 4.6.4 Feynman Diagrams and Sum Frequency Generation . . . 204
- 4.7 Virtual States and Virtual Excitations 207
- 4.8 Broadening Mechanisms 210
- 4.9 Introduction to Density Matrices 213
  - 4.9.1 Phenomenological Model of Damping 218
- 4.10 Calculating the Polarizability with the Density Matrix 220
- 4.11 Parity 223
- 4.12 Sum Rules 227
- 4.13 Local Fields 229
- 4.14 Surface Plasmons in Nanospheres 241

5 Using the OKE to Determine Mechanisms **247**

- 5.1 Intensity Dependent Refractive Index 247
  - 5.1.1 Mechanisms of $\chi^{(3)}$ 249
- 5.2 Tensor Nature of $\chi^{(3)}_{ijkl}$ in Liquids 254
  - 5.2.1 Inversion Symmetry 254
  - 5.2.2 90° Rotation Symmetry 256
  - 5.2.3 45° Rotation Symmetry 256
  - 5.2.4 Intrinsic Permutation Symmetry 257
- 5.3 Measurements of the Intensity-Dependent Refractive Index . . . 259
  - 5.3.1 Plane Waves 261
- 5.4 Intensity-Dependent Refractive Index Mechanisms 264
  - 5.4.1 Electronic Response 264

5.5	Molecular Reorientation	269
	5.5.1 Zero Electric Field	272
	5.5.2 Non-Zero Electric field	273
	5.5.3 General Case	275

6 Applications   277

6.1	Optical Phase Conjugation and Time Reversal	277
6.2	Phase Conjugate Mirror	281
	6.2.1 A Physical Picture	281
	6.2.2 Solving for the Intensities	282
6.3	Self Focusing	286
6.4	Nonlinear Optics at Negative Temperature	289
6.5	Optical Bistability and Multistability	292
	6.5.1 Reflections from an interface	293
	6.5.2 Reciprocity	294
	6.5.3 The Fabry-Perot Interferometer	294
	6.5.4 Multistability of a Fabry-Perot Interferometer with End Reflectors	296
	6.5.5 Graphical Solution to Transcendental Equations	303
6.6	Evaluating Materials for Applications and Devices	304
	6.6.1 Figure of Merit	304
	6.6.2 Epsilon Near Zero	306
6.7	Single Molecule Nonlinear Optics	308

Chapter 1

Introduction to Nonlinear Optics

The traditional core physics classes include electrodynamics, classical mechanics, quantum mechanics, and thermal physics. Add nonlinearity to these subjects, and the richness and complexity of all phenomena grow exponentially.

Nonlinear optics is concerned with understanding the behavior of light-matter interactions when the material's response is a nonlinear function of the applied electromagnetic field. In this book, we focus on building a fundamental understanding of wave propagation in a nonlinear medium, and the phenomena that result. Such an understanding requires both an understanding of the nonlinear Maxwell Equations as well as the mechanisms of the nonlinear response of the material at the quantum level.

We begin this chapter with a brief history of nonlinear optics, examples of some of the more common phenomena, and a non-rigorous but physically intuitive treatment of the nonlinear response.

1.1 History

1.1.1 Kerr Effect

The birth of nonlinear optics is often associated with J. Kerr, who observed the change in the refractive index of organic liquids and glasses in the presence of an electric field. [1, 2, 3] Figure 1.1 shows a diagram of the experiment.

Kerr collimated sunlight and passed it through a prism, essentially mak-

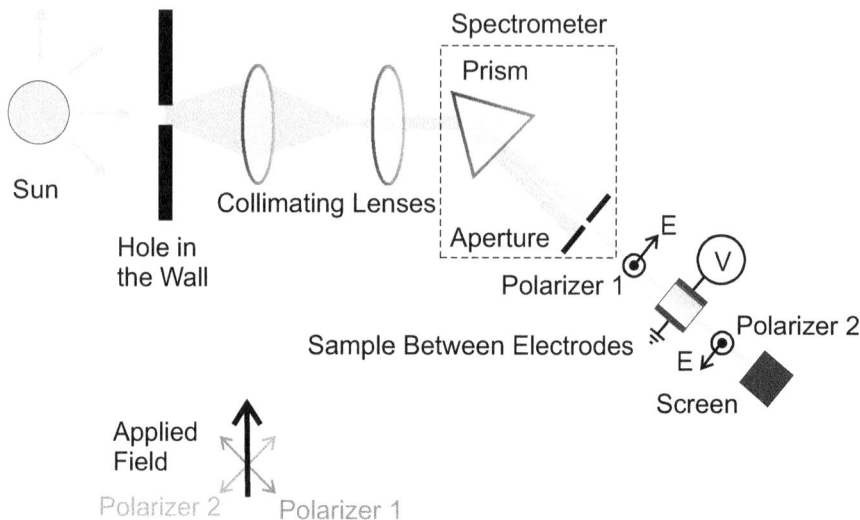

Figure 1.1: Kerr observed the change in transmittance through a sample between crossed polarizers due to an applied voltage. Inset at the bottom left shows the orientations of the polarizers and the applied electric field due to the static voltage.

ing a spectrometer that could be used to vary the color of the light incident on the sample. He placed an isotropic sample between crossed polarizers (i.e. the polarizer's axes are perpendicular to each other), so that no light makes it to the screen. A static electric field is applied 45° to the axis of each polarizer as shown in the bottom left portion of Figure 1.1. He found that the transmitted intensity is a quadratic function of the applied voltage.

This phenomena is called the Kerr Effect, or the quadratic electrooptic effect; and, originates in a birefringence induced by the electric field. The invention of the laser in 1960[4] provided light sources with high-enough electric field strengths to induce the Kerr effect with a second laser that replaces the applied voltage. This latter phenomena is called the Optical Kerr Effect (OKE). Since the intensity is proportional to the square of the electric field, the OKE is sometimes called the intensity dependent refractive index,

$$n = n_0 + n_2 I, \tag{1.1}$$

where n_0 is the linear refractive index, I is the intensity and n_2 the material-dependent Kerr coefficient.

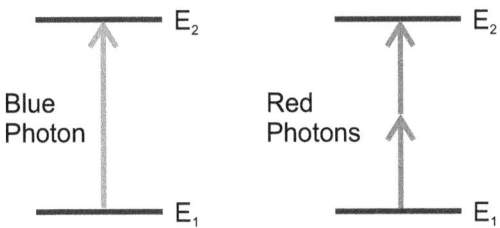

Figure 1.2: (left) A system is excited by a photon if its energy matches the difference in energies between two states. (right) Two-photon absorption results when two photons, each of energy $(E_2 - E_1)/2$, are sequentially absorbed.

1.1.2 Two-Photon Absorption

Two Photon Absorption (TPA) was first predicted by Maria Goeppert-Meyer in 1931,[5] and is characterized by an intensity dependent absorption, of the form

$$\alpha = \alpha_0 + \alpha_2 I, \tag{1.2}$$

where α_0 is the linear absorption coefficient and α_2 is the two-photon absorption coefficient. Since the absorption coefficient is proportional to the imaginary part of the refractive index, a comparison of Equations 1.1 and 1.2 implies that α_2 is proportional to imaginary part of n_2. Thus, both the OKE and TPA are the real and imaginary parts of the same phenomena.

Figure 1.2 can be used as an aid to understand the quantum origin of TPA. A transition can be excited by a photon when the photon energy matches the energy difference between two states (sometimes called the transition energy), $\hbar\omega = E_2 - E_1$, where ω is the frequency of the light. Two-photon absorption results when $\hbar\omega = (E_2 - E_1)/2$, and two photons are sequentially absorbed even when there are no two states that meet the condition $\hbar\omega = E_2 - E_1$.

We can understand TPA as follows. Imagine that the first photon perturbs the electron cloud of a molecule (or any quantum system) so that the molecule is in a superposition of states $|\psi\rangle$ such that the expectation value of the energy is $\hbar\omega = \langle\psi|H|\psi\rangle$. This is called a virtual state, which will be more clearly understood when treated rigorously using perturbation theory . If a second photon interacts with this perturbed state, it can lead to a transition into an excited energy eigenstate . Since the process depends on one photon perturbing the system, and the second photon interacting with the perturbed system, the probability of a double absorption is proportional to the square of the number of photons (the probability of the first absorption times the

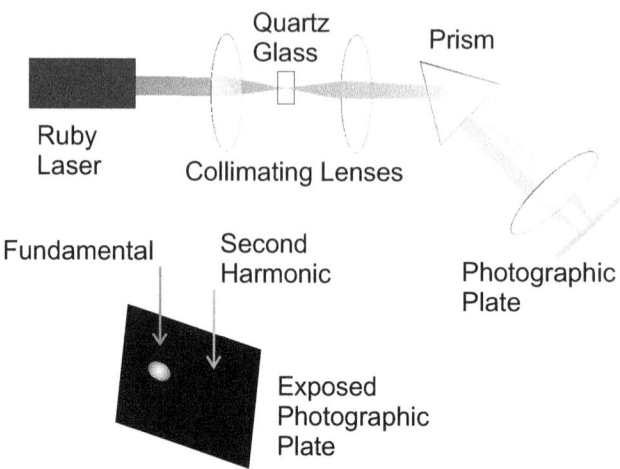

Figure 1.3: The experiment used by Franken and coworkers to demonstrate second harmonic generation. The inset shows an artistic rendition of the photograph that recorded the two beams.

probability of the second absorption), and thus the intensity.

1.1.3 Second Harmonic Generation

Second harmonic generation was the first nonlinear-optical phenomena whose discovery relied on the invention of the laser. P. A. Franken and coworkers focused a pulsed ruby laser into a quartz crystal and showed that light at twice the laser frequency was generated.[6] Figure 1.3 shows the experiment.

A prism was used to separate the the fundamental from the second harmonic, and the two beams were recorded on photographic film. The film's appearance is shown in the inset of Figure 1.3 . Since the second harmonic signal is weak, the spot on the film appeared as a tiny spec. The story is that someone in the publications office thought this spec was a piece of dirt, and air-brushed it away before the journal went to press. So, in the original paper, an arrow points to the location of the second harmonic spot, but the tiny spot is not visible.

1.1.4 Optical Kerr Effect

Three years later, the Optical Kerr Effect was observed by Mayer and Gires.[7] In their experiment, shown in Figure 1.4, a strong light pulse from a

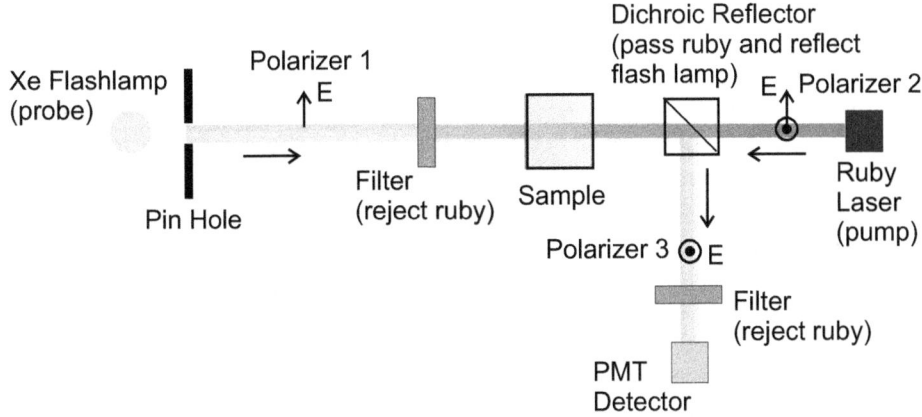

Figure 1.4: In the Optical Kerr Effect a strong pump laser is polarized 45° to the weak probe beam, causing a rotation of its polarization.

ruby laser and a weak probe beam from a xenon flashlamp counterpropagate in the sample. The electric field of the pump and probe beams are polarized 45° to each other. The pump beam induces a birefringence in the material that causes a rotation of the polarization of the probe beam. The probe beam is reflected from a dichroic mirror to the detector while allowing the pump beam to pass unaltered. The reflected probe beam passes through polarizer 3, which rejects the original polarization of the probe. Thus, the amount of light reaching the detector is related to the degree of depolarization of the probe beam, which in turn, is proportional to the pump intensity.

While the OKE is similar to the quadratic electrooptic effect, in that a strong electric field rotates the polarization of a weaker probe beam, OKE has many more diverse applications. Interestingly, two-photon absorption also contributes to polarization rotation, as you will show in the exercises below.

The Optical Kerr Effect can be modeled using solutions to the linear wave equation in an anisotropic medium. The problems that follow are based on using solutions to the wave equation and applying them to determine how the polarization of a linearly-polarized beam is affected by a birefringent material (one in which the refractive index is different for two orthogonal polarizations). The underlying physics needed to solve the problem: (1) the two polarizations along the principle axes propagate independently; and, (2) a polarizer passes only the electric field along the polarizer's axis. You may want to thumb through a basic optics book if these concepts are not familiar to you.

Problem 1.1-1(a): A monochromatic plane wave passes through a quartz plate of thickness d with its Poynting Vector perpendicular to the plate. The k-vector is given by $\mathbf{k} = k\hat{z}$ and the refractive indices along $\hat{x}$ and $\hat{y}$ are n_x and n_y. If the polarization vector of the beam when it enters the sample makes an angle θ with respect to $\hat{y}$, calculate the fraction of the beam intensity that passes through a polarizer that is perpendicular to the light's original polarization.

(b): Now assume that the refractive index is complex so that $n_x \rightarrow n_x^R + in_x^I$ and $n_y \rightarrow n_y^R + in_y^I$. Recalculate the transmittance (transmitted/incident intensity) through the crossed polarizers.

(c): Next assume that the material is isotropic but that a strong pump beam of intensity I_p is incident at normal incidence and polarized along $\hat{y}$. If $n_y^R = n_0^R + n_{2y}^R I$ and $n_y^I = n_0^I + n_{2y}^I I$, where the refractive index changes only along the direction of the pump beam's polarization, calculate the transmittance of probe beam through the crossed polarizer as a function of the pump intensity.

Homework Hint

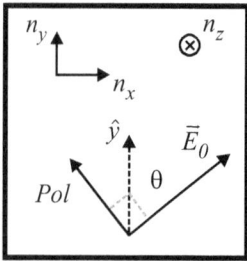 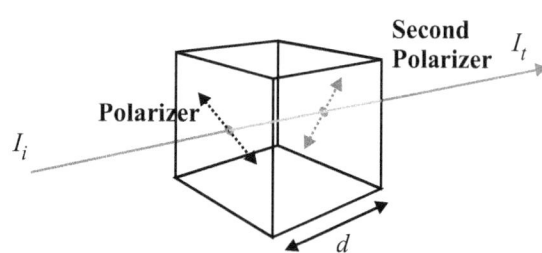

Figure 1.5: Polarizer set perpendicular to the original wave propagation direction (the first defines the polarization of the incident wave.)

Consider a polarizer that is set perpendicular to the polarization of the incident electric field $\mathbf{E}_0$ as is shown in Figure 1.5. The transmittance is defined as

$$T = \frac{I_t}{I_i}, \tag{1.3}$$

where the initial intensity is calculated from the electric field,

$$I_i \propto |\mathbf{E}(z=0)|^2. \tag{1.4}$$

The intensity transmitted through the polarizer is proportional to the square of the transmitted electric field, or

$$I_f \propto |\mathbf{E}(z=d) \cdot (\cos\theta \hat{x} - \sin\theta \hat{y})|^2, \tag{1.5}$$

where the constant of proportionality is the same for each intensity calculation. The electric field of the wave at any point in the material is given by

$$\mathbf{E} = E_0 \left(\sin\theta \hat{x} e^{i(kn_x z - \omega t)} + \cos\theta \hat{y} e^{i(kn_y z - \omega t)} \right), \tag{1.6}$$

where $k = \omega/c = 2\pi/\lambda$.

If the refractive index acting on the monochromatic plane wave has both a real and an imaginary component, it can be expressed as

$$n = n_R + i n_I, \tag{1.7}$$

where n_R and n_I are the real and imaginary parts of the refractive index. For a pump along $\hat{y}$, assume that

$$n_x = n_{0x} \tag{1.8}$$
$$n_y = n_{0y} + n_{2y} I, \tag{1.9}$$

where Equations 1.8 and 1.9 are substituted into Equation 1.6 with the understanding that all quantities are complex.

1.2 Units

Most of the scientific literature in nonlinear optics has accepted the use of SI units (i.e. meter, kilogram, second, coulomb). However, they lead to ugly equations and mask the beauty of the unification of electric and magnetic fields.

For example, consider Faraday's Law in Gaussian units,

$$\nabla \times \mathbf{E} = -\frac{1}{c}\frac{\partial \mathbf{B}}{\partial t}. \tag{1.10}$$

The units of the electric and magnetic fields are the same, placing them on an equal footing.

Coulomb's Law, on the other hand, is given by

$$\nabla \cdot \mathbf{E} = 4\pi \rho, \tag{1.11}$$

which for a point charge leads to

$$\mathbf{E} = \frac{q}{r^2}\hat{r}. \tag{1.12}$$

The electric potential of a point charge q and the energy of two interacting charges q and Q are given by

$$V = \frac{q}{r} \quad \text{and} \quad U = \frac{qQ}{r}. \tag{1.13}$$

Note that ϵ_0 and 4π are absent. Also, ϵ_0 does not appear in the induced polarization, which is simply given by

$$P_i = \chi_{ij}^{(1)} E_j, \tag{1.14}$$

where summation convention is implied (i.e. double indices are summed).

It is instructive to compare Coulomb's law in SI and Gaussian units. For two electrons separated by an Ångstrom, in SI units the electron charge is $e = 1.6 \times 10^{-19}$C, the radius of orbit in a typical atom is about $r = 1\text{Å} = 10^{-10}$m, and $k = \frac{1}{4\pi\epsilon_0} = 9 \times 10^9 \frac{\text{N·m}^2}{\text{C}}$, the resulting force is

$$|F| = 2.304 \times 10^{-8} \text{ N},$$

which is expressed in Newtons.

In Gaussian units the charge is expressed in statcoulombs (usually called "stat coul" or "esu"), as $e = 4.8 \times 10^{-18}$esu, and the orbital radius is $r = 10^{-8}cm$. In Gaussian units, Equation 1.12 with $\mathbf{F} = q\mathbf{E}$ yields

$$|F| = 2.304 \times 10^{-3} \text{ dyn},$$

where the force is expressed in dynes. Applying the conversion factor 1 dyn = 10^{-5} N, we confirm that the two forces are indeed the same, as expected.

There are many conventions for defining the various quantities of nonlinear optics, and this is often a source of much confusion. For example. Boyd's book, "Nonlinear Optics," uses SI units and defines a sinusoidal linearly polarized electric field to be of the form

$$\mathbf{E}(t) = E\hat{x}\exp(-i\omega t) + \text{c.c.}. \tag{1.15}$$

I prefer the fields to be of the form $\cos\omega t$ rather than $2\cos\omega t$, so I will use the convention,

$$\mathbf{E}(t) = \frac{1}{2}E\hat{x}\exp(-i\omega t) + \text{c.c.} \tag{1.16}$$

Other issues of convention will be explained when they are needed.

1.3 Example: Second Order Susceptibility

It is instructive to give a simple example of the ramifications of a nonlinear polarization before using the full rigor of tensor analysis and nonlocal response. We begin by illustrating the source of second harmonic generation.

Charges within a material are displaced under the influence of an electromagnetic field. The displacement of charges can be represented as a series of electric moments (dipole, quadrupole, etc). Currents can be similarly represented as a series of magnetic moments. The electric dipole field of bound charges is usually (but not always) the largest contribution to the electromagnetic fields. Next in order of importance is the electric quadrupole and magnetic dipole. We use the electric dipole approximation for most of this book and ignore these higher order terms.

In the *linear* dipole approximation, the induced dipole moment per unit volume of material – called the polarization – is proportional to the electric field. The constant of proportionality that relates the applied electric field to the polarization is called the linear susceptibility. If the induced polarization, P, is in the same direction as the applied electric field, E, the linear susceptibility, χ, is a scalar, and we have,

$$P = \chi^{(1)} E. \tag{1.17}$$

It should be stressed that Equation 1.17 is not a fundamental relationship, but rather a model of how charges react to an applied electric field. In the case of Equation 1.17, the model makes an intuitively reasonable assumption that the charges are displaced along the direction of the applied electric field. As we will see later, such models can become more complex when applied to real systems.

If the applied field is sinusoidal, as is the case for an optical field, the induced dipole will oscillate and reradiate the light. In the linear case, the dipole moment oscillates at the same frequency as the incident light.

Now consider a nonlinear material. The polarization will be given by

$$P = \chi^{(1)} E + \chi^{(2)} E^2 + \ldots, \tag{1.18}$$

where $\chi^{(2)}$ is called the second-order susceptibility. The source of the nonlinear terms originates in charges bound by an anharmonic potential, which we will study later. If we substitute the scalar form of Equation 1.15 into the second term of Equation 1.18, we get

$$\begin{aligned} E^2(t) &= \frac{1}{4}E^2(\exp[-i\omega t]+\exp[+i\omega t])^2 \\ &= \frac{1}{2}E^2\left(\frac{\exp[-i2\omega t]+\exp[+i2\omega t]}{2}+1\right). \end{aligned} \quad (1.19)$$

Equation 1.19 when substituted into Equation 1.18, clearly shows that the polarization oscillates at frequency 2ω, which acts as a source that radiates light at the second harmonic frequency. Also note that the constant offset (the "+1" term) is the source of the phenomenon called optical rectification, which results in a static polarization, and therefore a static electric field.

The origin of all the nonlinear-optical processes described in Section 1.1 can be explained using the nonlinear polarization.

1.4 Maxwell's Equations

In this section, we introduce the origin of interactions between light and matter. This requires that we incorporate into Maxwell's equation the effect of light on matter. Motivated by the fact that electric fields separate charges in a neutral material, we develop the standard model of describing the charge distribution as a series of moments. These various moments are affected by light depending on how the electric field field varies from point to point in the material: A uniform electric field induces a dipole moment, a field gradient the quadrupole, etc. This interaction is formulated using the concept of a response function.

We begin by introducing conventions and motivate the use of Gaussian units when expressing Maxwell's Equations due to the symmetry between the electric and magnetic fields that results. We assume non-magnetic materials with no free charges and no free currents. In particular no free charges nor free currents implies the absence of sources such as static charge or currents due to batteries. We shall assume that the materials are nonmagnetic, so $\mu = 1$ and $\chi_m = 0$.

When light interacts with a material it separates the charges from their equilibrium positions within the constituent molecules or within the lattice of a crystal. Since these charges are intimately associated with the material, they are called *bound charges*, and are "swept under the rug" by the introduction of the electric displacement vector, **D**, which has only free electrons as its source.

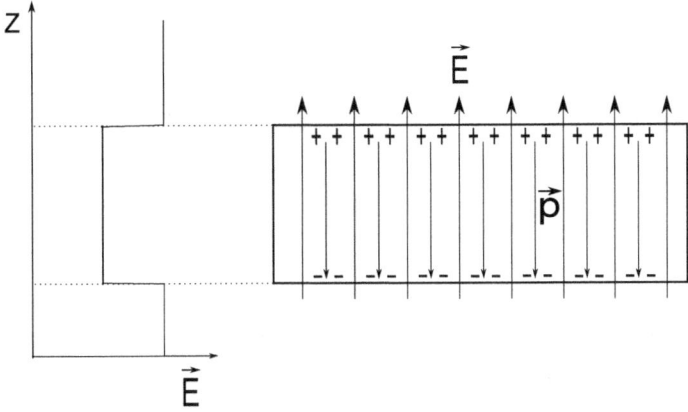

Figure 1.6: Plot of the electric field across the interface of a polarized material and a diagram of the field, **E**, bound charges, and polarization, **P**.

In Gaussian units, Maxwell's Equations in the absence of free charges and currents; and, for a nonmagnetic material (**H** = **B**) are,

$$\nabla \times \mathbf{E} = -\frac{1}{c}\frac{\partial \mathbf{B}}{\partial t}, \tag{1.20}$$

$$\nabla \times \mathbf{B} = \frac{1}{c}\frac{\partial \mathbf{D}}{\partial t}, \tag{1.21}$$

$$\nabla \cdot \mathbf{D} = 0, \tag{1.22}$$

$$\nabla \cdot \mathbf{B} = 0. \tag{1.23}$$

Note that in Gaussian units **E** and **B** have the same units, and are therefore considered on an equal footing.

In SI units, ϵ_0, the permittivity of free space, and μ_0, the permeability, yield the speed of light, $c\sqrt{\epsilon_0 \mu_0} = 1$. Indeed, this relationship suggests that light is an electromagnetic wave. However, ϵ_0 and μ_0 are absent in Gaussian units. Instead, the speed of light appears explicitly.

1.4.1 Electric displacement

The electric field in a material is the superposition of the applied electric field, the field due to free charge, and the fields due to all of the displaced charges. The electric field due to the bound charges can be expressed as a series of the moments. Ignoring terms higher order than the dipole term, the electric displacement is given by

$$\mathbf{D} = \mathbf{E} + 4\pi \mathbf{P} + \ldots \tag{1.24}$$

The higher-order terms will be discussed later. In Equation 1.24, **E** is the total field of all charges(bound and free), **D** is the electric displacement due to free charges and **P** is the electric dipole moment per unit volume, and accounts for the bound charge.

Figure 1.6 shows a material with an applied electric field and the induced charge. The electric field polarizes the material, causing the positive and negative charges to move in opposite directions until they encounter the surface. The net result is a dipole moment. Note that the polarization has the same units as the electric field. The electric field due to the polarization opposes the applied electric field, thus reducing the field strength inside the dielectric.

An electric field gradient results in a different magnitude of force on two neighboring electric charges of the same sign. Thus, two identical neighboring charges will be displaced by differing amounts, resulting in a quadrupole moment. Similarly, higher-order field gradients will couple to higher-order moments. To take this into account, the electric displacement can be expressed as a series of moments of the form

$$D_i = E_i + 4\pi P_i + 4\pi \frac{\partial}{\partial x_j} Q_{ij} + 4\pi \frac{\partial^2}{\partial x_j \partial x_k} O_{ijk} + \ldots, \qquad (1.25)$$

where Q_{ij} is the quadrupole moment and O_{ijk} is the octupole moment, etc.[1]

A vast number of nonlinear-optical processes can be approximated by the dipole term in Equation 1.25. However, we must be keep in mind that the dipole approximation does not always hold, especially when there are large field gradients, as can be found at interfaces between two materials. Figure 1.6 shows a plot of the electric field as a function of the surface-normal coordinate z. The abrupt drop at the interface due to the surface charge leads to a large surface gradient, so higher order terms need to be taken into account.

1.4.2 The Polarization

The electric polarization, **P**, is the dipole moment per unit volume, and is a function of the electric field. When the applied electric field strength is much smaller than the electric fields that hold an atom or molecule together, the

---

[1]Here we use **Summation Convention** where,

$$A_{ij}V_j = \sum_{j=1}^{3} A_{ij}V_j.$$

An index appearing once represents a vector component; an index appearing twice indicates the component that is to be summed over the three Cartesian components.

polarization **P(E)** can be expanded as a series in the electric field,

$$\mathbf{P}(\mathbf{E}) = \mathbf{P}^{(0)} + \mathbf{P}^{(1)}(\mathbf{E}) + \mathbf{P}^{(2)}(\mathbf{E}) + \ldots, \quad (1.26)$$

where the first two terms constitute linear optics and the subsequent terms are the subject of nonlinear optics.

It is instructive to consider the units of the polarization to confirm that they are the same as the units of the electric field. Doing a dimensional analysis, we get

$$[P] = \left[\frac{\text{dipole}}{\text{volume}}\right] = \left[\frac{\text{charge} \cdot \text{length}}{\text{length}^3}\right] = \left[\frac{\text{charge}}{\text{length}^2}\right].$$

Comparing Equation 1.4.2 with the electric field of a point charge given by Equation 1.12 confirms that the polarization is of the same units as the electric and magnetic field. Doing a similar analysis of Equation 1.25 of the higher-order terms, we can determine the units of the quadrupole, octupole, and higher-order moments.

1.5 Interaction of Light with Matter

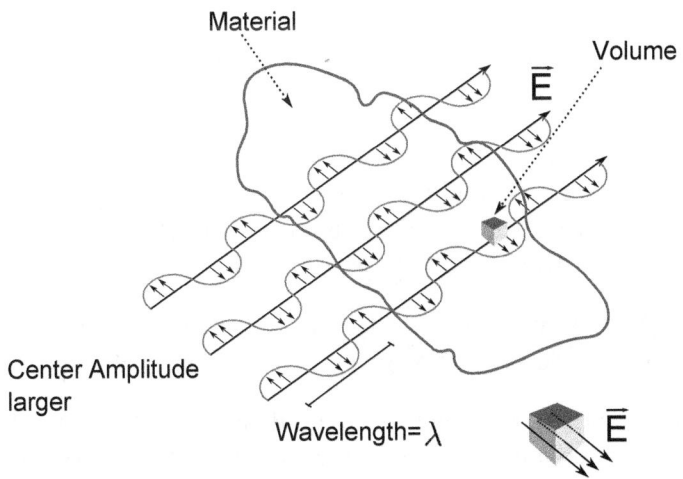

Figure 1.7: Light inducing a polarization inside a material. The light source is a monochromatic plane wave.

We will consider the most general case of light interacting with matter in the dipole approximation. Figure 1.7 shows the material and the beam of light. We then select a volume, V, that is large enough to contain so many molecules that the material appears homogeneous, but much smaller than a wavelength, so that the electric field is approximately uniform over the volume. The long wavelength approximation that $\lambda \gg \sqrt[3]{V}$ will generally hold in the visible part of the electromagnetic spectrum since light has a wavelength on the order of 1 μm and an atom's size is on the order of 10^{-4} μm.

Now we are ready to formulate the most general theory. Figure 1.8 shows an expanded view of the small volume element in Figure 1.7. Over this volume the electric field, **E**, is uniform but varying with time. In this case the most general response has the following properties:

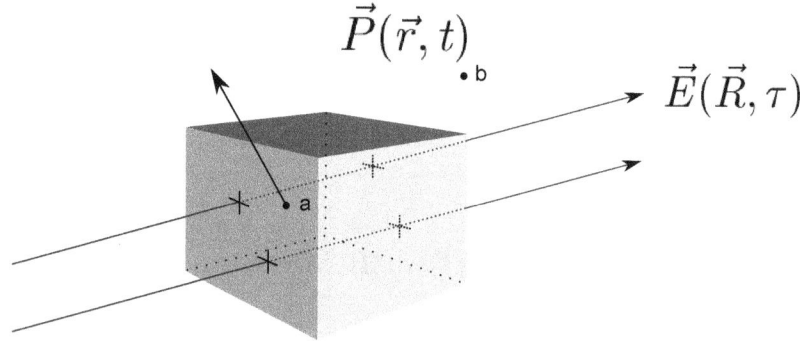

Figure 1.8: A small volume element of material in an electric field

1. **P** is not parallel to **E**: We assume that the polarization, **P**, can point in any direction, depending on the direction of the electric field and on the material's anisotropy. To account for this, the susceptibility (which relates **P** and **E**) must be a tensor.

2. **P(E)** = <u>Series</u>: When the electric field is small, we approximate **P** as a series of the electric field as given by Equation 1.26.

3. The polarization, **P**, at point "b" can depend on **E** at point "a": The response may be nonlocal if information about the electric field is transmitted to another point. For example, light absorbed at point "a" can excite an acoustical wave that propagates to "b" and leads to a polarization there. Thus, the polarizability at one point in the material can depend on the electric field everywhere in the material.

4. The polarization at a given time can depend on the electric field in the past: The polarization response can be delayed relative to the electric

field. For example, an impulse can start the oscillation of a spring that persists long after the impulse has subsided. Similarly, the charges in a material may come to rest long after the excitation pulse has left the material. Thus, the polarization at any time depends on the electromagnetic history of the material.

After the polarization is expanded as a series in the electric field,

$$\mathbf{P} = \sum_{n=0}^{\infty} \mathbf{P}^{(n)}(\mathbf{r},t), \qquad (1.27)$$

the most general form of the first-order polarization, $\mathbf{P}^{(1)}(\mathbf{r},t)$, is given by

$$P_i^{(1)}(\mathbf{r},t) = \int_{\substack{\text{all}\\ \text{space}}} d^3\mathbf{R} \int_{-\infty}^{\infty} d\tau\, T_{ij}(\mathbf{r},t;\mathbf{R},\tau) E_j(\mathbf{R},\tau), \qquad (1.28)$$

where $T_{ij}(\mathbf{r},t;\mathbf{R},\tau)$ is the response function, which carries all of the information that relates the applied electric field to the polarization. The tensor form of the response function accounts for anisotropy of the response, and relates the induced polarization along i to the field polarized along j. The non-locality of the response resides in the position dependence; that is, the polarization at coordinate $\mathbf{r}$ is related to the electric field at coordinate $\mathbf{R}$. The integral over all space thus accounts for the influence of every part of the material on the polarization. The polarization at time t is related by the response function to the field at time τ. The integral over time thus takes into account the influence of the the fields in the past.

Clearly, causality must be built into the temporal part of the response function as does the delay to the retarded potentials. Causality will be described later, while the small size of the sample will allow us to ignore retarded potentials.

The second-order polarization is of the form

$$P_i^{(2)}(\mathbf{r},t) = \int_{\substack{\text{all}\\ \text{space}}} d^3\mathbf{R}_1 \int_{\substack{\text{all}\\ \text{space}}} d^3\mathbf{R}_2 \int_{-\infty}^{\infty} d\tau_1 \int_{-\infty}^{\infty} d\tau_2 \qquad (1.29)$$
$$T_{ijk}\bigl(\mathbf{r},t;(\mathbf{R}_1,\tau_1),(\mathbf{R}_2,\tau_2)\bigr) E_j(\mathbf{R}_1,\tau_1) E_k(\mathbf{R}_2,\tau_2).$$

As in the linear case, the polarization at one point in the material is determined by the fields at other points in the material, thus the integration over all space. Again, the polarization at a given time depends on the electric field at previous times, requiring the integral over time. However, the polarization depends on the square of the electric field, so the induced dipole vector depends on the directions of the two electric fields, which is taken into account by the sum over all possible polarizations of both fields. The response function is therefore a third-rank tensor. Boyd described symmetry arguments that can be used to reduce the number of independent tensor components of the response function,[8] thus reducing significantly the complexity of the problem. Details of such symmetry arguments will not be discussed here.

Finally the most general form of the nth-order polarization is expressed as

$$P_i^{(n)}(\mathbf{r},t) = \int_{\substack{all \\ space}} d^3\mathbf{R_1} \ldots \int_{\substack{all \\ space}} d^3\mathbf{R_n} \int_{-\infty}^{\infty} d\tau_1 \ldots \int_{-\infty}^{\infty} d\tau_n \qquad (1.30)$$
$$T_{ijklm\ldots z} E_j(\mathbf{R_1},\tau_1) \ldots E_z(\mathbf{R_z},\tau_z).$$

Bulk Material

It is instructive to consider the static limit of a spatially uniform electric field. The polarization is then simply given by

$$P_i = P_i^{(0)} + \chi_{ij}^{(1)} E_j + \chi_{ijk}^{(2)} E_j E_k + \ldots, \qquad (1.31)$$

where P_i is the electric dipole moment per unit volume, the first two terms are the focus of linear optics, and $\chi^{(n)}$ is called the nth-order susceptibility.

Molecular

For a molecule, the quantum description of volume is not a well defined concept. Thus, it is more appropriate to characterize a molecule in a static electric field by its induced dipole moment, given by

$$p_i = \mu_i + \alpha_{ij} E_j + \beta_{ijk} E_j E_k + \gamma_{ijkl} E_j E_k E_l + \ldots, \qquad (1.32)$$

where p_i is the induced dipole moment of the molecule; μ is static dipole moment; α(units of volume) is the polarizability, or also known as the first-order molecular susceptibility; β is the hyperpolarizability, also known as the second-order molecular susceptibility; and γ is the second hyperpolarizability, or the third-order molecular susceptibility.

Calculating the Nonlinear Susceptibilities

In light of Equations 1.31 and 1.32, and the fact that $P_i = Np_i$ for N identical molecules, $\chi^{(n)}_{ijk...}$ can be calculated from a model of $\mathbf{p(E)}$ by differentiation,

$$\chi^{(n)}_{ijk...} = \frac{N}{n_x! n_y! n_z!} \frac{\partial^n p_i}{\partial E_j \partial E_k ...}\bigg|_{\mathbf{E}=0}, \tag{1.33}$$

where n_i represents the number of fields polarized along the ith Cartesian direction and N is the number density of molecules. A similar differentiation can be applied to the molecular dipole moment given by Equation 1.32 to get the molecular susceptibilities.

Example:

(a) A model of the polarization gives

$$P_x = aE_x^2. \tag{1.34}$$

Determine $\chi^{(1)}_{xx}$ and $\chi^{(2)}_{xxx}$.

(b) Given the polarization

$$P_x = aE_x + bE_y E_z^2, \tag{1.35}$$

find $\chi^{(1)}_{xy}$ and $\chi^{(3)}_{xyzz}$.

Solution:

(a) Taking the derivative with respect to the field and evaluating it at zero field

$$\boxed{\chi^{(1)}_{xx} = \frac{1}{1!} \frac{\partial P_x}{\partial E_x}\bigg|_0 = 2aE_x|_0 = 0.} \tag{1.36}$$

$$\boxed{\chi^{(2)}_{xxx} = \frac{1}{2} \frac{\partial^2 P_x}{\partial E_x^2}\bigg|_0 = a.} \tag{1.37}$$

(b) Here we apply the same method as in part (a)

$$\chi^{(1)}_{xy} = \frac{1}{1!} \frac{\partial P_x}{\partial E_y}\bigg|_{\mathbf{E}=0} = bE_z^2|_{\mathbf{E}=0} = 0, \tag{1.38}$$

so,

$$\boxed{\chi_{xy}^{(1)} = 0.} \tag{1.39}$$

For $\chi_{xyzz}^{(3)}$ we can apply the same method, getting

$$\boxed{\chi_{xyzz}^{(3)} = \frac{1}{1!}\frac{1}{2!}\frac{\partial^3 P_x}{\partial E_y \partial E_z \partial E_z}\bigg|_{\mathbf{E}=0} = b.} \tag{1.40}$$

The student is encouraged to invent problems in which the dipole moment is a nonlinear function of the electric field. The problems that follow can be solved using simple ideas from statistical mechanics, quantum mechanics, and electrostatics. Note that these problems might be difficult if your background in quantum mechanics and statistical physics is weak or rusty. For some of you, it may be best to move on and come back to these problems after reading further through this book.

Problem 1.5-1: A particle of charge q and mass m is confined to a ring made of a thin wire of radius a that is in thermal equilibrium with a bath of temperature T. What is α, β, and γ for a uniform static electric field applied in the plane of the ring?

Problem 1.5-2: Consider a particle of charge q in an infinite potential well defined by:

$$U = 0 \qquad -\frac{a}{2} < x < \frac{a}{2}$$
$$U = \infty \qquad x > \frac{a}{2}; \quad x < -\frac{a}{2}.$$

Assuming that the particle is in thermal equilibrium with a bath at temperature T, use classical means to calculate the static molecular susceptibilities α, β, and γ.

Problem 1.5-3: An electric dipole (at zero temperature) is embedded in an electret that has a uniform internal static field $\mathbf{E}_0 = E_0 \hat{z}$. If the dipole can only be oriented along $+z$ or $-z$ such that its Hamiltonian is of the form:

$$H_0 = \begin{pmatrix} -\mu E_0 & 0 \\ 0 & +\mu E_0 \end{pmatrix}, \quad (1.41)$$

with eigenstates

$$|\psi_+\rangle = \begin{pmatrix} 1 \\ 0 \end{pmatrix} \quad |\psi_-\rangle = \begin{pmatrix} 0 \\ 1 \end{pmatrix}, \quad (1.42)$$

and dipole operator

$$p = \begin{pmatrix} \mu & 0 \\ 0 & -\mu \end{pmatrix}, \quad (1.43)$$

calculate the static susceptibilities μ_0, α, β, and γ for an externally applied electric field $\mathbf{E}$ if this electric field changes the Hamiltonian to $H = H_0 + H'$ with

$$H' = \begin{pmatrix} 0 & +\epsilon E \\ +\epsilon E & 0 \end{pmatrix}. \quad (1.44)$$

Problem 1.5-4: Calculate the static susceptibilities μ_0, α, β, and γ for a freely rotating rigid molecule with dipole moment μ if it is in thermal equilibrium with a bath at temperature T.

1.6 Magnitudes of Nonlinear susceptibilities

Here we review the units for the susceptibilities in Gaussian units. They are:

$$p = [\text{esu} \cdot \text{cm}] \qquad P = \left[\frac{\text{esu}}{\text{cm}^2}\right] = \left[\frac{\text{statvolt}}{\text{cm}}\right],$$

$$\alpha = [\text{cm}^3] \qquad \chi^{(1)} = [\text{dimensionless}],$$

$$\beta = \left[\frac{\text{cm}^5}{\text{esu}}\right] \qquad \chi^{(2)} = \left[\frac{\text{cm}}{\text{statvolt}}\right] = \left[\frac{\text{cm}^2}{\text{esu}}\right] = \left[\frac{\text{cm}^3}{\text{erg}}\right]^{\frac{1}{2}},$$

$$\gamma = \left[\frac{\text{cm}^7}{\text{esu}^2}\right] = \left[\frac{\text{cm}^6}{\text{erg}}\right] \qquad \chi^{(3)} = \left[\frac{\text{cm}^3}{\text{erg}}\right].$$

We remind the reader that 'esu' is shorthand for 'stat coul' – the unit of charge. This usage is sometimes frowned upon. A more egregious infraction is the use of 'esu' to signify the units of β, a practice that was common in the past.

In SI units, the units of the dipole moments and the nonlinear susceptibilities are:

$$p = [\text{C} \cdot \text{m}] \qquad P = \left[\frac{\text{C}}{\text{m}^2}\right],$$

$$\alpha = [\text{m}^3] \qquad \chi^{(1)} = [\text{dimensionless}],$$

$$\beta = \left[\frac{\text{m}^3}{\text{V}}\right] \qquad \chi^{(2)} = \left[\frac{\text{m}}{\text{V}}\right],$$

$$\gamma = \left[\frac{\text{m}^5}{\text{V}^2}\right] \qquad \chi^{(3)} = \left[\frac{\text{m}^2}{\text{V}^2}\right].$$

Order of magnitude ranges for typical molecular susceptibilities are:

$$\alpha \approx \left(10^{-21} - 10^{-22}\right) \text{cm}^3,$$

$$\beta \approx \left(10^{-27} - 10^{-30}\right) \frac{\text{cm}^5}{\text{esu}},$$

$$\gamma \approx \left(10^{-33} - 10^{-36}\right) \frac{\text{cm}^6}{\text{erg}}.$$

Note that molecular size is on the order of $\alpha^{1/3}$. Thus, the range of typical α above gives a size range of 5Å to 10Å.

In order to observe these nonlinear effects, the field strengths must be large enough to induce a displacement of charge but small enough so that the molecules do not ionize. For example, using typical field strengths inside a hydrogen atom, which bind the electron to the proton, we get

$$E \ll 10^{11} \cdot \frac{\text{V}}{\text{m}},$$

$$E \ll 10^{7} \cdot \frac{\text{stat volt}}{\text{cm}}.$$

1.7 Nonlinear Optics at the Quantum Level

In the classical limit, light is represented by a beam and the material as a continuous substance. The quantum mechanical analogue of a nonlinear process is represented by photons interacting with a molecule. The correspondence between the two is usually straightforward, with beams being replaced by

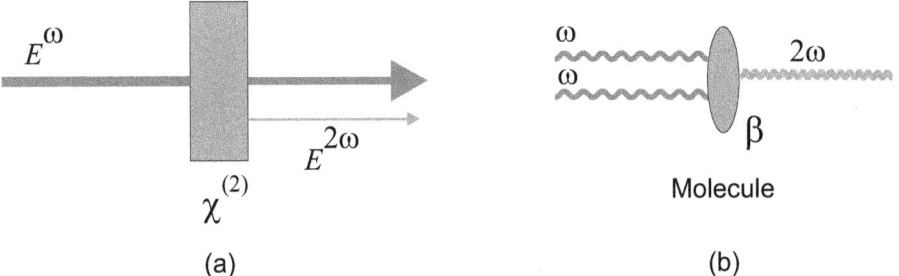

Figure 1.9: (a) Macroscopic view of second harmonic generation; and, (b) the quantum view.

photons in the low intensity regime, and molecules coming into focus at high magnification.

Consider as an example second harmonic generation. The macroscopic view is embodied by Equation 1.19, where a beam at frequency ω leads to the generation of a beam at 2ω due to an oscillating polarization field, as represented in Figure 1.9a. Some of the energy in the fundamental beam is converted into the second harmonic.

In the quantum limit, the beam intensity is turned down until only two quanta of light are incident on an individual molecule, as depicted in Figure 1.9b. The two photons, each at frequency ω, merge to yield a photon of frequency 2ω. Energy is conserved in the process, with $\hbar\omega + \hbar\omega = 2\hbar\omega$. Since both photons need to be present for the process to take place, the probability of a 2ω photon being emitted is proportional to the product of the probabilities of finding each photon at the molecular site. This is consistent with the macroscopic result – that the second harmonic field is proportional to the square of the fundamental field, or that the intensity of the second harmonic is proportional to the square of the fundamental intensity.

Now consider difference frequency generation. With two beams incident on the sample, of wavelengths ω_1 and ω_2, the polarization, which is of the form

$$\begin{aligned}P &= \chi^{(2)}\left(\frac{1}{2}E^{\omega_1}\exp[-i\omega_1 t] + \frac{1}{2}E^{\omega_2}\exp[-i\omega_2 t] + \text{c.c.}\right) \\ &= \frac{1}{4}E^{\omega_1}E^{\omega_2}\exp[-i(\omega_1 - \omega_2)t] + \ldots \end{aligned} \quad (1.45)$$

leads to light at the difference frequency, as depicted in Figure 1.10a. The low intensity limit, where individual photons interact with a molecule, is shown in Figure 1.10b. This process clearly does not conserve energy since

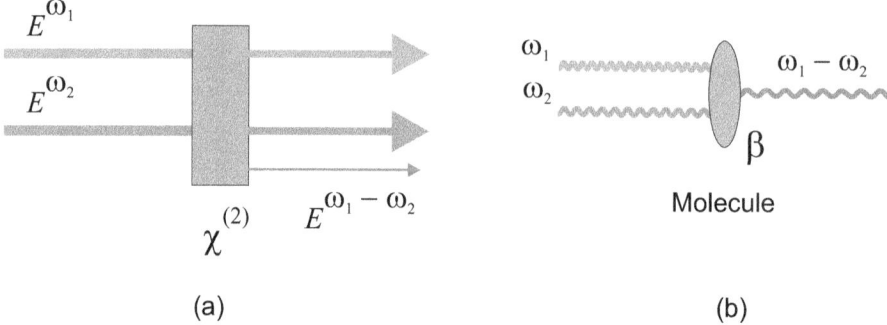

Figure 1.10: (a) Macroscopic view of difference frequency generation; and, (b) the quantum view.

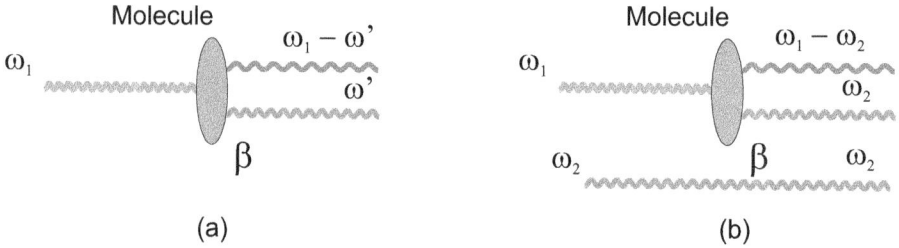

Figure 1.11: (a) Quantum view of reverse sum frequency generation; and, (b) difference frequency generation as a stimulated process of reverse sum frequency generation.

$\hbar\omega_1 + \hbar\omega_2 \neq \hbar(\omega_1 - \omega_2)$.

This problem can be reconciled by first considering time-reversed sum frequency generation as shown in Figure 1.11a, in which one photon splits into two. In this case, energy is conserved, with $\hbar\omega_1 = (\hbar\omega_1 - \hbar\omega') + \hbar\omega'$. In this process, ω' can be any frequency in the range $0 \leq \omega' \leq \omega_1$, and all colors would be equally likely to be emitted if β were frequency independent. As such, the probability of getting any particular color is exceedingly small.

Figure 1.11b shows the true difference frequency generation process, which calls upon the phenomenon of stimulated emission. When a molecule is in an excited state, the probability of a transition between states of energy difference $\hbar\omega$ will be enhanced in proportion to the number of photons at frequency ω. In difference frequency generation, the photon at ω_1 — which is in general not on resonance with any transitions in the molecule, will excite the system into a superposition of states with expectation energy $\hbar\omega_1$. An incident photon at frequency ω_2 will stimulate an emission at frequency ω_2, forcing the other photon to be emitted at frequency $\omega_1 - \omega_2$.

Without the stimulating beam, the probability of emitting two photons is proportional to the number of incident photons at frequency ω_1. The probability of the stimulated process is in proportion to the number of photons at frequency ω_2, thus the probability of producing a difference frequency photon is proportional to the product of the number of photons at each frequency, in line with the macroscopic process, which is proportional to the product of the two intensities.

These quantum processes will be more rigourously in Chapter 4.

1.8 Numerical Simulations using Python

Traditional physics textbooks tend to assign problems with analytical solutions or approximations that can be expressed in closed form. Given that many programming languages are free and easily accessible, there is no reason for students to not use them early in their careers. I find it pedagogically useful for a student to solve problems numerically both to elucidate the underlying physics and to give students practice in simple numerical techniques. Being freed from the burden of finding problems with analytical solutions, there are a broader range of representative problems from which to choose.

A strong understanding of the simple fundamentals will give the student the tools needed to deal with the more complex topics. In the spirit of this approach, we introduce numerical techniques using Python. I have programmed in Basic, Fortran, and C++ and more recently have learned Python. Since Python is free, easy to use, and boasts a steadily-growing library of scientific modules, it provides the ideal platform for introducing numerical techniques.

Rather than lecture on how one can use Python to solve useful problems, we start with a simple example, as follows.

Problem: A laser beam passes through a material whose absorbance, $\alpha(z)$, depends on the depth within the sample, z. The intensity, $I(z)$, is given by

$$\frac{dI(z)}{dz} = -\alpha(z)I(z). \tag{1.46}$$

(a) With $\alpha(z)$ as an input, write code that calculates the intensity profile inside the material under the condition that $I(0) = I_0$.

(b) If the loss is given by $\alpha(z) = \alpha_0 + \sin z$ (note that negative loss is gain, and is perfectly acceptable), calculate $I(z)$ in the range $0 \le z \le 6\pi$ and plot your results for $\alpha(z) = 0.1, 0.2, 0.3, 0.4,$ and 0.5.

(c) Determine the function $\alpha(z)$ that gives an oscillating intensity $I(z) = \cos^2 z$. Do this problem both analytically and numerically and compare your results.

While simple problems can be coded without much thought, it is most efficient to use a disciplined approach to help you think through the process without confusion. The steps are,

1. Draw a diagram

2. Discretize the variables and express the problem in terms of relationships between vectors and tensors

3. Make a flowchart

4. Write and debug the code

5. Run the code

6. Plot the results

We apply this approach to the example problem, below.

Parts (a) and (b)

Diagram

Figure 1.12 shows a diagram of the material, which is divided into N cells. Labeled are the front and back coordinates of each cell, the intensities at each boundary, the input boundary condition, and the loss in each cell. The number of cells is labeled and related to the macroscopic length of the space.

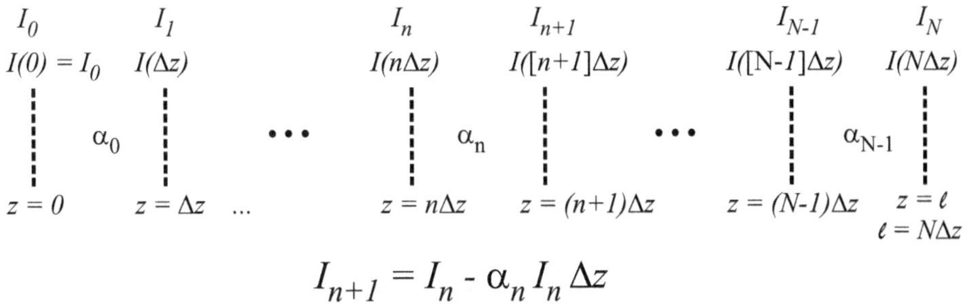

Nonlinear Optics: A student's perspective

I_0 I_1 I_n I_{n+1} I_{N-1} I_N
$I(0) = I_0$ $I(\Delta z)$ $I(n\Delta z)$ $I([n+1]\Delta z)$ $I([N-1]\Delta z)$ $I(N\Delta z)$

    α_0 ... α_n ... α_{N-1}

$z = 0$ $z = \Delta z$ ... $z = n\Delta z$ $z = (n+1)\Delta z$ $z = (N-1)\Delta z$ $z = \ell$
                                                                                            $\ell = N\Delta z$

$$I_{n+1} = I_n - \alpha_n I_n \Delta z$$

Figure 1.12: The intensity as a function of position on N cells and the discretized variables.

Discretization

Equation 1.46 is made discrete by expressing the change in intensity, $\Delta I(z)$, at point z in the material in terms of the loss and thickness of the cell,

$$\Delta I(z) = -\alpha(z) \cdot I(z) \cdot \Delta z. \qquad (1.47)$$

With $z = n\Delta z$ for a material of thickness $\ell = N\Delta z$, Equation 1.47 becomes

$$I([n+1]\Delta z) - I(n\Delta z) = -\alpha(n\Delta z) \cdot I(n\Delta z) \cdot \Delta z. \qquad (1.48)$$

Equation 1.47 can be reexpressed as

$$I_{n+1} = I_n - \alpha_n \cdot I_n \cdot \Delta z = I_n(1 - \alpha_n \Delta z), \qquad (1.49)$$

where we have used the simpler notation $I_n \equiv I(n\Delta z)$ and $\alpha(n\Delta z) \equiv \alpha_n$. Being in vector form, Equation 1.49 can be directly converted into computer code.

It is prudent to anticipate potential problems that can arise in the flow of the code under special circumstances. An obvious problem in Equation 1.49 is that the intensity is negative when $\alpha_n \Delta z \geq 1$ – a clear violation of basic physics.

There are several options for dealing with this problem. One is to test for the condition $\alpha_n \Delta z \geq 1$ and halt execution if it holds. An alternative is to recast the problem in a way that avoids the issue. The approach we take is to define a more robust algorithm.

Consider the special case when α is a constant. Equation 1.49 implies that

$$I_n = I_0(1 - \alpha_n \Delta z)^n, \qquad (1.50)$$

which in the limit of small Δz gives

$$\lim_{\Delta z \to 0} I_n = I_0 \lim_{\Delta z \to 0} (1 - \alpha \Delta z)^n = I_0 \lim_{n \to \infty} \left(1 - \alpha \frac{z}{n}\right)^n, \qquad (1.51)$$

where the last equality uses the definition of Δz, that is, $n\Delta z = z$. Equation 1.51 yields

$$I_n = I(z) = I_0 \lim_{n\to\infty} \left(1 - \alpha\frac{z}{n}\right)^n = I_0 \exp(-\alpha z). \tag{1.52}$$

This result is not surprising given that Equation 1.52 is just the solution of Equation 1.46 for constant α.

The problem of negative intensity in Equation 1.49 can be fixed by recognizing that the term in parentheses is an approximation to an exponential, that is, $\exp(-\alpha\Delta z) \approx 1 - \alpha\Delta z$. When $\alpha\Delta z$ becomes appreciable compared with unity, this approximation breaks down. There are two approaches to fixing this problem. One is to subdivide the offending cell to decrease Δz, then using Equation 1.50 to determine I_n. Alternatively, we can apply the exact form

$$I_{n+1} = I_n \exp(-\alpha_n \Delta z), \tag{1.53}$$

to all cells. Note that Equation 1.53 applied even when α depends on z provided that α changes negligibly over the step size Δz. We choose to use Equation 1.53 when implementing the code below.

Flowchart

The flowchart for calculating the intensity as a function of position is given by

1. **Input parameters**:

    - incident intensity I_0 (scalar)
    - position-dependent loss α (vector)
    - sample thickness ℓ (scalar).

2. Determine N, the number of cells, from the size of α.

3. Initialize the intensity vector of size $N+1$ and set $I[0] = I_0$.

4. Initialize the position vector of size $N+1$ and set $z[0] = 0$.

5. Loop n from 0 to $N-1$, calculating the intensity element by element using Equation 1.49 or 1.53 and z using $z = n\Delta z$.

    - At each step, check that $\alpha_n \Delta z < 1$. If condition not met, define strategy to deal with it or avoid the problem using the exponential form.

6. **Output**:
    - intensity profile I (vector)
    - position z (vector).

The Code

When writing code, use lots of comments to make clear your thought process to your future selves as well as to others.

The code for the example above is as follows:

```python
### This module calculates the intensity vector.
### Input parameters are the incident intensity I0, the loss
### vector, alphaZ, which gives the loss as a function of
### depth, and Lmaterial, the length of the material. The
### output is the vector z, the depth vector, and Intensity,
### the intensity as a function of depth.

def Intensity(I0,alphaZ,Lmaterial):

    import numpy as np    # use numpy package for vector/matrix
                          # functions
    import math           # import math functions

    N = len(alphaZ)       # determine number of elements
                          # in alphaZ

    DeltaZ = Lmaterial/N
                          # determine DeltaZ

    Intensity = np.zeros(N+1)
                          # create the intensity vector
                          # filled with zeros using numpy
    Intensity[0] = I0     # set first intensity to I0

    z = np.zeros(N+1)     # make a vector with zeros
    z[0] = 0.             # input face is defined as z=0

    for i in range(0,N):
                          # i will take values of 0 to N-1
```

```
            Intensity[i+1] = (math.exp(-1. * alphaZ[i] *   DeltaZ)
            * Intensity[i])
                        # apply difference equation

            z[i+1] = z[i] + DeltaZ
                        # calculate depth

    return Intensity, z
```

Note that it is inefficient to import numpy and math within a function definition, as we have done here. It is far more efficient to import modules prior to running a particular function that will be called many times. The method above is self contained and works fine as long as execution time is not an issue.

Running the Code

Run your code from the Python Shell. The Python Shell is a convenient environment for building projects from the modules that you have defined. In this example, there is only one module, called "Intensity." The Canopy Development environment uses IPython.

When IPython is ready for your input, it displays the prompt,

```
In [1]:
```

To make a vector V with 1000 elements and setting them to zero requires the numpy package. To do so requires the following steps,

```
In [1]: import numpy as np
In [2]: V = np.zeros(1000)
In [3]:
```

IPython in Canopy allows the abbreviated procedure:

```
In [1]: V = zeros(1000)
In [2]:
```

Below is an example of input to the IPython interpreter and the results returned by IPython. Note that the first three lines came up when I opened the shell. Then I opened the Python module in the editor, and pressed the compile button with the green triangle. This returned the line with "%run D:/ClassModules.py" telling me that the module name is "ClassModules.py" and it is on drive D.

```
Welcome to Canopy's interactive data-analysis environment!
 with pylab-backend set to: qt
Type '?' for more information.

In [1]: %run D:/ClassModules.py
In [2]: alpha = zeros(1000)
In [3]: for i in range(0,1000):
   ...:     alpha[i] = 0.5 + sin(6.*pi*i/1000.)
   ...:
In [4]: I, z = Intensity(1.,alpha,6.*pi)
In [5]: plot(z, I)
Out[5]: [<matplotlib.lines.Line2D at 0x8d8a2e8>]
In [6]: for i in range(0,1000):
   ...:     alpha[i] = 0.4 + sin(6.*pi*i/1000.)
   ...:
In [7]: I, z = Intensity(1.,alpha,6.*pi)
In [8]: plot(z, I)
Out[8]: [<matplotlib.lines.Line2D at 0x8d91c18>]
In [9]: for i in range(0,1000):
   ...:     alpha[i] = 0.3 + sin(6.*pi*i/1000.)
   ...:
In [10]: I, z = Intensity(1.,alpha,6.*pi)
In [11]: plot(z, I)
Out[11]: [<matplotlib.lines.Line2D at 0x8d97f28>]
In [12]: for i in range(0,1000):
   ...:     alpha[i] = 0.2 + sin(6.*pi*i/1000.)
   ...:
In [13]: I, z = Intensity(1.,alpha,6.*pi)
In [14]: plot(z, I)
Out[14]: [<matplotlib.lines.Line2D at 0x8d99240>]
In [15]: for i in range(0,1000):
   ...:     alpha[i] = 0.1 + sin(6.*pi*i/1000.)
   ...:
In [16]: I, z = Intensity(1.,alpha,6.*pi)
In [17]: plot(z, I)
Out[17]: [<matplotlib.lines.Line2D at 0x8d99d30>]
In [18]:
```

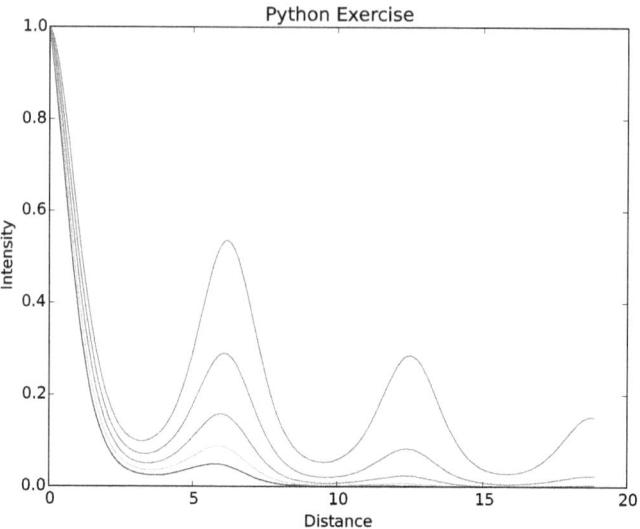

Figure 1.13: Plots made in Python.

Resulting plots

When running the commands in the IPython interpreter as shown above, a plot Window opens yielding the graph shown in Figure 1.13. The axes labels were added after the plot was generated using appropriate buttons on the plot window GUI.

Part (c)

Using Equation 1.46 with $I(z) = \cos^2 z$, we can solve for the loss,

$$\alpha(z) = 2\tan z. \tag{1.54}$$

The code can be used to check Equation 1.54. Does it work as expected? You may want to try simpler functions such as $I(z) = 1 - z^2$.

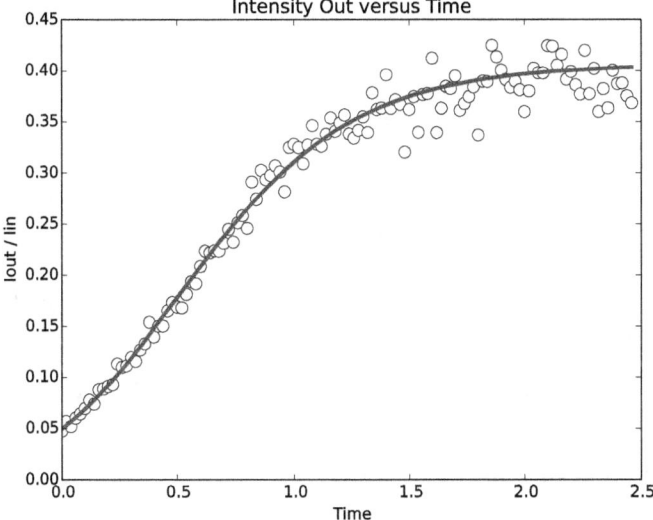

Figure 1.14: Ratio of transmitted to incident intensity as a function of time over a 2.5s data run (points). The smooth curve is a fit to the data.

Problem 1.8-1: Optical damage can be viewed as a self action effect where the light beam changes the optical absorbance of the material according to

$$\frac{\partial \alpha(z,t)}{\partial t} = -b(\alpha(z,t) - \alpha_\infty)I(z,t), \qquad (1.55)$$

where α_∞ and b are constants and the intensity inside the material varies according to Equation 1.46,

$$\frac{\partial I(z,t)}{\partial z} = -\alpha(z,t)I(z,t). \qquad (1.56)$$

In this problem, then, the intensities and losses are functions of space and time. If the beam is not too intense, at $t = 0$ the intensity distribution $I(z,0)$ is calculated by numerically integrating Equation 1.56 with $\alpha(z,t) = \alpha(z,0)$. $\alpha(z,\Delta t)$ can then be determined from Equation 1.55 using $\alpha(z,0)$ and $I(z,0)$. Next, $I(z,\Delta t)$ is calculated from Equation 1.56 using $\alpha(z,\Delta t)$, etc. The process is repeated, incrementing by Δt and Δz until you get the absorbance and intensity at each point z in the material at all times. You are to calculate $\alpha(z,t)$ and $I(z,t)$ for an incident beam of intensity I_0 and an initially uniform material so that $\alpha(z,0)$ is a constant.

You are to submit this project using a file created with LaTeX. Be sure to do the following,

(a) Draw a diagram that will help you discretize the problem.

(b) Discretize Equations 1.55 and 1.56.

(c) Prepare a flowchart of how you plan to code the problem.

(d) Write your code

(e) Present your results in the forms of pretty plots and discuss if they make sense in the various limiting cases.

Problem 1.8-2: Figure 1.14 shows the measured ratio of transmitted to incident intensity in an experiment that seeks to study photodegradation. The sample is $3\mu m$ thick and the incident intensity is $1MW/cm^2$ over a data run spanning $2.5s$. The sample is prepared in a way that makes it uniform (isotropic and homogeneous). The curve is a fit to the data. A text file with the data can be found at www.nlosource.com/DegradationData.txt.

(a) Determine α_0, α_∞, and b (defined in problem 1.8-1) from the data using any method that suits your fancy.

1.9 Goals

Nonlinear optics encompasses a broad range of phenomena. The goal of this course is to build a fundamental understanding that can be applied to both the large body of what is known as well as to provide the tools for approaching the unknown. In your studies, keep the following questions in mind:

- Are the nonlinear optical process related to each each other; and if so, how?
- What are the underlying microscopic mechanisms?
- How are the microscopic mechanisms related to macroscopic observations?
- What governs wave propagation in the nonlinear regime?
- How can nonlinear optics be applied?

Chapter 2

Models of the NLO Response

In this chapter, we develop the classical nonlinear harmonic oscillator model of the nonlinear-optical response of a material, show how the nonlinear-optical susceptibility in the frequency domain is related to the response function, and show how symmetry arguments can be used to reduce the number of independent tensor components.

2.1 Harmonic Oscillator

In this section, we begin by developing a model of the static response of a linear harmonic oscillator and generalize it to the nonlinear case to show how the nonlinear susceptibility is related to the linear and nonlinear spring constant. The model is subsequently applied to a nonlinear spring in a harmonic electric field to develop an understanding of the dependence of the nonlinear-optical susceptibility on the frequency of the electric field.

There are many nuances pertaining to the nonlinear harmonic oscillator that may be difficult for first-time learners to comprehend. Nevertheless, we discuss these in some detail because of their importance in developing deeper insights into the nonlinear response. Students who are more practically-minded can skim this section without disadvantage for the rest of this book.

2.1.1 Linear Harmonic Oscillator

Consider, as an example, a charge on a spring as shown in Figure 2.1. For the static one-dimensional harmonic oscillator model, we apply a static electric field **E**, which causes the spring to stretch.

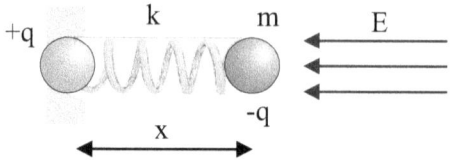

Figure 2.1: A charge on a spring with mass m and charge $-q$ at a distance x from the origin, which contains a fixed point charge $+q$. An electric field E is applied along $\hat{x}$.

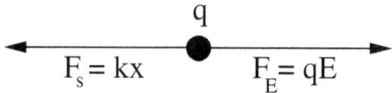

Figure 2.2: Free body diagram of the charge in Figure 2.1.

The applied electric field can be so large that the spring becomes unraveled, thus leading to anharmonic terms. We assume the field is small enough to keep the spring intact. This is analogous to a molecule where the applied field is large enough to strongly displace the electrons from their equilibrium positions, but not strong enough to ionize them. Figure 2.2 shows a free body diagram of a charge where the electric force is balanced by an opposing force due to the spring.

In equilibrium, the forces are equal to each other, that is,

$$F_s = F_E. \qquad (2.1)$$

Substituting the forces into Equation 2.1 yields

$$kx = -qE. \qquad (2.2)$$

Solving for x,

$$x = -\frac{q}{k}E, \qquad (2.3)$$

using the definition of the dipole moment

$$p = -q \cdot x + q \cdot 0 = -qx, \qquad (2.4)$$

and substitution of Equation 2.3 into Equation 2.4 yields

$$p = \frac{q^2}{k}E. \qquad (2.5)$$

Note that in an atom, the electrons are displaced and the nuclei remain approximately stationary. Thus, by picking the origin of the coordinate system at the center of positive charge, only electron charge will contribute to the dipole moment. For the remainder of this book, we will ignore the positive charges under the assumption that they are stationary and define the origin of the coordinate system.

The static linear susceptibility is defined as

$$\alpha = \frac{\partial p}{\partial E}\bigg|_{E=0}, \qquad (2.6)$$

which applied to Equation 2.5 yields the linear susceptibility

$$\alpha = \frac{q^2}{k}. \qquad (2.7)$$

When k is small, the charge is loosely bound and large displacements result from small applied fields. When q is large, the linear susceptibility, or polarizability, is large because the applied force scales in proportion to q^2.

For N springs per unit volume we get

$$\chi^{(1)} = N\alpha. \qquad (2.8)$$

Substituting Equation 2.7 into Equation 2.8 yields

$$\chi^{(1)} = N\frac{q^2}{k}. \qquad (2.9)$$

Note that Equation 2.9 explicitly assumes that the springs are non-interacting so that each unit acts independently. This approximation is often called the dilute gas approximation. When atoms or molecules get closer, interactions can be taken into account using mean-field methods, as we will describe later. Even then, as long as the charges cannot hop from one unit to the other, the result will be similar to Equation 2.9 with only a minor modification.

The susceptibility is a linear function of N for low concentrations. However, when N is sufficiently large, the susceptibility deviates from linearity (as depicted in Figure 2.3). In materials such as dye-doped polymers, deviations from linearity is a test of molecular aggregation.

2.1.2 Nonlinear Harmonic Oscillator

We model the nonlinear harmonic oscillator by adding a quadratic term to the force – a good approximation to real systems for sufficiently large displacements, as shown in Figure 2.4. Then, the spring force is given by

$$F_s = -k^{(1)}x - k^{(2)}x^2. \qquad (2.10)$$

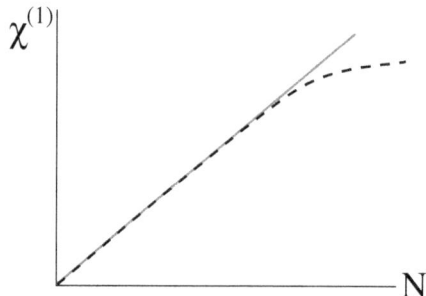

Figure 2.3: The susceptibility as a function of the number density of molecules, N. The susceptibility deviates from linearity when molecules aggregate at large N.

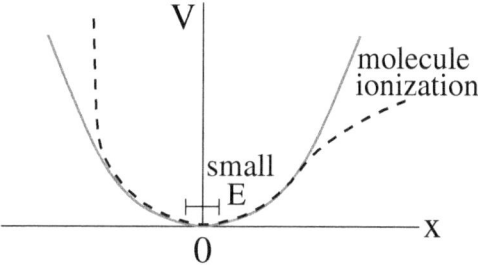

Figure 2.4: The electric potential as a function displacement. For large-enough x, the potential deviates from a quadratic function. The molecule will ionize beyond a critical displacement.

In equilibrium, the spring and electric forces balance so that

$$k^{(1)}x + k^{(2)}x^2 = qE. \tag{2.11}$$

Rearranging Equation 2.11 yields

$$x^2 + \frac{k^{(1)}}{k^{(2)}}x - \frac{q}{k^{(2)}}E = 0. \tag{2.12}$$

Solving for x, we get

$$x = \frac{-k^{(1)}}{2k^{(2)}} \pm \frac{1}{2}\sqrt{\left(\frac{k^{(1)}}{k^{(2)}}\right)^2 + \frac{4q}{k^{(2)}}E}, \tag{2.13}$$

Nonlinear Optics: A student's perspective

and factoring out $\frac{k^{(1)}}{k^{(2)}}$,

$$x = \frac{-k^{(1)}}{2k^{(2)}}\left[1 \mp \sqrt{1 + \frac{4q}{k^{(2)}}\left(\frac{k^{(2)}}{k^{(1)}}\right)^2 E}\right]. \tag{2.14}$$

For small electric field, a series expansion yields

$$\sqrt{1 + \frac{4q}{k^{(2)}}\left(\frac{k^{(2)}}{k^{(1)}}\right)^2 E} = 1 + \frac{1}{2}\frac{4q}{k^{(2)}}\left(\frac{k^{(2)}}{k^{(1)}}\right)^2 E + \ldots. \tag{2.15}$$

Note that the negative root yields the correct linear result, i.e.

$$\lim_{E \to 0} x = 0,$$

so we drop the positive one, yielding

$$\begin{aligned} x &= \frac{k^{(1)}}{2k^{(2)}}\frac{1}{2}\frac{4q}{k^{(2)}}\left(\frac{k^{(2)}}{k^{(1)}}\right)^2 E + \ldots \\ &= \frac{q}{k^{(1)}}E + \ldots. \end{aligned} \tag{2.16}$$

Then,

$$\chi^{(1)} = N\alpha = N\frac{q^2}{k^{(1)}}. \tag{2.17}$$

Note that this is the linear result previously obtained.

Applying Equation 1.33 to Equation 2.14 with the negative root, we get the second-order susceptibility

$$\chi^{(2)} = \frac{1}{2}\frac{\partial^2 P}{\partial E^2}\bigg|_{E=0} = \frac{-Nq^3}{(k^{(1)})^2}\frac{k^{(2)}}{k^{(1)}}, \tag{2.18}$$

and the third-order susceptibility

$$\chi^{(3)} = \frac{1}{6}\frac{\partial^3 P}{\partial E^3}\bigg|_{E=0} = \frac{2Nq^4}{(k^{(1)})^3}\left(\frac{k^{(2)}}{k^{(1)}}\right)^2. \tag{2.19}$$

all nonlinear susceptibilities vanish when $k^{(2)} \to 0$, as they should.

2.1.3 Non-Static Harmonic Oscillator

The equation of motion of a one-dimensional harmonic oscillator is given by Newton's first law,

$$F = m\ddot{x} = -m\omega_0^2 x - 2m\Gamma\dot{x} - e\sum_j E_{0j}\cos(\omega_j t) - m\left(\xi^{(2)}x^2 + \xi^{(3)}x^3 + \ldots\right), \quad (2.20)$$

where $-m\omega_0^2 x$ is the linear restoring force of the spring, $-2m\Gamma\dot{x}$ is the damping force, and $-e\sum_j E_{0j}\cos(\omega t)$ is the driving force on the electron by a sum of time-harmonic electric fields. The nonlinear restoring force is given by the power series in the parentheses. We will eventually use complex displacements to simplify the problem, but for now, consider all quantities in Equation 2.20 to be real.

Equation 2.20 is a classical model of a molecule in which the nonlinear response arises from the nonlinearity of the spring. However, there are other ways to make the harmonic oscillator nonlinear, such as expanding the damping force as a power series of the velocity. The model that most closely fits the data gives insights into the origins of the nonlinear response based on the physics used to build the model. For the remainder of this treatment, we assume that the nonlinearity originates in the spring's restoring force. Budding theorists should think about other models that yield a nonlinearity as practice for the real world.

First-Order Susceptibility

We start with the linear harmonic oscillator in a single time harmonic field. The electric field is more conveniently expressed in complex form

$$E = \frac{E_0}{2}e^{-i\omega t} + \text{c.c.}, \quad (2.21)$$

where the electric field is given by

$$E = \text{Re}\left[E_0 e^{-i\omega t}\right]. \quad (2.22)$$

Once transients have dissipated long after the fields are turned on, the electron will oscillate at the driving frequency, so we assume that

$$x = \frac{A}{2}e^{-i\omega t} + \frac{A^*}{2}e^{i\omega t}. \quad (2.23)$$

As later discussed, the complex amplitude in Equation 2.23 accommodates a phase difference between the applied electric field and the displacement of the electron. Substituting the ansatz given by Equation 2.23 into the linear equation of motion given by Equation 2.20 with $\xi^{(n)} = 0$, collecting all terms proportional to $e^{-i\omega t}$, and solving for A, we find

$$A = \frac{-eE_0}{m} \frac{1}{\omega_0^2 - 2i\Gamma\omega - \omega^2}. \tag{2.24}$$

The polarization is given by

$$P^{(1)} = -eNx^{(1)}. \tag{2.25}$$

In anticipation of the more general calculation to come, we have used the notation that $x^{(1)}$ is the solution to the first-order equation, which describes the linear spring. Substituting Equation 2.24 into 2.23 and substituting the result into Equation 2.25 gives

$$P^{(1)} = \frac{e^2 E_0 N}{2m} \frac{1}{D(\omega)} e^{-i\omega t} + \text{c.c}, \tag{2.26}$$

where

$$D(\omega) = \omega_0^2 - 2i\Gamma\omega - \omega^2. \tag{2.27}$$

The first-order polarization at frequency ω in complex form is

$$P^{(1)} = \frac{P^{(1)}_\omega}{2} e^{-i\omega t} + \frac{P^{(1)}_{-\omega}}{2} e^{i\omega t}, \tag{2.28}$$

where $P^{(1)*}_\omega = P^{(1)}_{-\omega}$ is complex. These are the Fourier components at ω and $-\omega$. The constitutive equation for the polarization is

$$P^{(1)} = \text{Re}\left[\chi^{(1)} E_0 e^{-i\omega t}\right] = \frac{\chi^{(1)}}{2} E_0 e^{-i\omega t} + \text{c.c.} \tag{2.29}$$

Equating the righthand sides of Equations 2.26 and 2.29 and comparing the coefficients of terms with $e^{-i\omega t}$ (equivalent to taking the ω Fourier component), we get

$$\chi^{(1)} = \frac{e^2 N}{m} \frac{(\omega_0^2 - \omega^2) + 2i\Gamma\omega}{(\omega_0^2 - \omega^2)^2 + 4\Gamma^2\omega^2} \equiv \frac{e^2 N}{m} \frac{1}{D(\omega)}, \tag{2.30}$$

where we have multiplied the numerator and denominator in Equation 2.24 by $\left((\omega_0^2 - \omega^2) + 2i\Gamma\omega\right)$ to arrive at a susceptibility that is the sum of a real and imaginary part.

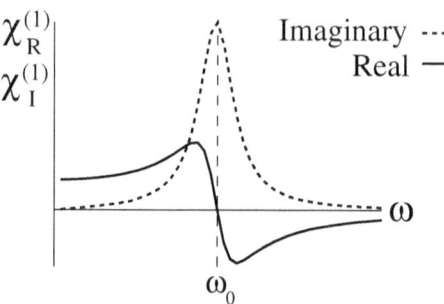

Figure 2.5: The real and imaginary parts of the susceptibility of a harmonic oscillator.

Figure 2.5 plots the real and imaginary parts of $\chi^{(1)}$ as a function of frequency. The real part represents the in-phase component while the imaginary part is the $\frac{\pi}{2}$ out of phase component of the induced dipole moment relative to the driving field. This complex form of the susceptibility, which is proportional to the displacement $x^{(1)}$, can be related to the purely real form, as follows.

Consider the position of the electron expressed in real form

$$x = x_0 \cos(\omega t + \phi). \tag{2.31}$$

Using the trigonometric identity for the cosine of a sum, we get

$$x = x_0 \left(\cos(\omega t) \cos(\phi) - \sin(\omega t) \sin(\phi) \right). \tag{2.32}$$

Our goal is to show that the time dependence of the position can be equally well represented by A_R and A_I, where

$$A = A_R + i A_I. \tag{2.33}$$

It is clear that $x_0 = |A|$ and ϕ from a comparison between Equations 2.32 and 2.33.

The first order solution for the displacement is given by the real part of the complex solution, or

$$\begin{aligned} x^{(1)} &= \mathrm{Re}\left[A e^{-i\omega t} \right] & (2.34) \\ &= \mathrm{Re}[(A_R + i A_I)(\cos(\omega t) - i \sin(\omega t))] & (2.35) \\ &= \mathrm{Re}[A_R \cos(\omega t) + A_I \sin(\omega t) + i \times \text{stuff}] & (2.36) \\ &= A_R \cos(\omega t) + A_I \sin(\omega t), & (2.37) \end{aligned}$$

where $A_R\cos(\omega t)$ is in phase with the field and $A_I\sin(\omega t)$ is $\frac{\pi}{2}$ out of phase with the field. Solving for the components using Equations 2.32 and 2.37, we get

$$A_R = x_0 \cos(\phi), \qquad (2.38)$$
$$A_I = -x_0 \sin(\phi), \qquad (2.39)$$
$$\phi = -\tan^{-1}\left(\frac{A_I}{A_R}\right), \qquad (2.40)$$

and

$$x_0 = \sqrt{(A_R^2 + A_I^2)}. \qquad (2.41)$$

Equations 2.38 through 2.41 define the transformation between (A_R, A_I) and (x_0, ϕ). Each representation conveys the same information, but, the complex form is more computationally expedient.

We are now ready to proceed to the nonlinear calculation using the complex representation of the displacement and calling upon the linear results as a guide for generalizing it to the nonlinear case. As we will see shortly, the approach is generally applicable to solving nonlinear equations provided that the solutions can be approximated by a converging series. Indeed, we will apply the same method to solve the nonlinear wave equation, so it is important for the student to understand this method before moving on.

To review, we started with a differential equation based on Newton's first law, which represents all the forces acting on the material. This starting point requires a model of the material, in this example a charge with a restoring force that opposes displacement and a viscous force that opposes motion. The nonlinearity enters through the restoring force, which is a power series in the displacement under the assumption of small displacements. The nonlinearity could equally well have been added to the viscous term, making it a power series in the velocity. Then, we added the force between the time harmonic electric field and the charge.

To solve the equation of motion to lowest-order, we ignored the nonlinear terms. Then, in essence, we solved the lowest-order equation using perturbation theory and a trial function that was chosen based on our experience that the displacement will oscillate at the frequency of the applied electric field once transients dissipate.

We start with the general nonlinear equations of motion given by Equation 2.20. To keep it simple for now, we only include the quadratic term, so $\xi^{(2)} \neq 0$ and $\xi^{(n)} = 0$ for $n > 2$. To keep track of the magnitude of the displacement, we express it as

$$x = \lambda x^{(1)} + \lambda^2 x^{(2)} + \ldots, \qquad (2.42)$$

where the exponent of λ tracks the order of each term. We will also assume that only one electric field is applied, and since the linear displacement is proportional to the applied electric field, both are of the same order so

$$E = \lambda E_0 \cos(\omega t). \tag{2.43}$$

Substituting Equations 2.42 and 2.43 into Equation 2.20 and keeping only terms to first-order in λ, we get

$$\ddot{x}^{(1)} + 2\Gamma \dot{x}^{(1)} + \omega_0 x^{(1)} = \boxed{-\frac{eE_0}{2m}} e^{-i\omega t} + \text{c.c.}, \tag{2.44}$$

which we solve to first order using a trial function that oscillates at the driving field's frequency and then projecting out the ω Fourier component, yielding

$$x^{(1)}(t) = \boxed{-\frac{eE_0}{2m}} \frac{1}{D(\omega)} e^{-i\omega t} + \text{c.c}, \tag{2.45}$$

where Equations 2.23, 2.24 and 2.27 were used. Note that the amplitude of Equation 2.45 is the product of the amplitude of the driving force – the boxed term, and the energy denominator term evaluated at the driving frequency.

To summarize, the solution to the equation

$$\ddot{x} + 2\Gamma \dot{x} + \omega_0 x = B e^{-i\omega t} + \text{c.c.} \tag{2.46}$$

is

$$x(t) = \frac{B}{D(\omega)} e^{-i\omega t} + \text{c.c.} \tag{2.47}$$

We will call upon Equation 2.47 multiple times in latter sections.

Second-Order Susceptibility

The differential equation for $x^{(2)}$ is calculated by keeping terms of order λ^2 in Equation 2.20, yielding

$$\ddot{x}^{(2)} + 2\Gamma \dot{x}^{(2)} + \omega_0^2 x^{(2)} = -\xi^{(2)} (x^{(1)})^2. \tag{2.48}$$

Equation 2.48 is identical in form to Equation 2.46 where $\xi^{(2)}(x^{(1)})^2$ is the time-dependent driving term. Since $x^{(1)}$ varies sinusoidally with time, $\xi^{(2)}(x^{(1)})^2$ will have two pieces; one that oscillates at the second harmonic and one static field. To remind ourselves why this is so, recall that $\left(e^{-i\omega t} + e^{+i\omega t}\right)^2 = e^{-2i\omega t} + e^{+2i\omega t} + 2$.

Since the source term contains two frequencies (2ω and 0), the trial solution for $x^{(2)}(t)$ must also contain both, so it is of the form

$$x^{(2)}(t) = \left(\frac{A_2}{2}e^{-2i\omega t} + \text{c.c.}\right) + A_0 \equiv x_0^{(2)}(t) + x_{2\omega}^{(2)}(t). \quad (2.49)$$

After substituting Equation 2.49 into Equation 2.48, we get the second harmonic equation by collecting terms that are multiplied by $e^{-2i\omega t}$ while the constant terms determine the static second order susceptibility, as follows.

The source term at frequency 2ω is of the form

$$\xi^{(2)}\left(x^{(1)}(t)\right)^2\bigg|_{2\omega} = \boxed{\xi^{(2)}\left[\frac{eE_0}{2m}\frac{1}{D(\omega)}\right]^2} e^{-2i\omega t} + \text{c.c.} \quad (2.50)$$

Equation 2.48 is of the same form as Equation 2.46 with B given by the boxed expression in Equation 2.50, so the solution is given by 2.47, yielding

$$x_{2\omega}^{(2)}(t) = \boxed{\xi^{(2)}\left[\frac{eE_0}{2mD(\omega)}\right]^2} \frac{1}{D(2\omega)} e^{-2i\omega t} + \text{c.c.} \quad (2.51)$$

The static source term is given by

$$\xi^{(2)}\left(x^{(1)}(t)\right)^2\bigg|_0 = \boxed{\frac{\xi^{(2)}}{2}\left|\frac{eE_0}{m}\frac{1}{D(\omega)}\right|^2}. \quad (2.52)$$

Collecting all the terms that are constants of time in Equation 2.48, and using Equation 2.52, we get

$$x_0^{(2)} = \frac{1}{\omega_0^2}\boxed{\frac{\xi^{(2)}}{2}\left|\frac{eE_0}{m}\frac{1}{D(\omega)}\right|^2}. \quad (2.53)$$

To calculate the second-order susceptibilities, we first need the second-order polarization, which in general can be expressed as

$$P^{(2)}(t) = P_0^{(2)} + \frac{P_{2\omega}^{(2)}}{2}e^{-2i\omega t} + \frac{P_{-2\omega}^{(2)}}{2}e^{2i\omega t} \quad (2.54)$$

to reflect the fact that the second-order process with one driving frequency ω can excite a polarization at frequency 0 and 2ω. Applying the definition of the polarization $P^{(2)}(t) = -Nex^{(2)}$ to Equations 2.51 and 2.53, and comparing the results to Equation 2.54 yields

$$P_0^{(2)} = -\xi^{(2)} \cdot \frac{Ne^3}{2m^2} \cdot \frac{1}{D(0)|D(\omega)|^2} E_0^2 \quad (2.55)$$

and
$$P^{(2)}_{2\omega} = -\zeta^{(2)} \cdot \frac{Ne^3}{2m^2} \cdot \frac{1}{D(2\omega)D^2(\omega)} E_0^2, \qquad (2.56)$$

where we have used $D(0) = \omega_0^2$.

To obtain the constitutive equation between the polarization and the electric field, we first need to express E^2 in complex form, which from Equation 2.21 is given by

$$E^2 = \left(\frac{E_0}{2}e^{-i\omega t} + \text{c.c.}\right)^2 = \frac{E_0^2}{4}e^{-2i\omega t} + \frac{E_0^2}{4} + \text{c.c.}. \qquad (2.57)$$

The constitutive equation is then given by

$$\begin{aligned}P^{(2)}(t) &= \text{Re}\left[\chi^{(2)}\left(\frac{E_0^2}{2}e^{-2i\omega t} + \frac{E_0^2}{2}\right)\right] \\ &= \frac{\chi^{(2)}}{4}E_0^2 e^{-2i\omega t} + \frac{\chi^{(2)*}}{4}E_0^2 e^{+2i\omega t} + \text{Re}\left[\chi^{(2)}\right]\frac{E_0^2}{2}. \end{aligned} \qquad (2.58)$$

Setting the righthand sides of Equation 2.54 and 2.58 equal to each other, then collecting terms proportional to $e^{-2i\omega t}$ and using Equation 2.56, we get the second-order susceptibility at frequency 2ω,

$$\chi^{(2)}(2\omega) = -\frac{Ne^3}{2m^2} \cdot \zeta^{(2)} \cdot \frac{1}{D^2(\omega)D(2\omega)}. \qquad (2.59)$$

Doing the same with the zero-frequency terms, we get

$$\chi^{(2)}(0) = -\frac{Ne^3}{2m^2} \cdot \zeta^{(2)} \cdot \frac{1}{D(0)|D(\omega)|^2}. \qquad (2.60)$$

The most general second-order process is the interaction between two fields of arbitrary frequency. Even if more fields are present, the interactions can be reduced to the sum of pairwise interactions. Thus, let's express the field as

$$E = \frac{E^{\omega_1}}{2}e^{-i\omega_1 t} + \frac{E^{\omega_2}}{2}e^{-i\omega_2 t} + \text{c.c.} \qquad (2.61)$$

The first-order approximation to the nonlinear spring will respond with a displacement that contains all possible frequencies present in E, that is

$$x^{(1)} = x^{(1)}_{\omega_1} + x^{(1)}_{\omega_2} \qquad (2.62)$$

and to second order, we have the displacement

$$x^{(2)} = x^{(2)}_{2\omega_1} + x^{(2)}_{2\omega_2} + x^{(2)}_{\omega_1+\omega_2} + x^{(2)}_{\omega_1-\omega_2} + x^{(2)}_{\omega_1-\omega_1} + x^{(2)}_{\omega_2-\omega_2}, \qquad (2.63)$$

where we have kept the frequencies in the last two terms to remind us from which terms these zero-frequency components originate.

The first-order solutions are fully decoupled, so can be treated independently and give the same results as for one incident field. Similarly, the second harmonic and static terms also give the same results as before. However, the sum and difference terms include both frequencies as a product of the two fields. This is the term we describe here.

Starting with

$$x^{(1)} = \frac{A_1}{2}e^{-i\omega_1 t} + \frac{A_1^*}{2}e^{i\omega_1 t} + \frac{A_2}{2}e^{-i\omega_2 t} + \frac{A_2^*}{2}e^{i\omega_2 t} \quad (2.64)$$

and substituting Equation 2.64 into Equation 2.48, we isolate the sum and difference terms, i.e. those with $e^{-i(\omega_1 \pm \omega_2)t}$ to solve for $x^{(2)}_{\omega_1 \pm \omega_2}(t)$, as we did in arriving at Equation 2.51 from Equation 2.50. In doing so, we get

$$x^{(2)}_{\omega_1 \pm \omega_2}(t) = \zeta^{(2)} \frac{e^2}{2m^2} \frac{E^{\omega_1} E^{\pm \omega_2}}{D(\omega_1)D(\pm \omega_2)D(\omega_1 \pm \omega_2)} e^{-2i\omega t} + \text{c.c.} \quad (2.65)$$

where $E^{\omega_2 *} = E^{-\omega_2}$

The general expression for the polarization in complex form of the sum and difference frequency terms is given by

$$\begin{aligned} P^{(2)}(t) &= \text{Re}\left[\chi^{(2)}\left(E^{\omega_1} E^{\omega_2} e^{-i(\omega_1 + \omega_2)t} + E^{\omega_1} E^{-\omega_2} e^{-i(\omega_1 - \omega_2)t}\right)\right] \\ &= \frac{\chi^{(2)}}{2} E^{\omega_1} E^{\omega_2} e^{-i(\omega_1 + \omega_2)t} + \frac{\chi^{(2)}}{2} E^{\omega_1} E^{-\omega_2} e^{-i(\omega_1 - \omega_2)t} + \text{c.c.} \end{aligned} \quad (2.66)$$

Using Equation 2.65 to calculate the polarization $P = -eNx^{(2)}_{\omega_1 \pm \omega_2}$, setting it equation to Equation 2.66 and collecting terms at the sum and difference frequencies, we get

$$\chi^{(2)}(\omega_1 \pm \omega_2) = -\frac{Ne^3}{m^2} \cdot \zeta^{(2)} \cdot \frac{1}{D(\omega_1 \pm \omega_2)D(\omega_1)D(\pm \omega_2)}. \quad (2.67)$$

Third-Order Susceptibility

Finally, let's consider the third-harmonic susceptibility $\chi^{(3)}(3\omega)$. There are often multiple ways to get a nonlinear susceptibility of a given order. For example, if $\zeta^{(3)}$ is the only nonzero nonlinear coefficient, then the differential equation for the third-order displacement, $x^{(3)}$, has a third-order driving term

$$\zeta^{(3)} \left(\lambda x^{(1)} + \lambda^2 x^{(2)} + \lambda^3 x^{(3)} \ldots\right)^3 \xrightarrow[\text{order}]{3^{rd}} \lambda^3 \left[x^{(1)}\right]^3. \quad (2.68)$$

Alternatively, if $\xi^{(2)}$ is the only nonzero nonlinear coefficient, then the differential equation for the third-order displacement has a third-order driving term,

$$\xi^{(2)}\left(\lambda x^{(1)} + \lambda^2 x^{(2)} + \lambda^3 x^{(3)} \ldots\right)^2 \xrightarrow[\text{order}]{3^{rd}} 2\lambda^3 x^{(1)} x^{(2)}. \tag{2.69}$$

The lesson here is that combinations of lower-order terms can yield higher-order effects.

Lets start by considering the purely third-order term as given by Equation 2.68. We first use the fact that the source term is given by

$$\xi^{(3)} \left(x^{(1)}(t)\right)^3 \bigg|_{3\omega} = \boxed{-\xi^{(3)} \left[\frac{eE_0}{2m} \frac{1}{D(\omega)}\right]^3} e^{-3i\omega t} + \text{c.c.} \tag{2.70}$$

so the solution will be given by Equation 2.47 with B given by the boxed expression Equation 2.70. It is only a couple steps removed from getting the final result

$$\chi^{(3)}(3\omega) = \frac{Ne^4}{4m^3} \cdot \xi^{(3)} \cdot \frac{1}{D^3(\omega)D(3\omega)}. \tag{2.71}$$

Next we apply one driving frequency, $\xi^{(2)} \neq 0$ and $\xi^{(n)} = 0$ for $n > 2$. Rather than solving the problem from scratch, we will use the patterns that have emerged in the other recautions for guidance.

First, we note that the driving term of interest is the third-order contribution to $\xi^{(2)}(\lambda x_1 + \lambda^2 x_2 + \ldots)^2 \to 2\lambda^3 x_1 x_2$. We have already calculated x_1 and x_2 for processes involving one color, and the results are expressed by Equations 2.45 and 2.51, so the source term is

$$2\xi^{(2)} x_1 x_2 = \boxed{-2\xi^{(2)} \left\{\frac{eE_0}{2m} \frac{1}{D(\omega)}\right\} \cdot \left\{\xi^{(2)} \left[\frac{eE_0}{2mD(\omega)}\right]^2 \frac{1}{D(2\omega)}\right\}} e^{-3i\omega t} + \text{c.c.} \tag{2.72}$$

Equation 2.73 is of the same form as Equation 2.46 where the boxed expression is B, so the solution for $x^{(3)}_{3\omega}(t)$ is

$$x^{(3)}_{3\omega}(t) = -2\left(\xi^{(2)}\right)^2 \left[\frac{eE_0}{2m}\right]^3 \frac{1}{D^3(\omega)D(2\omega)D(3\omega)} e^{-3i\omega t} + \text{c.c.}, \tag{2.73}$$

which leads to the nonlinear susceptibility

$$\chi^{(3)}(3\omega) = \frac{Ne^4}{2m^3} \cdot \left(\xi^{(2)}\right)^2 \cdot \frac{1}{D^3(\omega)D(2\omega)D(3\omega)}. \tag{2.74}$$

Equations 2.71 and 2.74 have different dispersion, so a measurement of the third-harmonic signal as a function of wavelength would differentiate between the two. When using the dispersion to problem the mechanisms, one would first need to establish that a third-order process is at work by checking that the third-harmonic intensity is proportional to the third power of the incident intensity.

As illustrated for third harmonic generation above, $\xi^{(n)}$ can contribute to all orders of nonlinear susceptibilities larger than n. The technique introduced in this section applies to all such cases.

> **Problem 2.1-1:** Fill in the details to arrive at Equation 2.71.
>
> **Problem 2.1-2:** A nonlinear spring with $\xi^{(2)} \neq 0$ and $\xi^{(n)} = 0$ for $n > 2$ is driven with two fields of frequency ω_1 and ω_2. Determine the fourth-order susceptibility $\chi^{(4)}(3\omega_1 - \omega_2)$ in a form similar to Equation 2.67. What if the two nonzero nonlinear terms are $\xi^{(2)}$ and $\xi^{(4)}$? What if $\xi^{(3)}$ is the only nonzero term?
>
> **Problem 2.1-3a:** There is some regularity in the form of the nonlinear susceptibilities – for example there are energy denominators at the driving and response frequencies, the prefactor $\xi^{(n)}$ as well as powers of the charge and the electron mass.
>
> (a) Use this pattern to write a Python module that will give you the correct expression for a nonlinearity of given order in terms of the driving frequencies, response frequencies, and nonlinearities $\xi^{(n)}$. You can output the code in LaTeX so that the equations are of professional typeset quality.
>
> b) Write a Python module that as output gives a Python module for computing the function $\chi^{(n)}(\ldots)$ for any combination of frequencies for an arbitrary number of driving fields and nonlinear coefficients.
>
> **Problem 2.1-4:** Redo the calculations in this section by expanding the damping term in Equation 2.20 in a power series of the velocity instead of the spring force. How do the results compare with the normal treatment where the spring force is expanded in a series of the position?

2.2 Macroscopic Propagation

In the previous sections, we developed intuition about the bulk nonlinear processes and then studied the nonlinear oscillator to learn about the microscopic models at the molecular level. Next we considered the macroscopic behavior of the non-interacting gas model, where we related $\chi^{(n)}$ to α, β and γ. Here we study the bulk response independent of its molecular origin. First we consider macroscopic propagation in a crude way to gain physical insights. Then, we study it in more detail using the full mathematical formalism.

Understanding $\chi^{(2)}(\omega_1, \omega_2)$ enables us to describe a number of processes such as second harmonic generation (SHG), sum frequency generation (SFG) and optical rectification (OR). The dispersion of the second-order susceptibility of a system provides complementary information about its characteristics, which $\chi^{(1)}(\omega)$ alone would miss.

Two frequencies uniquely determine $\chi^{(2)}$ because $P^{(2)}$ is proportional to the product of two fields $E^{\omega_1} E^{\omega_2}$, $E^{\omega_1} E^{-\omega_2}$, $E^{\omega_1} E^{\omega_1}$ and so forth. If a material is simultaneously illuminated with many colors, the second order susceptibility will mix two of them, generating a field at the sum or difference frequency. All pairs of interactions will act independently to yield all possible sum and difference frequencies of all pairs of colors in the material. This is analogous to the action of $\chi^{(1)}(\omega)$, which acts on each color independently. As new frequencies are generated, they too will interact with all other colors, generating even more colors. In all cases, only pairs of fields interact. All of these processes are described simply by the dispersion of $\chi^{(2)}(\omega_1, \omega_2)$, which describes the strength of the dipole moment induced by any pair of frequencies. In general, n frequencies describe the nth order nonlinear response.

As shown in Figure 2.6, at a resonance frequency, the real part of $\chi^{(1)}$ changes sign and the imaginary part peaks. The resonance frequencies determine the energy differences between the excited and ground states while the peak height determines the strength of the transition. When peaks overlap, each contribution can be determined using a multiple peak fit.

$\chi^{(2)}$ shows more complex resonance features and yields complimentary information to $\chi^{(1)}$ because of its sensitivity to symmetries not probed by $\chi^{(1)}$. In addition, $\chi^{(2)}$ probes transitions between excited states as well as transitions from the ground state to excited states that are forbidden in a linear process. Therefore, observed resonances in $\chi^{(2)}$ may indicate states that are not detected by linear absorption spectroscopy. As a result, second harmonic microscopy is commonly applied to biological systems to highlight orientationally-ordered regions which stand out in high contrast above background due to the fact that they are observed with different colors than those incident on the tissue.

To understand nonlinear resonance, it is instructive to first consider $\chi^{(1)}$. Figure 2.6 shows the imaginary and real parts of $\chi^{(1)}$ versus frequency ω. The natural (resonant) frequency of oscillation of the system is ω_0.

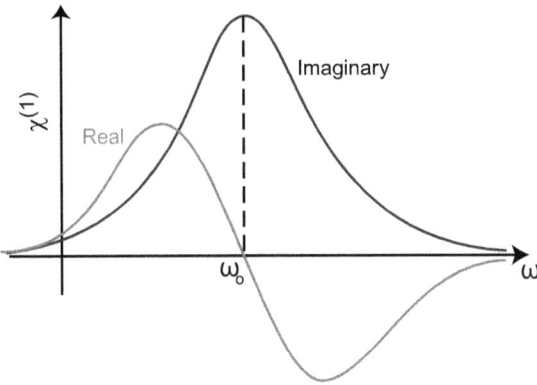

Figure 2.6: Plot of the real and imaginary parts of $\chi^{(1)}$. ω_0 is the resonant frequency of the system.

$\chi^{(2)}$ is calculated using

$$\chi^{(2)}(\omega_1,\omega_2) = \frac{1}{\mathscr{D}} \frac{\partial^2 (qNx_i)}{\partial E_j^{\omega_1} \partial E_k^{\omega_2}}\bigg|_{E=0}, \qquad (2.75)$$

where N is the number density of molecules, qx is the dipole moment, qNx is the polarization and $\mathscr{D}$ is the degeneracy factor, which should not be confused with the energy denominator $D(\omega_1,\omega_2)$. $\mathscr{D}$ depends on the components of the Fourier amplitudes of the electric fields. The dependence of $\chi^{(2)}$ on the field frequencies is complex because the energy denominators in $\chi^{(2)}$ include various combinations of frequencies.

The imaginary part of the linear susceptibility is a peaked function of wavelength while the real part is dispersive, as shown in Figure 2.6. Given that the second order susceptibility has products of complex denominators, its dependence on frequency is never simply a peaked or dispersive function. As such, a surface plot of $\chi^{(2)}(\omega_1,\omega_2)$ would be overly complex to view the resonant structure.

Figure 2.7 plots the positions of the resonances. The energy denominators are of the form $D(\omega_1)$, $D(\omega_2)$, $D(2\omega_1)$, $D(2\omega_2)$, and $D(|\omega_1 \pm \omega_2|)$ so the system has resonances when $\omega_1 = \omega_0$, $\omega_2 = \omega_0$, $\omega_1 = \omega_0/2$, $\omega_2 = \omega_0/2$, and $|\omega_1 \pm \omega_2| = \omega_0$. These resonances are shown in Figure 2.7, where the various lines shows the position of the resonances.

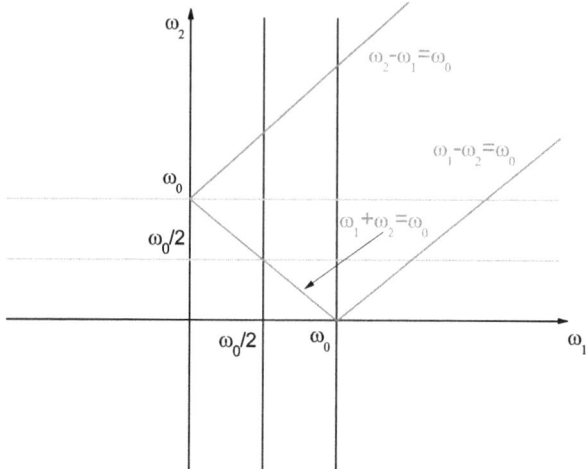

Figure 2.7: Resonances of $\chi^{(2)}$ in the ω_1–ω_2 plane

> **Problem 2.2-1:** Make a surface and contour plot of the real and imaginary parts of $\chi^{(2)}$ as a function of ω_1 and ω_2 to see that the resonances occur where we predict them to be in Figure 2.7.

The refractive index is a bulk material property that originates in contributions of molecules that radiate electromagnetic fields are a dipole pattern, but when taken as a superposition, yields a plane wave. Figure 2.8 shows the material (inside the box), as an array of radiating dipoles in response to the field incident from the left and shown as solid lines. The solid lines represent the planes of peak electric field with polarization upwards. The planes halfway between each pair of adjacent solid lines are not shown to reduce clutter, but correspond to the peak electric field polarized downward. We refer to this field as *external* by virtue of the fact that it is applied to the material from the outside. As drawn, this is the field that would be observed if the material were absent.

The external electric field, E^ω, induces an oscillating dipole in each molecule, which generates a dipole radiation field at frequency ω. The induced dipole moment is not, in general, in phase with the electric field, as shown by the shift in the planes of maximum field from the largest induced dipole moments, shown as arrows. The expanded view at the top portion of Figure 2.8 shows the incident field envelope and a larger sampling of the induced dipoles along a slice through the material as indicated by the line.

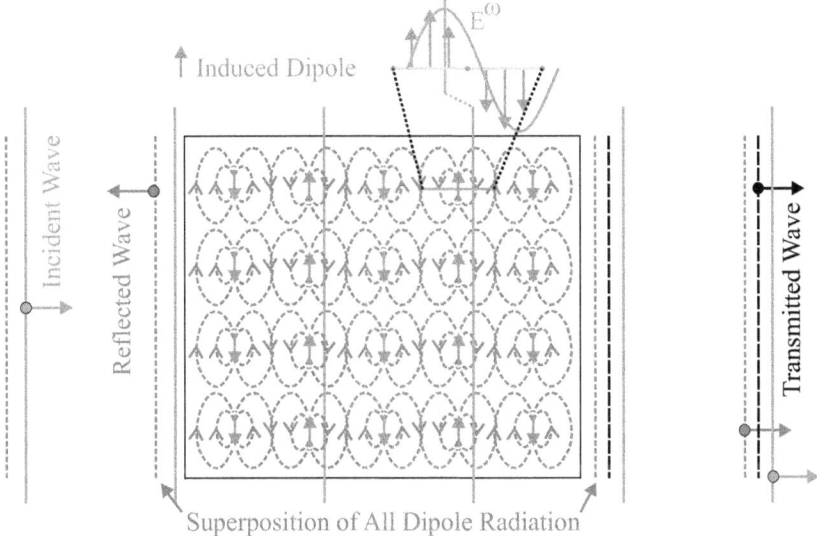

Figure 2.8: Illustration of the microscopic origin of the refractive index. The six solid vertical lines represent phase fronts of the external electric field in and outside the material and the shorter dashed lines represent the electric field generated by the induced dipoles fields (dotted curves) inside the sample. The superposition of the induced electric dipole fields and plane waves inside the sample leads to a plane wave with a decreased phase velocity, which defines the refractive index.

The superposition of the fields from all the radiating dipoles produces plane waves traveling to the left and right, as shown by the short-dash lines.

The total field is the superposition of the incident and radiated fields. The superposition of all fields due to the dipoles is the only field that travels to left and is therefore the reflected wave. The field due to the radiating dipoles that travels to the right adds to the incident field to yield the transmitted wave, shown by the long-dash vertical lines to the right.

The superposition of the incident and dipole fields inside the material leads to a phase shift relative to the incident wave and leads to the lower phase velocity in the sample. The refractive index quantifies the reduced phase velocity. As a result of the wave propagating more slowly, the peaks are closer together (not shown to reduce clutter). Outside, the net effect is a phase shift of the transmitted wave.

Reflection (or refraction) is often considered a surface effect, as one often hears the term 'surface reflection.' In the picture presented here, it is the full volume of material that creates the reflected wave. One may argue that this contradicts the observation that short pulses, which are spatially well localized, are known to reflect from a surface with a phase shift, but otherwise, with no time delay as one might expect if the pulse must interact with the whole material before reflecting. This seeming paradox can be resolved by noting that any pulse can be expressed as a superposition of planes waves of differing frequencies, which extend over all space. As such, before the peak of the pulse appears to enter the material, it is already interacting with all the underlying plane waves.

The student should keep this picture of radiating dipoles in mind when struggling to understand nonlinear wave propagation. At each point in the material, nonlinear interactions with molecules generate fields at other frequencies, which when added to the prorogating field, leads to the observed fields. Since the polarization is the induced dipole moment per unit volume, it is the macroscopic equivalent of radiating dipoles that contribute to the propagating waves.

Consider a nonlinear material with a second order susceptibility $\chi^{(2)} \neq 0$. A beam of light at frequency ω will induce the dipoles within to oscillate at frequency 2ω and will thus produce a beam at the second harmonic. A static field will also result. An incident field of the form $E^{\omega_1} + E^{\omega_2}$ will induce oscillations at 0, $2\omega_1$, $2\omega_2$, $\omega_1 + \omega_2$ and $|\omega_1 - \omega_2|$ and lead to beams at these frequencies. Solutions to the nonlinear Maxwell Equations give the amplitudes of the generated fields within the material and the beams that exit.

In Gaussian units, the electric displacement, **D**, is related to the electric field, **E**, and the polarization, **P**, according to

$$\mathbf{D} = \mathbf{E} + 4\pi\mathbf{P} = \varepsilon\mathbf{E}, \tag{2.76}$$

where ε is a rank two tensor. Note that in this section, we treat the linear and nonlinear susceptibilities as scalars since their tensor properties are not relevant to the discussion. When the nonlinear response vanishes,

$$\mathbf{P} = \chi^{(1)}\mathbf{E}, \tag{2.77}$$

then

$$\mathbf{D} = \mathbf{E} + 4\pi\chi^{(1)}\mathbf{E} = \left(1 + 4\pi\chi^{(1)}\right)\mathbf{E}, \tag{2.78}$$

whence the linear dielectric function, ϵ_0 (not to be confused with the permittivity of free space in SI units), becomes

$$\epsilon_0 = 1 + 4\pi\chi^{(1)}, \tag{2.79}$$

or in tensor form,

$$\epsilon_{ij} = \delta_{ij} + 4\pi\chi^{(1)}_{ij}. \tag{2.80}$$

Taking the curl of Faraday's law,

$$\nabla \times \mathbf{E} = -\frac{1}{c}\frac{\partial \mathbf{B}}{\partial t}, \tag{2.81}$$

leads to

$$\nabla \times \nabla \times \mathbf{E} = -\frac{1}{c}\frac{\partial}{\partial t}\nabla \times \mathbf{B}. \tag{2.82}$$

When free currents are absent, Ampere's law is

$$\nabla \times \mathbf{B} = \frac{1}{c}\frac{\partial \mathbf{D}}{\partial t}, \tag{2.83}$$

where **H** is equivalent to **B** when the material is nonmagnetic, i.e. when the magnetic susceptibility is given by $\mu = 1$. Inserting Equation 2.83 into Equation 2.82 and using Equation 2.78 yields

$$\nabla \times \nabla \times \mathbf{E} + \frac{\epsilon_0}{c^2}\frac{\partial^2 \mathbf{E}}{\partial t^2} = -\frac{4\pi}{c^2}\frac{\partial^2 \mathbf{P}_{NL}}{\partial t^2}, \tag{2.84}$$

where the linear polarization, $4\pi\chi^{(1)}\mathbf{E}$, is absorbed into the second term of Equation 2.84 with the help of Equation 2.79 for ϵ_0. The nonlinear polarization, $\mathbf{P}_{NL}$, is given by

$$\mathbf{P}_{NL}(\mathbf{E}) = \chi^{(2)}\mathbf{E}^2 + \chi^{(3)}\mathbf{E}^3 + \cdots = \mathbf{P}^{(2)} + \mathbf{P}^{(3)} + \cdots. \tag{2.85}$$

Equations 2.84 and 2.85 together constitute the nonlinear wave equation, which we consider through third order in the electric field. Chapter 3 deals with the rigourous solution to the nonlinear wave equation using the slowly varying amplitude approximation. In this section, we use semi-quantitative arguments to show how various beams are formed and how we can interpret certain nonlinear processes as an electric-field-dependent refractive index.

In the following treatment, we assume that the material is local in space and instantaneous in time so that the polarization at one point or time in the material is not affected by the field at any other point or time. These restrictions will be relaxed in Section 2.3, but for the discussion at hand, this feature is unimportant and will be neglected.

For incident fields of frequencies ω_1 and ω_2, the total electric field, in scalar form for simplicity, is given by

$$E = \frac{E^{\omega_1}}{2}e^{-i\omega_1 t} + \frac{E^{\omega_2}}{2}e^{-i\omega_2 t} + \text{c.c.}, \tag{2.86}$$

and the polarization, P, is given by

$$\begin{aligned} P &= \left[\frac{P^{2\omega_1}}{2}e^{-2i\omega_1 t} + \frac{P^{2\omega_2}}{2}e^{-2i\omega_2 t} + \frac{P^{\omega_1+\omega_2}}{2}e^{-i(\omega_1+\omega_2)t} + \cdots + \text{c.c.}\right] \\ &+ P_0 E^{\omega_1} E^{\omega_2} + \ldots \end{aligned} \tag{2.87}$$

With

$$P_{NL} = \chi^{(2)} E^2, \tag{2.88}$$

substituted into Equation 2.84, we pick the Fourier component of interest in the nonlinear wave equation. To this end, we first determine the Fourier components of E^2 in Equation 2.88 using Equation 2.86,

$$\begin{aligned} E^2 &= \frac{(E^{\omega_1})^2}{4}e^{-2i\omega_1 t} + \frac{(E^{\omega_2})^2}{4}e^{-2i\omega_2 t} + \frac{E^{\omega_1}E^{-\omega_1}}{2} + \\ &\quad \frac{E^{\omega_1}E^{-\omega_2}}{4}e^{-i(\omega_1-\omega_2)t} + \cdots + \text{c.c.} \end{aligned} \tag{2.89}$$

The $2\omega_1$ Fourier component is given by

$$\int_{-\infty}^{\infty} e^{-2i\omega_1 t}\left(P = \chi^{(2)} E^2\right) dt, \tag{2.90}$$

which yields

$$P^{2\omega_1} = \frac{1}{2}\chi^{(2)}\left(E^{\omega_1}\right)^2. \tag{2.91}$$

The coefficient 1/2 in Equation 2.91 comes from squaring the field given by Equation 2.86. The coefficient for other processes to n^{th} order is calculated in the same way by taking the input fields to the nth power and picking out the desired Fourier component.

A zero frequency polarization, P^0, arises from the cross terms,

$$P^0 = \frac{1}{2}\chi^{(2)} E^{-\omega_1} E^{\omega_1}, \qquad (2.92)$$

and the sum frequency field at $\omega_1 + \omega_2$ arises from cross terms when the sum of the two incident fields are squared,

$$P^{\omega_1+\omega_2} = \chi^{(2)} E^{\omega_1} E^{\omega_2}. \qquad (2.93)$$

If the two incident fields are E^ω and E^0, the total incident field is given by

$$E = \left(\frac{E^\omega}{2} e^{-i\omega t} + \text{c.c.}\right) + E^0. \qquad (2.94)$$

Squaring Equation 2.94,

$$E^2 = E^\omega E^0 e^{-i\omega t} + \frac{(E^\omega)^2}{4} e^{-2i\omega t} + \cdots, \qquad (2.95)$$

whence

$$P^{(2)} = 2\chi^{(2)} E^\omega E^0. \qquad (2.96)$$

To understand how this leads to a voltage-dependent refractive index called the electro-optic effect, consider a beam with field E^ω that travels through a material between capacitor plates, which apply a static electric field E^0. The electric displacement is given by

$$\begin{aligned} \mathbf{D}^\omega &= \mathbf{E}^\omega + 4\pi\chi^{(1)}\mathbf{E}^\omega + 8\pi\chi^{(2)} E^0 \mathbf{E}^\omega \\ &= \left(1 + 4\pi\chi^{(1)} + 8\pi\chi^{(2)} E^0\right) \mathbf{E}^\omega \\ &\equiv \epsilon \mathbf{E}^\omega, \end{aligned} \qquad (2.97)$$

which defines the effective dielectric function ϵ. With the linear dielectric function given by $\epsilon_0 = 1 + 4\pi\chi^{(1)}$,

$$\epsilon = \epsilon_0 + 8\pi\chi^{(2)} E^0. \qquad (2.98)$$

As such, the applied voltage has the effect of inducing a phase shift in the optical beam.

The refractive index n is calculated from ϵ using

$$\begin{aligned} n &= \sqrt{\epsilon} \\ &= \epsilon_0^{1/2}\sqrt{1 + \frac{8\pi\chi^{(2)}E^0}{\epsilon_0}}, \end{aligned} \qquad (2.99)$$

which for small applied static field yields

$$n = n_0\left(1 + \frac{4\pi\chi^{(2)}}{n_0^2}E^0\right). \qquad (2.100)$$

Equation 2.100 is expressed in standard form as

$$n = n_0 + \frac{4\pi\chi^{(2)}}{n_0}E^0 \equiv n_1 E_0, \qquad (2.101)$$

where $n_0 = \epsilon_0^2$ and n_1 is called the electro-optic coefficient and given by

$$n_1 = \frac{4\pi\chi^{(2)}}{n_0}. \qquad (2.102)$$

The general features of $\chi^{(3)}$ can be understood by considering three incident fields at frequencies ω_1, ω_2 and ω_3. For the sake of simplicity, we start with one field of frequency ω,

$$E = \frac{E^\omega}{2}e^{-i\omega t} + \text{c.c.} \qquad (2.103)$$

From Equation 2.103

$$E^3 = \frac{(E^\omega)^3}{8}e^{-3i\omega t} + \frac{3}{8}\left(E^\omega\right)^2 E^{-\omega}e^{-i\omega t} + \cdots + \text{c.c.}, \qquad (2.104)$$

whence

$$P^{3\omega} = \frac{1}{4}\left(E^\omega\right)^3 \chi^{(3)} \qquad (2.105)$$

and

$$P^\omega = \frac{3}{4}\chi^{(3)} E^\omega E^{-\omega} E^\omega, \qquad (2.106)$$

where $E^\omega E^{-\omega} = |E^\omega|^2$ (Recall that $E^{\omega*} = E^{-\omega}$), and is thus proportional to the intensity.

The polarization $P^{3\omega}$ leads to third harmonic generation. The third-order polarization at frequency ω when combined with the linear polarization is the given by

$$P^\omega = \chi^{(1)} E^\omega + \frac{3}{4}\chi^{(3)} |E^\omega|^2 E^\omega, \qquad (2.107)$$

and leads to the intensity-dependent refractive index as described below.

Using the same approach as leading to Equation 2.97, ϵ is given by

$$\epsilon = 1 + 4\pi\chi^{(1)} + \frac{3}{4}(4\pi)\chi^{(3)}|E^\omega|^2, \tag{2.108}$$

where we have neglected $\chi^{(2)}$. Equation 2.108 is exact for a centrosymmetric material such as a liquid. Section 2.3.9 describes why $\chi^{(2)}$ vanishes in a centrosymemtric medium. In analogy to Equation 2.99, the refractive index for the third-order response is

$$n = n_0 + \frac{3\pi\chi^{(3)}}{2n_0}|E^\omega|^2. \tag{2.109}$$

The Optical Kerr Effect coefficient, n_2, is given by

$$n_2 = \frac{3\pi\chi^{(3)}}{2n_0}. \tag{2.110}$$

Since the intensity, I, is easy to measure, the experimentalist finds it most convenient to express the intensity-dependent refractive index in the form

$$n = n_0 + n_2'I. \tag{2.111}$$

Since the intensity is related to the electric field according to,

$$I = \frac{n_0 c}{8\pi}|E|^2, \tag{2.112}$$

then

$$n_2' = \frac{12\pi^2\chi^{(3)}}{n_0^2 c}. \tag{2.113}$$

n_1 dominates in a non-centrosymmetric material such as in a molecular crystal or a poled dye-doped polymer while in a centrosymmetric medium, n_2 is the lowest-order non-vanishing term.

Problem 2.2-2: Intensity dependent absorbance is defined as $\alpha = \alpha_0 + \alpha_2 I$ where α is the absorbance, i.e.

$$\frac{I}{I_0} = \exp[-\alpha x],$$

where I is the intensity of light transmitted through the nonlinear material of thickness x and I_0 the incident intensity. This effect is related to two-photon absorption and occurs when the susceptibilities are complex. Relate α_2 to the real and imaginary parts of the refractive index and third-order susceptibility. (Hint: $k = n\frac{\omega}{c}$)

2.3 Response Functions

The nonlinear-optical properties of a material originate in the redistribution of charges in response to light. As we have seen in previous sections, the charge distribution can be described by the polarization, **P**. The problem at hand is to relate the polarization to the applied field, **E**. This relationship is called the constitutive relation and is formally described by the response function. The goal of this section is to determine the most general constitutive relation that is applicable to the broadest range of phenomena.

To first order in the electric field, the most general form of the polarization is given by

$$P_i^{(1)}(t) = \int_{-\infty}^{\infty} T_{ij}^{(1)}(t;\tau) E_j(\tau) d\tau, \quad (2.114)$$

where we have assumed that the response is spatially local, and the response function is T_{ij}. This form for the polarization takes into account how fields at different points in time, τ, will contribute to the material's response at time t. To build an intuition for response functions, we begin with the first order case, and then generalize the approach to higher orders.

2.3.1 Time Invariance

One of the most important axioms describing nature is that the laws of physics are invariant under time translation. So if we let $t \to t + t_0$ in $P_i^{(1)}$ and in E_j, the laws of physics should remain unchanged, thus the response function should not change. This symmetry is reflected in the fact that the constitutive equation remains unchanged if the polarization and electric field are advanced together by $t \to t + t_0$ but keeping the response function fixed in Equation 2.114, yielding

$$P_i^{(1)}(t + t_0) = \int_{-\infty}^{+\infty} T_{ij}^{(1)}(t;\tau) E_j(\tau + t_0) d\tau. \quad (2.115)$$

Next, we do just math, advancing the time variable t in Equation 2.114 according to $t \to t + t_0$, yielding:

$$P_i^{(1)}(t + t_0) = \int_{-\infty}^{+\infty} T_{ij}^{(1)}(t + t_0;\tau) E_j(\tau) d\tau. \quad (2.116)$$

Making the substitution $\tau \to \tau'$ in Equation 2.116 and $\tau = \tau' - t_0$ in Equation 2.115, then setting Equations 2.116 and 2.115 equal to each other yields

$$\int_{-\infty}^{+\infty} T_{ij}^{(1)}(t + t_0;\tau') E_j(\tau') d\tau' = \int_{-\infty}^{+\infty} T_{ij}^{(1)}(t;\tau' - t_0) E_j(\tau') d\tau'. \quad (2.117)$$

If Equation 2.117 is to hold for any arbitrary function $E_j(\tau')$, then it follows that

$$T_{ij}^{(1)}(t+t_0;\tau') = T_{ij}^{(1)}(t;\tau'-t_0). \tag{2.118}$$

With $t = 0$ and $t_0 \to t'$ we get

$$T_{ij}^{(1)}(t';\tau') = T_{ij}^{(1)}(0;\tau'-t'). \tag{2.119}$$

Equation 2.119 shows that the response depends only on one variable, $\tau' - t'$. We can thus re-express $T(t,\tau)$ (a function of two variables) in terms of $R(\tau - t)$ (a function of one variable). Equation 2.114 can then be re-written as

$$P_i^{(1)}(t) = \int_{-\infty}^{+\infty} R_{ij}^{(1)}(t-\tau)E_j(\tau)d\tau. \tag{2.120}$$

If we enforce causality by setting $R_{ij}^{(1)}(t-\tau) = 0$ for $\tau > t$ (i.e. ensuring that the induced polarization does not precede the application of a field) and defining $\tau' = t - \tau$, we get

$$P_i^{(1)}(t) = \int_0^{\infty} R_{ij}^{(1)}(\tau')E_j(t-\tau')d\tau'. \tag{2.121}$$

Equation 2.121 shows that the polarization of a material at time t is induced by all fields acting in the past. The response function is a weighting function that describes the strength of the electric field's past influence as a function of time prior to the appearance of the polarization.

Note that we can express the bottom limit of integration in Equation 2.121 as $\tau' = 0$, as we have done here, or as $\tau' = -\infty$ with the understanding that $R_{ij}^{(1)}(t) = 0$ for $t < 0$.

2.3.2 Fourier Transforms of Response Functions and Electric Susceptibilities

In the previous section we derived the polarization as a function of time due to an electric field using a response function that characterizes the material. Often it is more convenient to consider the polarization in frequency space. Our goal here is to Fourier transform the electric field, polarization and response function to re-express Equation 2.121 in the frequency domain.

The Fourier transform of the electric field is given by

$$\tilde{E}_j(\omega) = \frac{1}{2\pi}\int_{-\infty}^{+\infty} dt\, E_j(t)e^{-i\omega t}, \tag{2.122}$$

and the reverse transform by

$$E_j(t) = \int_{-\infty}^{+\infty} d\omega \tilde{E}_j(\omega) e^{+i\omega t}. \tag{2.123}$$

Substituting Equation 2.123 into Equation 2.121 yields

$$P_i^{(1)}(t) = \int_{-\infty}^{+\infty} d\tau' R_{ij}^{(1)}(\tau') \int_{-\infty}^{+\infty} d\omega \tilde{E}_j(\omega) e^{i\omega(t-\tau')}, \tag{2.124}$$

which we can regroup into the form,

$$P_i^{(1)}(t) = \int_{-\infty}^{+\infty} \left(\int_{-\infty}^{+\infty} R_{ij}^{(1)}(\tau') e^{-i\omega\tau'} d\tau' \right) \tilde{E}_j(\omega) e^{i\omega t} d\omega. \tag{2.125}$$

The term in parentheses in Equation 2.125 we call the first order electric susceptibility, which is given by

$$\chi_{ij}^{(1)}(\omega) = \int_{-\infty}^{+\infty} R_{ij}^{(1)}(\tau') e^{-i\omega\tau'} d\tau'. \tag{2.126}$$

Aside from the missing factor of $1/2\pi$, the susceptibility is the Fourier transform of the response function. Substituting Equation 2.126 into Equation 2.125, we find

$$P_i^{(1)}(t) = \int_{-\infty}^{+\infty} \chi_{ij}^{(1)}(\omega) \tilde{E}_j(\omega) e^{i\omega t} d\omega. \tag{2.127}$$

Next, we transform the polarization into frequency space according to,

$$\tilde{P}_i^{(1)}(\omega_\sigma) = \frac{1}{2\pi} \int_{-\infty}^{+\infty} P_i^{(1)}(t) e^{-i\omega_\sigma t} dt. \tag{2.128}$$

Substituting Equation 2.127 into Equation 2.128 yields

$$\tilde{P}_i^{(1)}(\omega_\sigma) = \frac{1}{2\pi} \int_{-\infty}^{+\infty} dt \int_{-\infty}^{+\infty} d\omega \chi_{ij}^{(1)}(\omega) \tilde{E}_j(\omega) e^{i\omega t} e^{-i\omega_\sigma t}. \tag{2.129}$$

Recalling that the Dirac delta function can be expressed as

$$\delta(\omega - \omega_\sigma) = \frac{1}{2\pi} \int_{-\infty}^{+\infty} e^{it(\omega-\omega_\sigma)} dt, \tag{2.130}$$

we can rewrite Equation 2.129 as

$$\tilde{P}_i^{(1)}(\omega_\sigma) = \int_{-\infty}^{+\infty} \chi_{ij}^{(1)}(\omega) \tilde{E}_j(\omega) \delta(\omega - \omega_\sigma) d\omega. \tag{2.131}$$

Performing the integration yields

$$\tilde{P}_i^{(1)}(\omega_\sigma) = \chi_{ij}^{(1)}(\omega_\sigma) \tilde{E}_j(\omega_\sigma). \tag{2.132}$$

In the frequency representation, there are no integrals. The polarization at frequency ω_σ is simply the product of the electric field at frequency ω_σ and the electric susceptibility, at frequency ω_σ. The utility of such a simple expression is obvious.

A Matter of Convention

There are two additional points that need to be addressed. First is a matter of convention that will become more apparent when we consider the nonlinear polarization. Energy conservation is built into Equation 2.132 by virtue of the Dirac delta function in Equation 2.131. In the linear case, this simply means that the polarization oscillates at the same frequency as the driving field. We express this fact by re-writing Equation 2.132 as

$$\tilde{P}_i^{(1)}(\omega_\sigma) = \chi_{ij}^{(1)}(-\omega_\sigma;\omega_\sigma)\tilde{E}_j(\omega_\sigma), \tag{2.133}$$

where the first argument of $\chi_{ij}^{(1)}(-\omega_\sigma;\omega_\sigma)$ is the frequency of the polarization and the argument to the right of the semicolon is the field frequency. The minus sign signifies that the oscillating polarization generates a radiating field. Energy conservation demands that the sums of all of the arguments vanish. In the case of nonlinear susceptibilities, multiple colors of light are incident on the material, generating multiple output colors.

Monochromatic Fields

The second point pertains to the meaning of the Fourier transformed functions. They are often erroneously thought of as the amplitude of the field at a given frequency, but this interpretation is not quite right. Note that $\tilde{E}_j$ cannot be a field amplitude because the units are not right. The Fourier transform of an electric field, according to Equation 2.122 has units of field multiplied by time.

The issue can be reconciled by considering a monochromatic electric field of the form

$$E_j(t) = \frac{1}{2}\left(E_j^{\omega_1}e^{-i\omega_1 t} + E_j^{-\omega_1}e^{+i\omega_1 t}\right), \tag{2.134}$$

where E_j^ω is the complex field amplitude and its complex conjugate is given by $E_j^{\omega*} = E_j^{-\omega}$. Taking the Fourier transform of Equation 2.134 using Equation 2.122 with the help of Equation 2.130 yields

$$\tilde{E}_j(\omega) = \frac{1}{2}\left(E_j^{\omega_1}\delta(\omega-\omega_1) + E_j^{-\omega_1}\delta(\omega+\omega_1)\right). \tag{2.135}$$

$\tilde{E}_j(\omega)$ in Equation 2.135 is thus a function of ω that peaks at $\pm\omega_1$ and has units of electric field multiplied by time.

For the linear case, the polarization frequency is the same as the electric field frequency, and there are no other frequency components, so the polarization is also of the form

$$\tilde{P}_i^{(1)}(\omega) = \frac{1}{2}\left(P_i^{\omega_1}\delta(\omega-\omega_1) + P_i^{-\omega_1}\delta(\omega+\omega_1)\right). \tag{2.136}$$

Substituting Equations 2.135 and 2.136 into Equation 2.133 and integrating around $\omega_\sigma = \omega_1$ yields

$$P_i^{\omega_1} = \chi_{ij}^{(1)}(-\omega_1;\omega_1)E_j^{\omega_1}. \tag{2.137}$$

Equation 2.133 has the same mathematical form as Equation 2.137, but the former is expressed in terms of the Fourier transform functions and the latter in terms of electric field and polarization amplitude of time-harmonic functions.

2.3.3 Second-Order Polarization and Susceptibility

For the second-order case, we begin with the fundamental relationship between the polarization $P^{(2)}(t)$ and response function $R_{ijk}^{(2)}(\tau_1,\tau_2)$

$$P_i^{(2)}(t) = \int_{-\infty}^{+\infty} \int_{-\infty}^{+\infty} R_{ijk}^{(2)}(\tau_1,\tau_2) E_j(t-\tau_1) E_k(t-\tau_2) d\tau_1 d\tau_2, \tag{2.138}$$

where $R_{ijk}^{(2)}(\tau_1,\tau_2) = 0$ when $\tau_1 < 0$ or $\tau_2 < 0$. Substituting Equation 2.123 into Equation 2.138 and rearranging terms yields

$$\begin{aligned}P_i^{(2)}(t) &= \int_{-\infty}^{+\infty} d\omega_1 \int_{-\infty}^{+\infty} d\omega_2 \tilde{E}_j(\omega_1)\tilde{E}_k(\omega_2) e^{i\omega_1 t} e^{i\omega_2 t} \\ &\times \left[\int_{-\infty}^{+\infty} d\tau_1 \int_{-\infty}^{+\infty} d\tau_2 R_{ijk}^{(2)}(\tau_1,\tau_2) e^{-i\omega_1 \tau_1} e^{-i\omega_2 \tau_2}\right]\end{aligned} \tag{2.139}$$

Defining the second order susceptibility as

$$\chi_{ijk}^{(2)}(\omega_1,\omega_2) = \int_{-\infty}^{+\infty} d\tau_1 \int_{-\infty}^{+\infty} d\tau_2 R_{ijk}^{(2)}(\tau_1,\tau_2) e^{-i\omega_1 \tau_1} e^{-i\omega_2 \tau_2}, \tag{2.140}$$

the second-order polarization becomes

$$P_i^{(2)}(t) = \int_{-\infty}^{+\infty} d\omega_1 \int_{-\infty}^{+\infty} d\omega_2 \chi_{ijk}^{(2)}(\omega_1,\omega_2) \tilde{E}_j(\omega_1) \tilde{E}_k(\omega_2) e^{i\omega_1 t} e^{i\omega_2 t}. \tag{2.141}$$

Using the inverse Fourier transform of $P_i^{(2)}(t)$,

$$\tilde{P}_i^{(2)}(\omega) = \frac{1}{2\pi} \int_{-\infty}^{\infty} P_i^{(2)}(t) e^{-i\omega t} dt, \tag{2.142}$$

and substituting Equation 2.141 into Equation 2.142 yields

$$\tilde{P}_i^{(2)}(\omega) = \int_{-\infty}^{\infty} \int_{-\infty}^{\infty} d\omega_1 d\omega_2 \chi_{ijk}^{(2)}(\omega_1,\omega_2) \tilde{E}_j(\omega_1) \tilde{E}_k(\omega_2) \delta(\omega_1+\omega_2-\omega). \tag{2.143}$$

Nonlinear Optics: A student's perspective

Upon integration, we get

$$\tilde{P}_i^{(2)}(\omega) = \int_{-\infty}^{\infty} d\omega_1 \chi_{ijk}^{(2)}(-\omega; \omega_1, \omega - \omega_1) \tilde{E}_j(\omega_1) \tilde{E}_k(\omega - \omega_1). \quad (2.144)$$

Note the Dirac delta function in Equation 2.143. As we saw in the linear case, by convention, the sum of the arguments of the second-order susceptibility in Equation 2.144 must vanish, which imposes the energy conservation constraint set by the delta function. The first argument of $\chi^{(2)}$ represents the polarization frequency and the second two represent the two incoming waves.

Monochromatic Fields

Textbooks typically stop at equation 2.144, and with some hand-waving, apply it to processes that mix a finite number of monochromatic waves. This section fills in the details by describing how monochromatic fields are formally treated, clearing up certain confusion that arises when hands wave. Readers who are interested in the final result are referred to Equation 2.153.

Monochromatic fields are most conveniently treated by starting with Equation 2.141. For illustration, let's assume that the light beam is composed of two monochromatic waves of frequency ω' and ω'', so the electric field is expressed as[1]

$$E_j(t) = \frac{1}{2}\left(E_j^{\omega'} e^{-i\omega' t} + E_j^{-\omega'} e^{+i\omega' t} + E_j^{\omega''} e^{-i\omega'' t} + E_j^{-\omega''} e^{+i\omega'' t}\right), \quad (2.145)$$

yielding the Fourier transform

$$\tilde{E}_j(\omega) = \frac{1}{2}\left(E_j^{\omega'} \delta(\omega - \omega') + E_j^{-\omega'} \delta(\omega + \omega') + E_j^{\omega''} \delta(\omega - \omega'') + E_j^{-\omega''} \delta(\omega + \omega'')\right). \quad (2.146)$$

Substituting Equation 2.146 into Equation 2.141 requires first that we calculate $\tilde{E}_j(\omega_1)\tilde{E}_k(\omega_2)$. Since each field is the sum of four terms, this product has 16 terms – an expression too complex to fit reasonably in these pages. However, evaluating a single term will give us a taste of how the rest of the terms are to be treated.

Using Equation 2.146 the product $\tilde{E}_j(\omega_1)\tilde{E}_k(\omega_2)$ is given by

$$\begin{aligned}\tilde{E}_j(\omega_1)\tilde{E}_k(\omega_2) &= \frac{1}{4}\Big(E_j^{\omega'}\delta(\omega_1 - \omega')E_k^{\omega''}\delta(\omega_2 - \omega'') + E_j^{\omega''}\delta(\omega_1 - \omega'')E_k^{\omega'}\delta(\omega_2 - \omega') \\ &+ E_j^{\omega'}\delta(\omega_1 - \omega')E_k^{-\omega''}\delta(\omega_2 + \omega'') + E_j^{-\omega''}\delta(\omega_1 + \omega'')E_k^{\omega'}\delta(\omega_2 - \omega')\Big) \\ &+ \ldots \end{aligned} \quad (2.147)$$

---

[1] Contrary to one's first impression, two incident fields is the most general case for a second-order process. If more colors are present, only pairs of photons interact through $\chi^{(2)}$.

For fun, let's consider only the two terms in the second row, and substitute it into Equation 2.141, which yields

$$\begin{aligned}P_i^{(2)}(t) &= \frac{1}{4}\int_{-\infty}^{+\infty}d\omega_1\int_{-\infty}^{+\infty}d\omega_2\,\chi_{ijk}^{(2)}(\omega_1,\omega_2)\\ &\times \left(E_j^{\omega'}\delta(\omega_1-\omega')E_k^{-\omega''}\delta(\omega_2+\omega'')+E_j^{\omega-\omega''}\delta(\omega_1+\omega'')E_k^{\omega'}\delta(\omega_2-\omega')\right)\\ &\times e^{+i\omega_1 t}e^{+i\omega_2 t}.\end{aligned} \quad (2.148)$$

It is straightforward to evaluate the integrals in Equation 2.148, yielding

$$P_i^{(2)}(t) = \frac{1}{4}\left(\chi_{ijk}^{(2)}(\omega',-\omega'')E_j^{\omega'}E_k^{-\omega''} + \chi_{ijk}^{(2)}(-\omega'',\omega')E_j^{-\omega''}E_k^{\omega'}\right)e^{i(\omega'-\omega'')t}. \quad (2.149)$$

Taking the Fourier transform of Equation 2.149 yields

$$\begin{aligned}\tilde{P}_i^{(2)}(\omega) &= \frac{1}{2\pi}\int_{-\infty}^{+\infty}P_i^{(2)}(t)e^{-i\omega t}dt = \frac{1}{4}\left(\chi_{ijk}^{(2)}(\omega',-\omega'')E_j^{\omega'}E_k^{-\omega''}\right.\\ &\quad \left.+\chi_{ijk}^{(2)}(-\omega'',\omega')E_j^{-\omega''}E_k^{\omega'}\right)\times\frac{1}{2\pi}\int_{-\infty}^{+\infty}dt\,e^{-i\omega t}e^{i(\omega'-\omega''-\omega)t}\\ &= \frac{1}{4}\left(\chi_{ijk}^{(2)}(\omega',-\omega'')E_j^{\omega'}E_k^{-\omega''}+\chi_{ijk}^{(2)}(-\omega'',\omega')E_j^{-\omega''}E_k^{\omega'}\right)\\ &\quad \times \delta(\omega'-\omega''-\omega).\end{aligned} \quad (2.150)$$

For monochromatic input beams, the polarization will be a sum of terms of discrete frequencies given by all the input frequencies and all combinations of sums and differences of all possible pairs of incident frequencies, which we express as

$$\tilde{P}_i^{(2)}(\omega) = \frac{1}{2}\sum_{\omega_\mu}P_i^{\omega_\mu}\delta(\omega-\omega_\mu), \quad (2.151)$$

where the sum includes a negative frequency for each positive one. Setting Equations 2.150 and 2.151 equal to each other yields

$$\begin{aligned}\frac{1}{2}\sum_{\omega_\mu}P_i^{(2)\omega_\mu}\delta(\omega-\omega_\mu) &= \frac{1}{4}\left(\chi_{ijk}^{(2)}(\omega',-\omega'')E_j^{\omega'}E_k^{-\omega''}+\chi_{ijk}^{(2)}(-\omega'',\omega')E_j^{-\omega''}E_k^{\omega'}\right)\\ &\quad \times \delta(\omega'-\omega''-\omega)+\ldots\end{aligned} \quad (2.152)$$

Recall that in deriving Equation 2.152, we included only one term for illustration, but there are many other such terms corresponding to sum frequency generation, second harmonic generation of each frequency, etc.

The Dirac delta functions on the left- and right-hand sides of Equation 2.152 are simultaneously nonzero when $\omega = \pm\omega_\mu$ and $\pm\omega_\mu+\omega'-\omega'' = 0$, under

Nonlinear Optics: A student's perspective

which condition the solutions to Equation 2.152 are obtained by setting the coefficients of the delta functions equal to each other. We can see this more clearly by integrating both sides of Equation 2.152 in the vicinity of ω_σ — the frequency of interest that is being measured. For $\omega' > \omega''$ we have,

$$P_i^{(2)\omega_\sigma} = \frac{1}{2}\left(\chi_{ijk}^{(2)}(-\omega_\sigma;\omega',-\omega'')E_j^{\omega'}E_k^{-\omega''} + \chi_{ijk}^{(2)}(-\omega_\sigma;-\omega'',\omega')E_j^{-\omega''}E_k^{\omega'}\right). \quad (2.153)$$

Keep in mind that the derivation leading to Equation 2.153 assumes that all the frequencies are different. If any of them are the same or zero, an additional numerical factor precedes Equation 2.153, a topic that will be discussed in Section 2.3.5.

The arguments of the second order susceptibility indicate that light at frequency ω' and ω'' are incident on the material and the difference frequency $\omega' - \omega''$ is generated. As we saw in Section 1.7, the process appears macroscopically as difference frequency generation - or parametric mixing. In the microscopic interpretation, light is incident at frequency ω' and splits into two beams at frequencies ω'' and $\omega' - \omega''$. The difference frequency is the beam of interest so its frequency precedes the semicolon; but, the negative sign associated with ω'' is a reminder that a stimulated beam at ω' is also emitted.

2.3.4 n^{th} Order Polarization and Susceptibility

We can generalize the method from the previous section to find higher-order polarizations and susceptibilities. n^{th}-order susceptibility $\chi^{(n)}$ is calculated from the nth-order response function via

$$\chi_{ijk...}^{(n)}(-\omega;\omega_1,\omega_2,....\omega_n) = \int d\tau_1 \int d\tau_2.... \int d\tau_n R_{ijk...}^{(n)}(\tau_1,\tau_2.....\tau_n)$$
$$\times\; e^{-i\omega_1\tau_1}e^{-i\omega_2\tau_2}....e^{-i\omega_n\tau_n}. \quad (2.154)$$

For n incident monochromatic waves, the nth-order polarization is given by

$$P_i^{(n)\omega} = \left(\frac{1}{2}\right)^{n-1}\left(\chi_{ijkl...}^{(n)}(-\omega;\omega_1,\omega_2,...\omega_n)E_j^{\omega_1}E_k^{\omega_2}...E_\xi^{\omega_n} + [\omega_1 \rightleftharpoons \omega_2]+...\right) \quad (2.155)$$

As in the second-order case, the derivation leading to Equation 2.155 assumes that all the frequencies are different. If any of them are the same or zero, an additional numerical factor precedes it, as described below.

Problem 2.3.4-1: The calculations of the polarization in frequency space as presented in Section 2.3 assumes a spatially local but non-instantaneous response. For practice, repeat the derivations for a material with an instantaneous response but spatially nonlocal. Then do the calculation that includes both a non-instantaneous and nonlocal response. To make life simple, assume that the material is homogenous and infinite so that you can assume that the spatial part of the response function depends only on the vector separating the polarization point and the field point.

2.3.5 Tensor Symmetries of Response Functions

The general relationship between the polarization and the applied fields can be used to determine the symmetries of the response functions. Starting with the polarization

$$P_i^{(n)}(t) = \int d\tau_1 \int d\tau_2 \ldots R_{ijk\ldots}^{(n)}(t;\tau_1,\tau_2,\ldots\tau_n) E_j(\tau_1) E_k(\tau_2)\ldots, \quad (2.156)$$

interchanging the two time variables τ_1 and τ_2; and, interchanging the two indices associated with the two fields with these time variable, j and k, yields

$$P_i^{(n)}(t) = \int d\tau_2 \int d\tau_1 \ldots R_{ikj\ldots}^{(n)}(t;\tau_2,\tau_1\ldots\tau_n) E_k(\tau_2) E_j(\tau_1)\ldots \quad (2.157)$$

Equations 2.156 and 2.157 are equal to each other since exchanging the integration variables τ_1 and τ_2 and the dummy indices j and k leaves the polarizations unchanged. Because this interchange also leaves the field product unchanged, i.e. $E_j(\tau_1)E_k(\tau_2) = E_k(\tau_2)E_j(\tau_1)$, it follows that

$$R_{ijk\ldots}^{(n)}(t;\tau_1,\tau_2\ldots) = R_{ikj\ldots}^{(n)}(t;\tau_2,\tau_1\ldots). \quad (2.158)$$

This is called intrinsic permutation symmetry.

Intrinsic permutation symmetry is a restatement of the fact that the order of time integration is irrelevant, so that both fields are treated on an equal footing. Using Equation 2.154, intrinsic permutation symmetry in the frequency domain is expressed as

$$\chi_{ijk\ldots}^{(n)}(-\omega;\omega_1,\omega_2,\ldots\omega_n) = \chi_{ikj\ldots}^{(n)}(-\omega;\omega_2,\omega_1,\ldots\omega_n). \quad (2.159)$$

Intrinsic permutation symmetry applies to the exchange of any two indices excluding the first one.

The second-order polarization for sum-frequency generation can be expressed in a form similar to Equation 2.153, or

$$P_i^{(2)\omega_\sigma} = \frac{1}{2}\left(\chi_{ijk}^{(2)}(-\omega_\sigma;\omega_1,\omega_2)E_j^{\omega_1}E_k^{\omega_2} + \chi_{ijk}^{(2)}(-\omega_\sigma;\omega_2,\omega_1)E_j^{\omega_2}E_k^{\omega_1}\right). \quad (2.160)$$

Intrinsic permutation symmetry allows us to rewrite the second-order susceptibility in the second term of Equation 2.160 as

$$\chi_{ijk}^{(2)}(-\omega_\sigma;\omega_2,\omega_1) = \chi_{ikj}^{(2)}(-\omega_\sigma;\omega_1,\omega_2), \quad (2.161)$$

which we substitute into Equation 2.160, yielding

$$P_i^{(2)\omega_\sigma} = \frac{1}{2}\left(\chi_{ijk}^{(2)}(-\omega_\sigma;\omega_1,\omega_2)E_j^{\omega_1}E_k^{\omega_2} + \chi_{ikj}^{(2)}(-\omega_\sigma;\omega_1,\omega_2)E_j^{\omega_2}E_k^{\omega_1}\right). \quad (2.162)$$

Since the j and k subscripts are summed, we interchange them in the second term of Equation 2.162, which leads to

$$P_i^{(2)\omega_\sigma} = \chi_{ijk}^{(2)}(-\omega_\sigma;\omega_1,\omega_2)E_j^{\omega_1}E_k^{\omega_2}. \quad (2.163)$$

The higher-order polarizations can be transformed in the same way.

In passing, we note that the second-order second harmonic generation susceptibility $\chi_{ijk}^{(2)}(-2\omega;\omega,\omega)$ has degenerate incident frequencies so permutation symmetry implies that

$$\chi_{xyx}^{(2)}(-2\omega;\omega,\omega) = \chi_{xxy}^{(2)}(-2\omega;\omega,\omega) \neq \chi_{yxx}^{(2)}(\omega;-2\omega,\omega). \quad (2.164)$$

Permutation symmetry does not hold when the first index is involved and the degeneracy gives $\chi_{xyx}^{(2)} = \chi_{xxy}^{(2)}$ (the frequencies need not be specified).

As we saw in Section 2.2 the definitions of the fields leads to numerical prefactors in the relationship between the polarization and the monochromatic field amplitudes. In the most general case for the notation used here, the polarization is given by

$$\begin{aligned}P_\mu^{(n)\omega} &= \sum_{\alpha_1,\alpha_2\ldots} K(-\omega;\omega_1,\omega_2,\ldots\omega_n)\chi_{\mu,\alpha_1,\alpha_2\ldots}^{(n)}(-\omega;\omega_1,\omega_2,\ldots\omega_n)\\&\quad\times E_{\alpha_1}^{\omega_1}E_{\alpha_2}^{\omega_2}\ldots E_{\alpha_n}^{\omega_n},\end{aligned} \quad (2.165)$$

where $K(-\omega;\omega_1,\omega_2\ldots\omega_n)$ is defined in our convention to be

$$K(-\omega;\omega_1,\omega_2\ldots\omega_n) = 2^{l+m-n}\Pi. \quad (2.166)$$

Π is the distinct number of permutations of $\omega_1,\omega_2\ldots\omega_n$, n is the order of the nonlinearity, m is the number of times $\omega_i = 0$ appears, and $l = 1$ if $\omega \neq 0$ and $l = 0$ if $\omega = 0$.

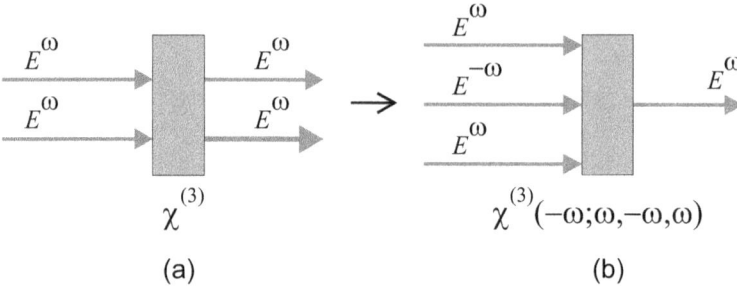

Figure 2.9: (a) In the optical Kerr effect two beams of frequency ω enter the sample and two leave the sample. The process can be viewed as one strong beam, called the pump, causing a change in the refractive index of the other beam. (b) Alternatively, the transmitted pump can be viewed as a negative frequency input beam.

Example: Optical Kerr Effect

In the Optical Kerr Effect, two photons of frequency ω are incident on the material and two photons are emitted, as diagramed by Figure 2.9. Its third order susceptibility tensor has a frequency dependence of the form $\chi^{(3)}_{ijkl}(-\omega;\omega,-\omega,\omega)$. To apply Equation 2.166, note that there are no zero frequencies so $m = 0$. Since the output frequency is nonzero, $l = 1$. The order of the nonlinearity is $n = 3$ and the number of distinct permutations of the frequencies $(\omega,-\omega,\omega)$ is $\Pi = 3$. Therefore

$$K = 2^{1+0-3}(3) = \frac{3}{4}, \tag{2.167}$$

which gives

$$P_x^{(3)\omega} = \frac{3}{4}\chi^{(3)}_{xxxx}(-\omega;\omega,-\omega,\omega)E_x^\omega E_x^{-\omega}E_x^\omega, \tag{2.168}$$

the result we got in Equation 2.106

Problem 2.3.5-1a: Derive Equation 2.166

Problem 2.3.5-2(a): Two light beams of frequency ω_1 and ω_2 impinge on a nonlinear material with a scalar and spatially local response ($\chi^{(n)}$ will not depend on **k**). Calculate the Fourier amplitude $P^{(4)}_{2\omega_1-2\omega_2}$ in terms of $\chi^{(4)}$, E^{ω_1} and E^{ω_2}. In other words, write an expression of the form $P^{(4)}_{2\omega_1-2\omega_2} = K\chi^{(4)}(-(2\omega_1-2\omega_2);\ldots)E^{\omega_1}\ldots$

(b): Two high-intensity light beams of frequency ω_1 and ω_2 are launched into a centrosymmetric material, in which even-order susceptibilities vanish. Light at frequency $\omega_1 - 2\omega_2$ is observed only when a static electric field is applied to the sample. Write an expression in a form as you did in (a) for the polarization at $\omega_1 - 2\omega_2$ for the lowest-order nonlinearity that could be responsible for the effect. Be sure to evaluate K. How does the intensity at frequency $\omega_1 - 2\omega_2$ depend on the applied static voltage?

Problem 2.3.5-3(a): An electric field pulse of the form $E(t) = t_0 E_0 \delta(t)$ is applied to a material. $\delta(t)$ is the Dirac Delta function, t_0 has units of time and E_0 has units of electric field strength. The polarization responds as a damped oscillator of the form $P^{(1)}(t) = P_0 e^{-t/\tau}\sin\omega_0 t$, where ω_0 is the frequency of oscillation and τ the damping time constant. Use this result to determine the response function $R^{(1)}(t)$ and from this determine the linear susceptibility $\chi^{(1)}(-\omega;\omega)$. Compare your answer with the result for the nonlinear spring.

(b): For a response function of the form

$$R^{(2)}(t_1,t_2) = R_0 \theta(t_1)\theta(t_2)e^{-t_1/\tau}\sin(\omega_0 t_1)e^{-t_2/\tau}\sin(\omega_0 t_2), \quad (2.169)$$

calculate $\chi^{(2)}(\omega_1,\omega_2)$ AND $P^{(2)}(t)$. To calculate $P^{(2)}(t)$, assume that the electric field is a step function of the form $E(t) = E_0 \theta(t)$.

(c): Consider a nonlinear spring with $k^{(1)} \neq 0$, $k^{(2)} \neq 0$, and $k^{(n)} = 0$ for $n > 2$. If $\Gamma = 0$ and there is no driving field, show that

$$x = Ae^{-i\omega_0 t} + \delta e^{-i2\omega_0 t}$$

is a solution of the equations of motion when $\delta \ll A$ and show how δ depends on A, $k^{(1)}$ and $k^{(2)}$. Do the results make physical sense in light of what you got in parts (a) and (b)?

2.3.6 Complex Susceptibilities

To understand the complex nonlinear susceptibility, we first consider the linear response, which through the response function leads to the polarization,

$$P^{(1)}(t) = \int_0^\infty R^{(1)}(\tau) E(t-\tau) d\tau. \tag{2.170}$$

Here $P^{(1)}(t)$ and $E(t-\tau)$ are observed quantities and are therefore real, which implies that $R^{(1)}(\tau)$ is also real.

The linear susceptibility $\chi^{(1)}(\omega)$ is a complex quantity with real and imaginary parts, which is given by Equation 2.126 to be

$$\chi^{(1)}(\omega) = \int_0^\infty R^{(1)}(\tau) e^{-i\omega\tau} d\tau. \tag{2.171}$$

Recall that $R^{(1)}(\tau)$ is the linear response function that relates the polarization at a time t to an applied electric field at some earlier time $t-\tau$. The lower limit of integration is set to zero to enforce causality. From Equation 2.171 and the fact that $R^{(1)}(\tau)$ is real we get

$$\chi^{(1)}(\omega_i)^* = \chi^{(1)}(-\omega_i). \tag{2.172}$$

The same approach can be applied to the nonlinear susceptibilities, in which case the complex conjugate operation flips the signs of all the frequencies.

2.3.7 Kramers-Kronig

The Kramers-Kronig equations come from evaluating the integral

$$\int_{-\infty}^\infty \frac{\chi^{(1)}(\omega')}{\omega' - \omega} d\omega'. \tag{2.173}$$

We will make ω complex as a computational convenience, with $\omega \to \omega + i\omega_I$. This allows the integral to be evaluated as part of a contour integral in the complex plane. Figure 2.10 shows the closed contour over which the integral can be evaluated using Cauchy's Residue Theorem. The contribution from the large semicircular path vanishes as $R \equiv |\omega| \to \infty$ because the denominator in Expression 2.173 scales in proportion to R as $R \to \infty$.

The principal value of an integral whose path passes through a singularity is defined as the part of the integral that excludes the singular point $\omega = \omega'$, and takes the form

$$\lim_{\delta \to 0} \left(\int_{-\infty}^{\omega'-\delta} \frac{\chi^{(1)}(\omega')}{\omega' - \omega} d\omega' + \int_{\omega'+\delta}^\infty \frac{\chi^{(1)}(\omega')}{\omega' - \omega} d\omega' \right). \tag{2.174}$$

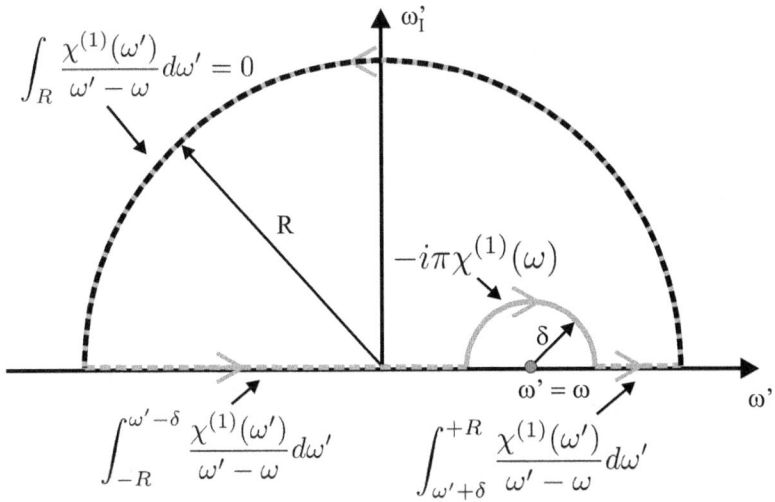

Figure 2.10: The contour integral is composed of three parts: (1) The semicircular part of radius $R \to \infty$; (2) the real axis, excluding the singularity at $\omega = \omega'$; and, (3) the semicircular part of radius $\delta \to 0$, which is integrated using Cauchy's Integral Formula.

The path of integration shown in Figure 2.10 excludes the pole, so Cauchy's Integral Formula yields

$$\oint_{-\infty}^{\infty} \frac{\chi^{(1)}(\omega')}{\omega' - \omega} d\omega' = 0. \tag{2.175}$$

This contour integral can be separated in the three paths as shown in Figure 2.10, as follows. The integral along the real axis that excludes the singularity is the principle value of the desired integral. The small semicircular path of radius $\delta \to 0$ circles the pole and the large semicircular path is of radius $R \to \infty$. Equation 2.175 is thus the sum of the three pieces, and are of the form

$$\int_{-\infty}^{\infty} \frac{\chi^{(1)}(\omega')}{\omega' - \omega} d\omega' - i\pi \chi^{(1)}(\omega) + \int_R \frac{\chi^{(1)}(\omega')}{\omega' - \omega} d\omega' = 0. \tag{2.176}$$

The middle term is half the residue of the result that would have been obtained for a closed circle.

The last term in Equation 2.176 vanishes as $R \equiv |\omega' - \omega| \to \infty$ since the numerator is a susceptibility that is finite in the limit of large frequency. Therefore, Equation 2.176 yields

$$\chi^{(1)}(\omega) = \frac{-i}{\pi} \int_{-\infty}^{\infty} \frac{\chi^{(1)}(\omega')}{\omega' - \omega} d\omega'. \tag{2.177}$$

Because Equation 2.177 spans the real axis, the frequencies are real. As such, only the linear susceptibility is complex.

Substituting $\chi^{(1)} = \chi_R^{(1)} + i\chi_I^{(1)}$ and $-i(\chi_R^{(1)} + i\chi_I^{(1)}) = -i\chi_R^{(1)} + \chi_I^{(1)}$ into Equation 2.177, and isolating the real and the imaginary parts of the result yields the Kramers-Kronig relations,

$$\text{Re}[\chi^{(1)}(\omega)] = \frac{1}{\pi} \int_{-\infty}^{\infty} \text{Im}\left[\frac{\chi^{(1)}(\omega')}{\omega' - \omega}\right] d\omega', \qquad (2.178)$$

and

$$\text{Im}[\chi^{(1)}(\omega)] = \frac{-1}{\pi} \int_{-\infty}^{\infty} \text{Re}\left[\frac{\chi^{(1)}(\omega')}{\omega' - \omega}\right] d\omega'. \qquad (2.179)$$

Equation 2.178 relates the real to the imaginary part of the susceptibility provided that the imaginary part is known over a broad enough range of angular frequencies, ω, to perform the required integration. The best results are obtained when the limits of integration span fully the domain where the susceptibility is nonzero – or equivalently, when the susceptibilities at the endpoints are small. The imaginary part can be determined from the real part in the same way. Since the dispersion of the imaginary part of $\chi^{(1)}$ is related to the linear absorption spectra, a property that is easily measured, Equation 2.178 can be used to predict the more-difficult-to-measure frequency dependence of the refractive index, which is related to the real part of $\chi^{(1)}$.

The Kramers-Kronig equations can be re-expressed in a more useful form using Equation 2.172, which requires that

$$\text{Re}[\chi^{(1)}(-\omega)] = \text{Re}[\chi^{(1)}(\omega)],$$
$$\text{Im}[\chi^{(1)}(-\omega)] = -\text{Im}[\chi^{(1)}(\omega)]. \qquad (2.180)$$

Rewriting Equation 2.179 as

$$\text{Im}[\chi^{(1)}(\omega)] = -\frac{1}{\pi} \int_{-\infty}^{0} \frac{\text{Re}[\chi^{(1)}(\omega')]}{\omega' - \omega} d\omega' - \frac{1}{\pi} \int_{0}^{\infty} \frac{\text{Re}[\chi^{(1)}(\omega')]}{\omega' - \omega} d\omega', \qquad (2.181)$$

and using Equations 2.180 yields

$$\text{Im}[\chi^{(1)}(\omega)] = \frac{1}{\pi} \int_{0}^{\infty} \frac{\text{Re}[\chi^{(1)}(\omega')]}{\omega' + \omega} d\omega' - \frac{1}{\pi} \int_{0}^{\infty} \frac{\text{Re}[\chi^{(1)}(\omega')]}{\omega' - \omega} d\omega'. \qquad (2.182)$$

Combining the two terms yields the more familiar form,

$$\text{Im}[\chi^{(1)}(\omega)] = -\frac{2\omega}{\pi} \int_{0}^{\infty} \frac{\text{Re}[\chi^{(1)}(\omega')]}{\omega'^2 - \omega^2} d\omega'. \qquad (2.183)$$

The real part can be massaged in the same way, yielding

$$\text{Re}[\chi^{(1)}(\omega)] = \frac{2}{\pi} \int_0^\infty \frac{\omega' \text{Im}[\chi^{(1)}(\omega')]}{\omega'^2 - \omega^2} d\omega'. \tag{2.184}$$

The Kramers-Kronig relations are a type of Hilbert transform filter, expressed as

$$H(\omega) = \int_{-\infty}^\infty e^{-i\omega t} h(t) dt = \int_{-\infty}^\infty (h(t)\cos(\omega t) - ih(t)\sin(\omega t)) dt. \tag{2.185}$$

The concept of causality is captured in the statement of the Kramers-Kronig relationships. Applying a frequency domain convolution amounts to sliding a kernel function, in this case projecting out the even and odd parts of the causal response of $\chi^{(1)}(\omega)$ which correspond to the real and imaginary pieces of the frequency response. For example,

$$\int_{-\infty}^\infty \chi^{(1)}(\omega - \omega') H(\omega) d\omega = \int_{-\infty}^\infty \chi^{(1)}(\omega - \omega')[\cos(\omega t) + i\sin(\omega t)] d\omega. \tag{2.186}$$

The Kramers-Kroning relationships can help determine when a response obeys causality simply by testing whether or not the real and imaginary parts of the Hilbert transforms are equal to each other.

Numerical Integration

Since optical absorption measurements are much simpler to implement than refractive index measurements, The Kramers-Kronig relations are commonly used to get the the refractive index spectrum from the absorption spectrum by numerically integrating the Kramers-Kronig relationships.

Numerical integration of a function $\phi(\omega)$ between the limits ω_1 and ω_2 using the trapezoidal method gives

$$\int_{\omega_1}^{\omega_2} \phi(\omega) d\omega \approx \sum_{j=0}^{N-1} \frac{\phi_j + \phi_{j+1}}{2} \Delta\omega = \left[\sum_{j=1}^{N-1} \phi_j + \left(\frac{\phi_0}{2} + \frac{\phi_N}{2} \right) \right] \Delta\omega, \tag{2.187}$$

where the interval $\omega_1 \to \omega_2$ is divided into N segments each of width $\Delta\omega = (\omega_1 - \omega_2)/N$ and $\phi_j \equiv \phi(\omega_1 + j\Delta\omega)$. The principle value of the integral excluding the point ω_0 for which $\phi(\omega_0) = \phi_i$ is calculated from the trapezoidal rule using

$$\begin{aligned}\mathcal{P} \int_{\omega_1}^{\omega_2} \phi(\omega) d\omega &\approx \left(\sum_{j=0}^{i-1} \frac{\phi_j + \phi_{j+1}}{2} + \sum_{j=i+1}^{N-1} \frac{\phi_j + \phi_{j+1}}{2} \right) \Delta\omega \\ &= \left[\sum_{j=1}^{i-2} \phi_j + \sum_{j=i+2}^{N-1} \phi_j + \left(\frac{\phi_0}{2} + \frac{\phi_{i-1}}{2} + \frac{\phi_{i+1}}{2} + \frac{\phi_N}{2} \right) \right] \Delta\omega.\end{aligned} \tag{2.188}$$

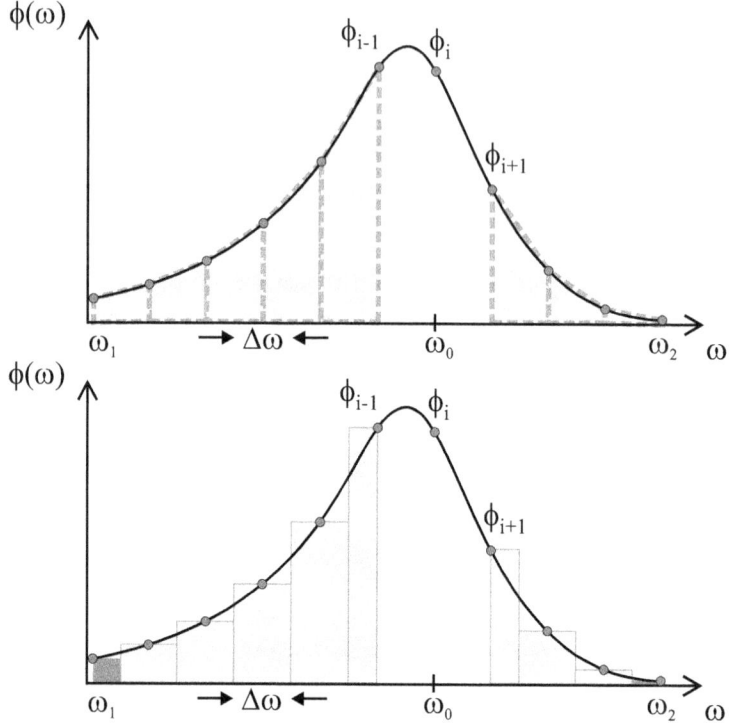

Figure 2.11: The principal value of an integral is calculated by adding the areas of the trapezoids (top) which is equivalent to adding the areas of the rectangles (bottom) when the interval $\Delta\omega$ is sufficiently small.

Figure 2.11 interprets Equation 2.188 as the sum of areas of the shaded trapezoids which is equivalent to adding the rectangles shown in the bottom part of the figure. Each rectangle is of width $\Delta\omega$ except for those at the ends of the interval and those straddling the point ω_0.

Problem 2.3.7-1(a): Using Equation 2.30 with $A \equiv e^2 N/m$, write a Python module that calculates the real and imaginary parts of $\chi^{(1)}$ that has as inputs the amplitude, A, the width, Γ, and the natural frequency of oscillation, ω_0. Also as inputs are the range of frequencies to be plotted, ω_1 and ω_2, and the number of intervals in that frequency range, N. Use this module to plot the real and imaginary parts of $\chi^{(1)}$ as a function of ω. Use $A = 1$, $\omega_0 = 5$, $\Gamma = 0.5$ $\omega_1 = 0$, $\omega_2 = 10$, and $N = 1000$ for every part of this problem.

(b): Use the Kramers-Kronig relation to determine the real part from the imaginary part and compare your result with the plot in part (a).

(c): Typical experiments are constrained by the light source and detector. Assuming that the light source works only in the range $\omega = 3 \rightarrow 5.5$, determine how well the true real part of $\chi^{(1)}$ is reproduced in this limited range.

(d): Use the data from part (c) but set to zero all the values outside this range to determine how this affects the results in the range $\omega = 0 \rightarrow 10$.

(e): Vary N to see how coarse you can make the frequency intervals in the full range $\omega = 0 \rightarrow 10$ while maintaining reasonable accuracy.

2.3.8 Permutation Symmetry

In general the nth-order susceptibility, $\chi^{(n)}$, is a tensor of rank $n+1$ and depends on a set of $n+1$ input and output frequencies that defines the nonlinear process. Negative frequencies (representing outgoing fields) are independent arguments that change the tensor elements of the nonlinear susceptibilities. As such, the susceptibility $\chi^{(n)}_{ijkl...}(-\omega;\omega_1,\omega_2,....\omega_n)$ defined in Equation 2.155 is distinct from $\chi^{(n)}_{ijkl...}(-\omega;\omega_1,-\omega_2,....\omega_n)$, and will thus have resonances at different combinations of frequencies and represent different processes. In the former case, where all frequencies to the right of the semicolon are positive, there are n absorbed photons and one emitted one. In the later case, $n-1$ photons are absorbed and two are emitted. Thus, for a fixed set of incident photons, energy conservation in the first process where n are absorbed photons demands an emission at one frequency. In the second process, energy conservation allows a spectrum of colors to be emitted as long as the sum of the energies of the two emitted photons matches the absorbed energy.

The dispersion of the process described by $\chi^{(n)}_{ijkl...}(-\omega;\omega_1,-\omega_2,....\omega_n)$ is determined by including all possible permutations of the distinct frequencies. Fortunately, we do not need to keep track of all these separate permutations by using symmetry considerations.

From the defining relationship expressing the polarization as a function of space and time, the time dependence of the sum-frequency polarization for two incident monochromatic waves of frequencies ω_1 and ω_2 is given by

$$P(t) = \frac{1}{2}\left(P_i^{\omega_1+\omega_2} e^{-i(\omega_1+\omega_2)t} + P_i^{-\omega_1-\omega_2} e^{i(\omega_1+\omega_2)t}\right). \tag{2.189}$$

The function $P(t)$ is real since it is an observable quantity, so using the fact that $P(t) = P^*(t)$, it is straightforward to show using Equation 2.189 that,

$$P_i^{-\omega_1-\omega_2} = P_i^{\omega_1+\omega_2 *}. \tag{2.190}$$

Since the electric fields are real, a similar argument as above leads to

$$E_j^{-\omega_1} = E_j^{\omega_1 *},$$
$$E_k^{-\omega_2} = E_k^{\omega_2 *}. \tag{2.191}$$

From Equations 2.190 and 2.191, with the help of Equation 2.155, we get

$$P_i^{\omega_1+\omega_2} = \chi_{ijk}^{(2)}(-\omega_1-\omega_2;\omega_1,\omega_2) E_j^{\omega_1} E_k^{\omega_2}, \tag{2.192}$$

and

$$P_i^{-\omega_1-\omega_2 *} = \chi_{ijk}^{(2)}(\omega_1+\omega_2;-\omega_1,-\omega_2)^* E_j^{-\omega_1 *} E_k^{-\omega_2 *}. \tag{2.193}$$

A comparison between Equations 2.192 and 2.193 leads us to conclude that the positive and negative frequency arguments of the nonlinear susceptibility tensor have the symmetry

$$\chi_{ijk}^{(2)}(-\omega_1-\omega_2;\omega_1,\omega_2) = \chi_{ijk}^{(2)}(\omega_1+\omega_2;-\omega_1,-\omega_2)^*, \tag{2.194}$$

which reduces the number of independent components needed to describe the susceptibility tensor χ.

Problem 2.3.8-1: Verify Equations 2.190, 2.191 and 2.194 by filling in the details.

Full Permutation Symmetry

In the regime where the field frequencies are far from material resonances, the susceptibilities are approximately independent of frequency. As such, pairs of certain indices of the nonlinear susceptibility tensor can be interchanged without affecting the polarization. For example, consider the second-order polarization $P_i = \sum_{jk} \chi_{ijk} E_j E_k$. Since multiplication commutes, we can exchange the two electric fields, yielding

$$\sum_{jk} \chi_{ijk} E_j E_k = \sum_{jk} \chi_{ijk} E_k E_j. \qquad (2.195)$$

Because the subscripts are summed, we can interchange them, i.e. $j \leftrightarrow k$. Doing so to the righthand side of Equation 2.195 yields

$$\sum_{jk} \chi_{ijk} E_j E_k = \sum_{jk} \chi_{ikj} E_j E_k. \qquad (2.196)$$

Equation 2.196 implies $\chi_{ijk} = \chi_{ikj}$.

In the regime where the second-order susceptibility depends on frequency, intrinsic permutation symmetry holds, as we have seen in Section 2.3.5, leading to

$$\chi^{(2)}_{ijk}(-\omega_n - \omega_m; \omega_n, \omega_m) = \chi^{(2)}_{ikj}(-\omega_n - \omega_m; \omega_m, \omega_n). \qquad (2.197)$$

The same procedure can be applied to higher-order susceptibilities, where intrinsic permutation symmetry demands that the nonlinear susceptibility to any order is unchanged when any pair of indices are interchanged along with the associated frequencies, excluding the first index.

For lossless media, where energy is conserved, the nonlinear susceptibility is guaranteed to be real and the interchange of any two indices along with the corresponding frequencies, including the first index, leaves the tensor unchanged. This is called full permutation symmetry and holds for nondispersive materials, as you will show in the following problems.

Problem 2.3.8-2: Show that the nth-order susceptibility is invariant when any two indices and their associated frequencies are interchanged, except for the first index. What makes the first index special?

Problem 2.3.8-3: Use the fact that when the energy density of a material in an electric field is given by $U = -\mathbf{P} \cdot \mathbf{E}$ to show that the nth-order susceptibility is invariant under full permutation symmetry.

Kleinman's Symmetry

The off-resonance regime is defined by the condition that the Bohr frequencies, defined by $\omega_{nm} = (E_n - E_m)/\hbar$ – where E_n and E_m at any two eigenenergies, of a material are much larger that the frequencies of all the photons in a nonlinear process. In this case, since the photon frequencies are negligibly small, they can be set to zero. Then, the interchange of any two frequencies leaves the nonlinear susceptibility unchanged. Full permutation symmetry then implies that any two indices can be interchanged. This is called Kleinman Symmetry. For the second-order susceptibility, full permutation symmetry yields

$$\chi^{(2)}_{ijk}(\omega_3 = \omega_1 + \omega_2) = \chi^{(2)}_{jki}(\omega_1 = -\omega_2 + \omega_3) = \chi^{(2)}_{kij}(\omega_2 = \omega_3 - \omega_1) =$$
$$\chi^{(2)}_{ikj}(\omega_3 = \omega_2 + \omega_1) = \chi^{(2)}_{kji}(\omega_2 = -\omega_1 + \omega_3) = \chi^{(2)}_{jik}(\omega_1 = \omega_3 - \omega_2)$$
(2.198)

and Kleinman symmetry gives

$$\chi^{(2)}_{ijk} = \chi^{(2)}_{jki} = \chi^{(2)}_{kij} = \chi^{(2)}_{ikj} = \chi^{(2)}_{kji} = \chi^{(2)}_{jik}. \tag{2.199}$$

Contracted Notation

The second harmonic tensor d_{ijk} can be simplified when Kleinman Symmetry holds or when the process is second-harmonic generation, which is fully symmetric in its second two indices since $\omega_n = \omega_m$). First, we note that for historic reasons, the second harmonic tensor is defined as

$$d_{ijk} = \frac{1}{2}\chi^{(2)}_{ijk}. \tag{2.200}$$

Under the above conditions, the polarization is invariant when the last two indices are interchanged, which allows us to define a 3 × 6 matrix that contains the same information as the second-order susceptibility. It is determined by:

- replacing (j,k)=(1,1) with 1
- replacing (j,k)=(2,2) with 2
- replacing (j,k)=(3,3) with 3
- replacing (j,k)=(1,2) or if (j,k)=(2,1) with 4
- replacing (j,k)=(1,3) or (j,k)=(3,1) with 5

- replacing (j,k)=(2,3) or (j,k)=(3,2) with 6.

The index i in the contracted tensor d_{il} spans from 1 to 3, and l spans 1 to 6. The matrix is thus given by

$$d_{il} = \begin{bmatrix} d_{11} & d_{12} & d_{13} & d_{14} & d_{15} & d_{16} \\ d_{16} & d_{22} & d_{23} & d_{24} & d_{14} & d_{12} \\ d_{15} & d_{24} & d_{33} & d_{23} & d_{13} & d_{14} \end{bmatrix}.$$

2.3.9 Symmetries

Symmetry is a powerful tool for simplifying the nonlinear susceptibility. Here we consider additional examples of symmetry and show how they can be applied to nonlinear optics.

Inversion Symmetry and Centrosymmetry

The inversion operation reverses the sign of each vector component of **v**, yielding $v_i \to -v_i$. If a coordinate system exists such that for every vector that defines the coordinate of an atom, inversion gives a coordinate of another atom of the same type, the structure is said to be centrosymmetric. More precisely, a material is centrosymmetric if there exists a coordinate origin such that the electron density at every point in space is the same as at the inversion point.

Figure 2.12 shows an example of two molecules. Figure 2.12a shows a pentagon, in which no center of inversion exists. In the xy coordinate frame, the vector shows the position of one of the atoms. After applying the inversion operation the dashed vector in the $x'y'$ coordinate frame results. No atom is found in this position.

The benzene molecule is shown in Figure 2.12b. Upon inversion, the dashed vector is found to represent the coordinate of another carbon atom. Since this condition holds for every atom in the molecule, it is centrosymmetric.

If a molecule or a material is centrosymmetric, then all physical properties are also unchanged upon inversion. We can use this fact to determine some general properties of the nonlinear susceptibility.

For example, consider the nonlinear polarization of an even-order susceptibility, given by

$$P_i^{(n)} = \chi_{ijk...}^{(n)} E_j E_k \ldots, \tag{2.201}$$

where n is even. The inversion operation yields

$$-P_i^{(n)} = \chi_{ijk...}^{(n)} (-E_j)(-E_k) \cdots = \chi_{ijk...}^{(n)} E_j E_k \ldots, \tag{2.202}$$

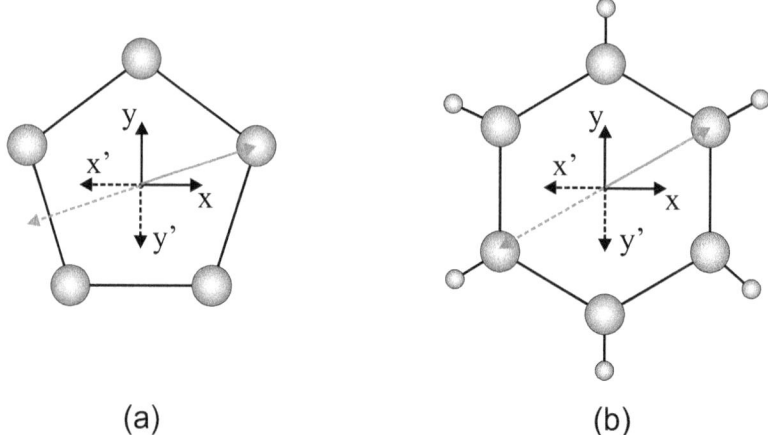

Figure 2.12: (a) A molecule made of 5 atoms in the form of a pentagon is not centrosymmetric because no center of inversion point exists. (b) A hexagon, such as the benzene molecule, which is made of 6 carbon atoms and six hydrogen atoms, is centrosymmetric because ever atomic coordinate represented by a vector (shown as a solid arrow) is also the coordinate of an atom when the inversion operation is applied (dashed arrow).

where the last equality reflects the fact that there are an even number of electric fields. Note that since the material remains unchanged upon inversion, this operation changes only the fields and the polarization, but the second-order susceptibility remains the same.

Adding Equations 2.201 and 2.202 yields

$$2\chi^{(n)}_{ijk...} E_j E_k \cdots = 0. \tag{2.203}$$

Since the electric fields are nonzero, this implies that $\chi^{(n)}_{ijk...} = 0$ for a centrosymmetric material.

Mirror Plane

A reflection changes the sign of the vector component that is perpendicular to the mirror plane. As an example of an application of mirror symmetry, consider a material that is invariant under reflection through the xz-plane. For second harmonic light that is emitted with a polarization along x for incident light at the fundamental frequency polarized along y, the nonlinear polarization is given by

$$P^{(2)}_x = \chi^{(2)}_{xyy} E_y E_y. \tag{2.204}$$

Reflection through the mirror plane gives

$$P_x^{(2)} = \chi_{xyy}^{(2)}(-E_y)(-E_y) = \chi_{xyy}^{(2)} E_y E_y, \qquad (2.205)$$

so $\chi_{xyy}^{(2)}$ is not required to vanish. Note that it might vanish due to other factors, but is allowed by mirror symmetry.

On the other hand, for the process described by

$$P_y^{(2)} = \chi_{yxx}^{(2)} E_x E_x, \qquad (2.206)$$

it is easy to show that $\chi_{yxx}^{(2)}$ vanishes for an xz mirror plane.

Problem 2.3.9-1: Show that $\chi_{yxx}^{(2)}$ vanishes for an xz mirror plane.

Rotations

The same ideas can be applied to rotations. Consider again the process described by Equation 2.204 and a material that is invariant to rotations about the z axis by $\pi/2$. This rotation transforms vectors according to $v_x \to v'_y$, $v_y \to -v'_x$, and $v_z \to v'_z$, yielding

$$P_y^{(2)} = \chi_{xyy}^{(2)}(-E_x)(-E_x) = \chi_{xyy}^{(2)} E_x E_x. \qquad (2.207)$$

But since by definition

$$P_y^{(2)} = \chi_{yxx}^{(2)} E_x E_x, \qquad (2.208)$$

a comparison between Equations 2.207 and 2.208 implies that $\chi_{yxx} = \chi_{xyy}$.

Finally, if the material is invariant to rotation of $\pi/2$, it is also invariant to rotation of π, which transforms the components by $v_x \to -v'_x$, $v_y \to -v'_y$, and $v_z \to v'_z$, leading to $\chi_{yxx} = \chi_{xyy} = 0$. Thus, rotational symmetries yield information about relationships between tensor components.

Problem 2.3.9-2: A material is isotropic in the x-y plane. If inversion symmetry is broken along the z axis, how many independent tensor components does the far far far off-resonance $\chi^{(2)}$ have? Explain why.

Problem 2.3.9-3(a): You set out to measure the $\chi^{(2)}$ and $\chi^{(3)}$ tensor of a material. Use the fact that the material is transparent at all wavelengths so that Kleinman symmetry holds in addition to permutation symmetry. If it takes one hour to measure each component, how much time will you save by applying Kleinman symmetry? Be sure to list all distinct tensor components and how they are related to each other.

(b): If the material is isotropic, how many measurements are required in each case? Hint: apply the inversion operation and any other symmetry that you think will be useful.

Problem 2.3.9-4: The charge distribution in an isosceles triangular molecule is shown below. Determine which tensor components vanish from the set $\chi^{(2)}_{xyy}$, $\chi^{(2)}_{yxy}$, $\chi^{(2)}_{xyx}$, $\chi^{(2)}_{yxx}$, $\chi^{(2)}_{xxx}$, and $\chi^{(2)}_{yyy}$.

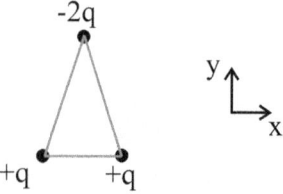

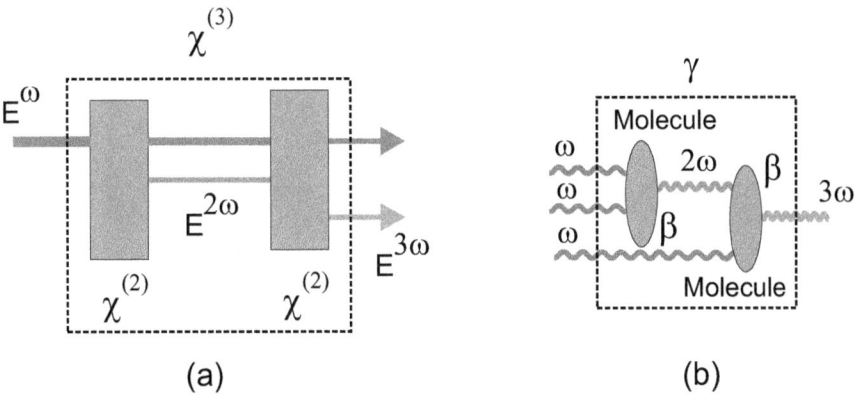

Figure 2.13: (a) Two crystals in series through a $\chi^{(2)}$ process mimic $\chi^{(3)}$. (b) Two molecules in series through a β process mimic γ.

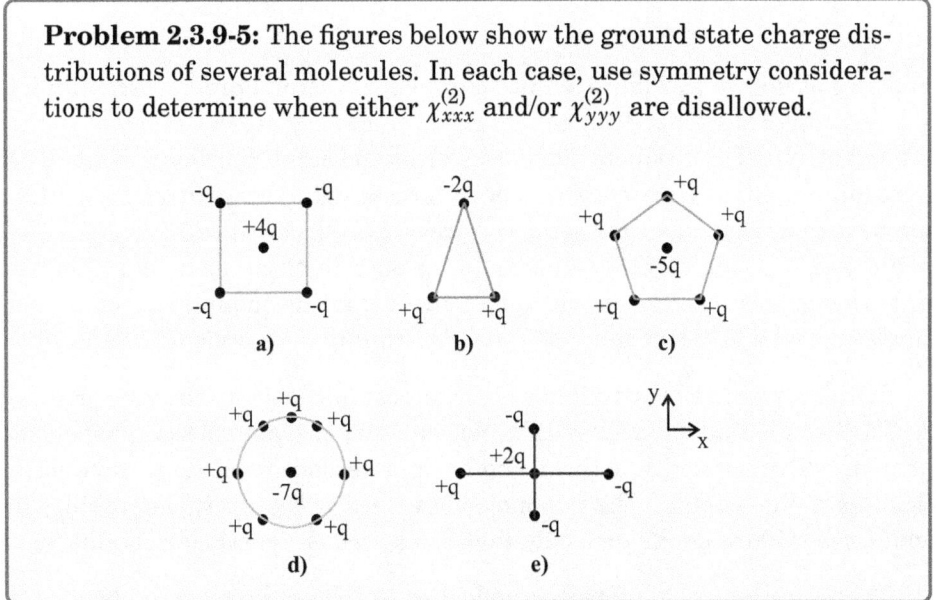

Problem 2.3.9-5: The figures below show the ground state charge distributions of several molecules. In each case, use symmetry considerations to determine when either $\chi^{(2)}_{xxx}$ and/or $\chi^{(2)}_{yyy}$ are disallowed.

2.4 Cascading

The nonlinear susceptibility of a material is a bulk property and the hyperpolarizabilities describe the nonlinearity at the molecular level. Materials are not so easily subdivided when light produced in one part of a material interacts with another part of the material. Consider as an example two crystals

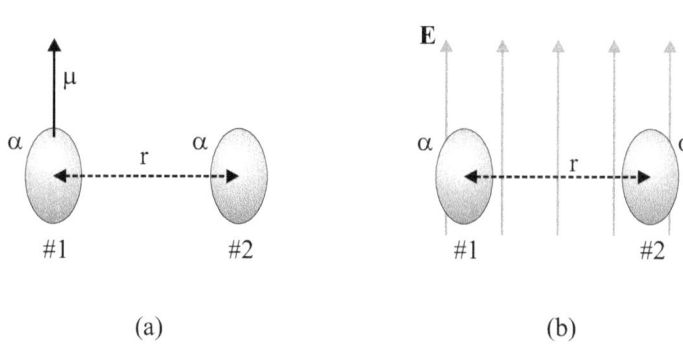

Figure 2.14: (a) Two side-by side molecules, each have polarizability α; but, only the left one has a permanent dipole moment μ. (b) Two molecules each with polarizability α in an electric field. The induced dipole moments and dipole fields are not shown.

in series as shown in Figure 2.13a. Through the second-order susceptibility $\chi^{(2)}$ (we ignore the tensor components here since they are not relevant to the argument), light at frequency ω is converted to second harmonic light. The remaining light at frequency ω then interacts with the light at 2ω within the next crystal, which through a $\chi^{(2)}$ process produces third harmonic. The two crystals acting together appear to an outside observer as a $\chi^{(3)}$ material. This process is more efficient than direct third harmonic generation in a single crystal and is a common method for tripling the frequency of a laser.

The same process is possible at the microscopic level as shown in Figure 2.13b. Two photons interact with a molecule to generate a second harmonic photon, which acts with a third photon in a second molecule to produce a third harmonic photon. The two molecules, each with hyperpolarizability β, act together like a single molecule that has a second hyperpolarizability γ.

There are many delightful ramifications of cascading, which lead to a riches of processes, especially in the quantum case. The reader is encouraged to think about how Figure 2.13b can be interpreted in light of quantum electrodynamics, where the 2ω photon can be emitted before the ω photons are absorbed. We discuss this process later using Feynman diagrams.

As a precursor to understanding cascading, discussions are limited to static fields for now. Figure 2.14 shows the two cases that we consider as illustrations: (1) two interacting molecules in the absence of an electric field (Figure 2.14a) and to molecules in a uniform electric field 2.14b).

Nonlinear Optics: A student's perspective

2.4.1 Two Interacting Molecules

As shown in Figure 2.14a, we start by considering two side-by-side molecules; one has a permanent dipole moment and both are polarizable. The dipole moment of Molecule #1 induces a dipole moment in Molecule #2, and this induced dipole moment acts on Molecule #1 to induce a dipole moment. The dipole moment of Molecule #1 is then the sum of the static and induced dipole moment, which then acts on Molecule #2, and so on.

Our goal is to find the dipole moment of this two-molecule system. There are at least three methods, which includes the self-consistent field method, the series method, and the derivative method. In the sections that follow, we illustrate two of these methods then apply the third method to the problem of two polarizable molecules in a static electric field. Note that since all electric fields and dipoles point either up or down, we express them as scalers.

While we are applying these methods to cascading, these techniques are more broadly applicable to other classes of problems. As such, they are useful to learn.

Self-Consistent Field

The self-consistent field method uses the local electric field F_i experienced by Molecule #i as an intermediate quantity that does not necessarily need to be calculated directly. Then, the dipole moment on Molecule #i, p_i, is expressed in terms of the local electric fields, yielding

$$p_1 = \mu + \alpha F_1, \tag{2.209}$$

and

$$p_2 = \alpha F_2. \tag{2.210}$$

But, the local electric field acting on Molecule #1 is just the dipole field due to Molecule #2, so,

$$F_1 = -\frac{p_2}{r^3} \quad \text{and} \quad F_2 = -\frac{p_1}{r^3}, \tag{2.211}$$

where the minus signs reflect the fact that for a dipole pointing upwards the electric field in its equatorial plane points downwards.

Substituting Equations 2.211 into Equations 2.209 and 2.210 yields

$$p_1 = \mu - \frac{\alpha}{r^3} p_2, \tag{2.212}$$

and

$$p_2 = -\frac{\alpha}{r^3} p_1. \tag{2.213}$$

Finally, substituting Equation 2.213 into Equation 2.212 and solving for p_1 yields

$$p_1 = \frac{\mu}{1-\left(\frac{\alpha}{r^3}\right)^2}, \qquad (2.214)$$

and solving for p_2 using Equation 2.213 yields

$$p_2 = -\frac{\alpha}{r^3} \cdot \frac{\mu}{1-\left(\frac{\alpha}{r^3}\right)^2}. \qquad (2.215)$$

The total dipole moment is the sum of the two dipoles given by Equations 2.214 and 2.215, yielding

$$p = p_1 + p_2 = \frac{\mu}{1+\frac{\alpha}{r^3}}. \qquad (2.216)$$

Note how Gaussian units units make this problem tidy, where the magnitude of the dipole field is simply p/r^3 and the dipole moment is given by $p = \mu + \alpha E + \beta E^2 + ...$ Also, since α has units of volume, the ratio α/r^3 is a dimensionless parameter.

Equation 2.216 deserves some explanation. First, we note that in the limit of infinite separation, the combined dipole moment is μ, which is reasonable since the molecules do not interact. In the limit of zero separation, the composite dipole moment vanishes, so the induced dipole moment cancels the permanent dipole moment.

It is telling to observe the individual dipole moments. At infinite separation, Molecule #1 has dipole moment μ while Molecule #2 has no dipole moment, as we would expect. When the separation between molecule's approaches a molecules size, given by $\alpha^{1/3}$, both dipole moments diverge. However, the individual dipole momets are of opposite sign, so that when they just "touch" each other, the infinities cancel and the dipole moment is $\mu/2$.

Nonlinear Optics: A student's perspective

Series Solution

In the series method, Equation 2.209, 2.210 and 2.211 are solved iteratively, by starting with no fields, as follows:

$$p_1 = \mu$$
$$p_2 = -\alpha \frac{p_1}{r^3} = -\alpha \frac{\mu}{r^3} = -\mu \frac{\alpha}{r^3}$$
$$p_1 = \mu - \alpha \frac{p_2}{r^3} = \mu - \alpha \frac{-\mu \frac{\alpha}{r^3}}{r^3} = \mu \left(1 + \left[\frac{\alpha}{r^3}\right]^2\right)$$
$$p_2 = -\alpha \frac{p_1}{r^3} = \cdots = -\mu \cdot \frac{\alpha}{r^3} \cdot \left(1 + \left[\frac{\alpha}{r^3}\right]^2\right)$$
$$p_1 = \mu - \alpha \frac{p_2}{r^3} = \cdots = \mu \left(1 + \left[\frac{\alpha}{r^3}\right]^2 + \left[\frac{\alpha}{r^3}\right]^4\right)$$
$$\cdots \quad (2.217)$$

Each iteration alternates between getting p_1 and p_2, using the previous result to update the next one, and leads to an additional term to a series. The sequence given by Equations 2.217 is clearly the series approximation to Equations 2.214 and 2.215.

The self consistent field technique is the most straightforward and elegant one provided that one gets a simple algebraic equation that is possible to solve for p_1. When the self-consistent field equations cannot be solved, the series approach is the appropriate one. Furthermore, it is amenable to numerical solutions.

2.4.2 Two Polarizable Molecules in a Static Electric Field

In this section, we calculate the polarizability of two molecules in a uniform electric field **E**, which induces a dipole moment in each one (see Figure 2.14b). The local field at each molecule is then the sum of the applied electric field and the dipole field due to the other molecule, or

$$F_1 = E - \frac{p_2}{r^3} \quad \text{and} \quad F_2 = E - \frac{p_1}{r^3}. \quad (2.218)$$

To get the polarizability, one first calculates $p(E) = p_1(E) + p_2(E)$ using any of the approaches discussed above, from which $\alpha^{\text{eff}} = \partial p/\partial E|_{E=0}$ gives the effective polarizability of the two together, as described below.

Derivative Method

The derivative method uses the fact that $\alpha_i = \partial p_i/\partial F$ defines the polarizability of Molecule i and $\alpha_i^{\text{eff}} = \partial p_i/\partial E$ is the effective polarizability that Molecule

i contributes to the aggregate material response. The linear or nonlinear response of a collection of molecules, then, is a sum over the effective responses of the individual ones.

Given Equations 2.218 for the local electric fields at each molecule, the induced dipoles are given by

$$p_1 = \alpha F_1 = \alpha \left(E - \frac{p_2}{r^3}\right) \qquad (2.219)$$

and

$$p_2 = \alpha F_2 = \alpha \left(E - \frac{p_1}{r^3}\right). \qquad (2.220)$$

With the definition

$$\alpha_i^{\text{eff}} = \left.\frac{\partial p_i}{\partial E}\right|_{E=0}, \qquad (2.221)$$

differentiating Equations 2.219 and 2.220 with respect to applied electric field and adding them yields

$$\alpha_1^{\text{eff}} + \alpha_2^{\text{eff}} = 2\alpha - (\alpha_1^{\text{eff}} + \alpha_2^{\text{eff}}) \cdot \frac{\alpha}{r^3}. \qquad (2.222)$$

Solving Equation 2.222 for $\alpha^{\text{eff}} = \alpha_1^{\text{eff}} + \alpha_2^{\text{eff}}$ yields

$$\alpha^{\text{eff}} = \alpha_1^{\text{eff}} + \alpha_2^{\text{eff}} = \frac{2\alpha}{1 + \frac{\alpha}{r^3}}. \qquad (2.223)$$

Applying the other methods to this problem is left as an exercise.

In the limit of infinite separation, Equation 2.223 says that the effective polarizability is the sum of individual polarizabilities, which is expected for non-interacting molecules. When the separation is $r = \alpha^{1/3}$, i.e. when the molecules "touch," the effective polarizability is α, and at $r = 0$, the induced dipole moments due to dipole interactions cancel the applied electric field, so the effective polarizability vanishes.

The derivative method become useful in calculating higher-order susceptibilities in cases where the self-consistent field method becomes messy.

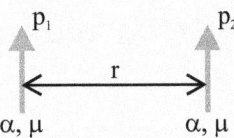

Problem 2.4-1(a): Two molecules are placed side by side a distance r apart. Each of them has a permanent dipole moment, μ, and a polarizability, α. No electric field is applied to the system. Using the self-consistent field method, calculate the dipole moment of the two-molecule system as a function of r when the two molecules are aligned.

(b): Calculate the local electric field at one of the molecules as a function of r.

(c): Repeat this problem for the case when the permanent dipoles point in opposite directions.

(d): Discuss the limiting cases when $r \to \infty$, $r \to \alpha^{1/3}$ and $r \to 0$. Argue why these limiting cases make sense.

(e): What if $\alpha < 0$? Is this ever the case?

Problem 2.4-2: Two molecules are placed side by side and aligned parallel to each other separated by a distance r as shown in Figure 2.14b.

(a) If the polarizability of each molecule is α, show that the effective polarizability of the two-molecule system is

$$\alpha^{\text{eff}} = \frac{2\alpha}{1 + \frac{\alpha}{r^3}}.$$

This problem was solved in Section 2.4.2 using the derivative technique. Apply the series and self-consistent fields techniques to solve the problem as a warmup for part b.

(b) Now assume that one molecule is linear with only a nonzero polarizability α while the second molecule is a purely second-order molecule with only nonzero β. Calculate the effective polarizability, hyperpolarizability and second hyperpolarizability of the two-molecule system. The self-consistent field method is a bit messy, but works. The derivative method is marginally simpler.

2.5 Magnetic and Quadrupole Contributions to Nonlinear Optics

Since the electric polarization describes the largest contributions to the nonlinear response, it is often the sole focus of nonlinear-optics textbooks. As such, material models of the nonlinear-optical response come down to determining how charges respond to an electric field to induce an electric dipole moment. This one simple approximation leads to a wealth of phenomena.

Nonlinear optics is complex enough to learn without adding additional complications such as magnetic contributions or higher-order moments. However, these higher-order terms are the subject of important research and therefore deserve mention. We will use an intuitive approach to help build an understanding of the origin of these effects.

In Gaussian units, the electric and magnetic fields have the same units, as do magnetic and electric dipoles. So, it is easy to generalize Maxwell's equations since every electric phenomenon has a magnetic analogue described by an equation of the same form. For example, the constitutive equation for the electric displacement given by Equation 2.76 and its consequences give the magnetic counterparts, i.e. the magnetic flux density $\mathbf{B}$ – which in the physics vernacular is often inaccurately identified as the magnetic field – is related to the magnetic field $\mathbf{H}$, and the magnetization, $\mathbf{M}$, according to,

$$\mathbf{B} = \mathbf{H} + 4\pi\mathbf{M} = \mu\mathbf{H}. \tag{2.224}$$

where μ is a dimensionless second rank tensor called the magnetic permeability. In the linear case, the dimensionless linear permittivity $\chi_m^{(1)}$ relates the magnetic field to the magnetization according to

$$\mathbf{M} = \chi_m^{(1)}\mathbf{H}. \tag{2.225}$$

Substituting Equation 2.225 in Equation 2.224 yields

$$\mathbf{B} = \mathbf{H} + 4\pi\chi_m^{(1)}\mathbf{H} = \left(1 + 4\pi\chi_m^{(1)}\right)\mathbf{H}, \tag{2.226}$$

whence the *linear magnetic permeability* μ_0 (not to be confused with the permeability of free space in SI units), is given by

$$\mu_0 = 1 + 4\pi\chi_m^{(1)}, \tag{2.227}$$

or in tensor form,

$$\mu_{ij} = \delta_{ij} + 4\pi\chi_{m_{ij}}^{(1)}. \tag{2.228}$$

The subscript m reminds us that these are magnetic properties. Note that μ_0, $\chi_m^{(1)}$, and μ are dimensionless, so $\mathbf{B}$, $\mathbf{M}$, and $\mathbf{H}$ have the same units. Thus, the rest of Maxwell's Equation can be easily generalized in this way.

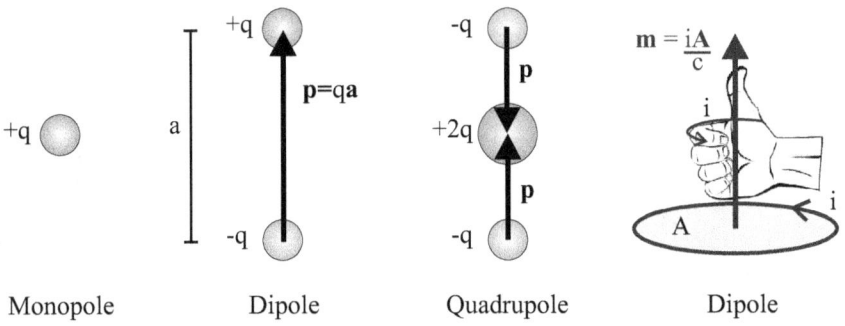

Figure 2.15: A monopole, dipole and quadrupole in one dimension. A current loop is also a magnetic dipole, where its direction is given by the righthand rule.

2.5.1 Multipole Moments

We begin with a general discussion of multipole moments. Figure 2.15 shows a monopole, made of a single charge, a dipole made by displacing a positive charge from a negative one, and a quadrupole, made by displacing two dipoles with opposing orientations. While mutipoles are generally three-dimensional objects, we will restrict the discussions that follow to point charges in one dimension to keep it simple.

The contributions of the higher-order moments are taken into account by their contributions to the electric displacement

$$D_i = E_i + 4\pi \left(P_i + \nabla_j Q_{ij} + \nabla_j \nabla_k O_{ijk} + \ldots \right) \tag{2.229}$$

and magnetic flux density

$$B_i = H_i + 4\pi \left(M_i + \nabla_j Q_{m_{ij}} + \nabla_j \nabla_k O_{m_{ijk}} + \ldots \right), \tag{2.230}$$

where summation convention applies. Q and O are the electric quadrupole and octupole moment densities, which we call the electric quadrupolarization and octupolarization; and, Q_m and O_m are the magnetic quadrupole and octupole moment densities, which we call the magnetic quadrupolarization and octupolarization. The macroscopic response of a material emerges from excitations of the various moments in the microscopic constituents in response to electric and magnetic fields.

To understand how the higher-order moments contribute to the fields, let's start by thinking about the polarization, which is more familiar to most of us, in a pictorial way. Figure 2.16a shows a sheet of aligned dipoles, which

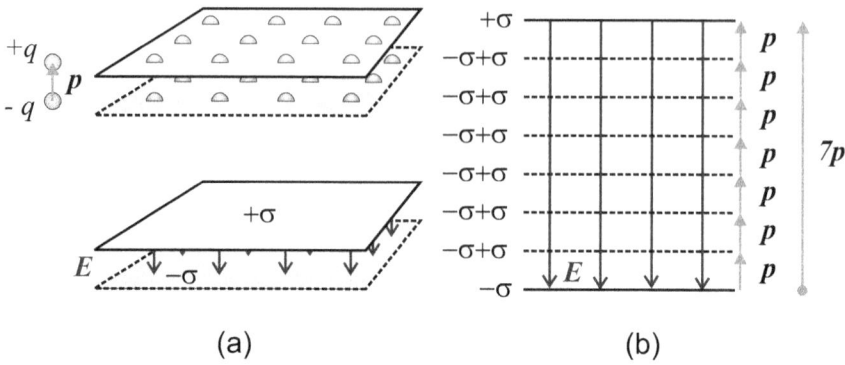

Figure 2.16: (a) A sheet of dipoles act as two sheets of opposite surface charge density, which for a small-enough gap, give a uniform field between them. (b) When such sheets are stacked – as shown here in side view, the charge densities cancel for all internal sheets, so the net field is due to the two outer ones.

act like two charged sheets of opposite surface charge density. A material is built by stacking such sheets. However, Figure 2.16b shows how the charges on all of the internal sheets cancel, so only the top and bottom surface sheets are charged. The far left part of the figure shows how the net dipole moment of the two outer chargers is equivalent to the sum over the individual dipoles.

The electric field in the material can be calculated using the fact that a planar sheet of surface charge density σ contributes a uniform electric field of magnitude $E = 2\pi\sigma$, so the two sheets of opposite charge density reinforce the electric field between them, where $E = 4\pi\sigma$; and, the fields outside cancel and vanish. Note that the electric field is in the direction opposite to the dipole moment.

For a cubic chunk of material of thickness d and cross-sectional area A that carries the net charge q, the polarization is given by $P = p/Ad$, where p is the total dipole moment. But $p = qd$, so the polarization is then given by

$$P = \frac{qd}{ad} = \frac{q}{a} = \sigma = -E/4\pi, \tag{2.231}$$

where the negative sign reflects the fact that the electric field is opposite to the dipole moment as illustrated in Figure 2.16b. Using equation 2.229 when the higher-order moments are absent, Equation 2.231 implies that $D = E + 4\pi P = 0$. This make sense since the charges reside inside dipoles, so they are bound charges. With no free charges, $D = 0$. Note that we know the charges are bound because they would not otherwise remain separated in

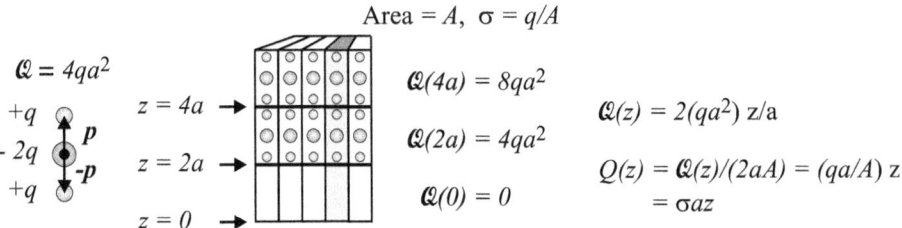

Figure 2.17: A vertical slice of material made of quadrupolar layers where the quadrupole moment increases linearly with each layer.

this way. Magnetic dipoles give the same result, so the magnetic field opposes the magnetization, so the flux density vanishes.

Next we consider the electric quadrupole moment $\mathcal{Q}$, which is a tensor. To avoid unnecessary complications, we consider only the diagonal component along z for a collection of point charges, or

$$\mathcal{Q} \equiv \mathcal{Q}_{zz} = 2\sum_i q_i z_i^2, \qquad (2.232)$$

where z_i is the coordinate of charge q_i.

We saw that the dipole moment is simply a vector sum of the dipole moments of the microscopic units. It is not obvious that this will also hold for the quadrupole moment, which is a tensor. To that end, we calculate the quadrupole moment for the arrangement of point charges shown in Figure 2.15, where we pick a coordinate system such that the bottom charge $-q$ is at coordinate z and the spacing between charges in a. From Equation 2.232 we get

$$\mathcal{Q} = 2\left(-qz^2 + 2q(x+a)^2 - q(x+2a)^2\right) = -4qa^2. \qquad (2.233)$$

Equation 2.233 is independent of coordinate system, so a collection of N quadrupoles, each of quadrupole moment $\mathcal{Q}$, will have a quadrupole moment $N\mathcal{Q}$. The quadrupolarization for a chunk of material of volume V with N dipoles is $Q = \frac{N}{V}\mathcal{Q}$.

Next we describe the need for the gradient in Equation 2.229. Figure 2.17 highlights one vertical slice of material that contains a single stack of quadrupoles whose quadrupole moments increases by one unit between successive layer according to

$$\mathcal{Q}(z) = 2qa^2 \cdot \frac{z}{a}. \qquad (2.234)$$

The cross-sectional area of each layer is A and each quadrupole occupies a hight $2a$ for a volume of $2aA$. Then, the electric quadrupolarization is calculated from Equation 2.234, yielding

$$Q(z) = \frac{\mathcal{Q}(z)}{2aA} = \frac{q}{a}z = \sigma z, \tag{2.235}$$

where we have used the fact that the surface charge density is given by $\sigma = q/A$. The gradient of Equation 2.235 gives

$$\frac{\partial Q(z)}{\partial z} = \sigma. \tag{2.236}$$

To see what this all means, substitute Equation 2.236 into Equation 2.229 for a system with no free charge and no polarization, so that the electric displacement and polarization vanish. This yields $E = -\nabla Q = -4\pi\sigma$, the same result as as Equation 2.231. As such, an electric quadrupolarization that increases linearly with distance is equivalent to a uniform polarization in the electric field that it creates.

The polarization P is usually the largest moment density. The electric quadrupolarization Q is typically of the same magnitude as the magnetization M, the electric octupolarization O is of the same order of magnitude as the magnetic quadrupolarization Q_m, etc. One can always design a material to force one of the moments to vanish so that these rough relationships no longer hold. For example, in a material that does not polarize, the magnetization can be larger than the polarization, but if each were optimized, the polarization would win.

2.5.2 Magnetic Monopoles

Magnetic monopoles have not been observed,[9] and their absence is reflected in the formulation of Maxwell's equations. However, given the symmetry of the equations in Gaussian units, if they are ever found to exist, magnetic monopoles can be easily added using the fact that magnetic monopoles and electric charges have the same units as do electric and magnetic currents.

Dirac showed that the existence of magnetic monopoles forces electric charge to be quantized,[9] and the existence of an electric monopole (the electron being an example) implies that magnetic charge is quantized. He determined the quantization relationship by considering how to stitch together the vector potential to avoid a singularity. Such a calculation is beyond the scope of this textbook; so instead, we state Dirac's quantization condition without proof as

$$q_m = n\frac{\hbar c}{2q_e}, \tag{2.237}$$

where n is an integer.

We can apply Equation 2.237 to all the derivations in nonlinear optics with electric charges to get analogous results for magnetic charges. Since Equation 2.237 implies that the magnetic monopole's magnetic charge is larger than the electron's charge, nonlinear susceptibilities in exotic materials made with magnetic monopoles would be much larger than electric susceptibilities provided that the monopole masses are comparable in magnitude.

In such a universe of complete symmetry, moving magnetic charges would constitute a magnetic current that would create an electric field in the magnetic analogy to Ampere's law and point magnetic charges would obey the magnetic equivalent of Coulombs Law. The Lorentz force would also be generalized in the same way.

> **Problem 2.5.2-1:** Verify that the smallest possible magnetic monopole moment is at least 68 times the electron charge.
>
> **Problem 2.5.2-2:** Generalize Maxwell's equations to include magnetic monopoles and magnetic current.

2.5.3 Magnetic Dipoles

The lowest-order moments observed in nature are the electric monopole and the magnetic dipole. The electron is an elementary particle that is both a point electric monopole and point magnetic dipole. A charged particle orbiting a center of force, such as an electron in a hydrogen atom, yields a physical magnetic dipole moment. Figure 2.15 shows that a current loop gives a magnetic dipole moment of magnitude

$$m = iA/c, \qquad (2.238)$$

where A is the area of the loop, i the current that it carries, and c the speed of light. The dipole vector is perpendicular to the loop and its direction is determined from the righthand rule. To get its direction, wrap your fingers around the loop pointing in the direction of the current, and your thumb will point along the dipole axis.

An electric field applied to a molecule induces a distortion in the electron cloud, leading to a redistribution of charge. The net separation between the electron cloud and the nuclei results in a change in the dipole moment. There is no magnetic counterpart in the absence of magnetic charges.

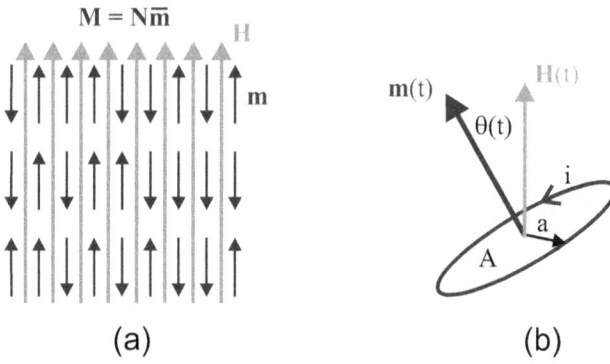

Figure 2.18: (a) Magnetization from an ensemble of magnetic dipoles at finite temperature and (b) induced dipole moment of a wire loop due to magnetic induction and subsequent rotation from magnetic torque.

The simplest way for a static magnetic field to induce a magnetization is for it to align existing magnetic dipoles. Alternatively, a changing electric field can induce a magnetic dipole moment. These two cases are treated below.

Linear and Nonlinear Permittivities of Permanent Dipoles

As an illustration, consider a dilute collection of noninteracting magnetic dipoles of number density N that can be aligned along or opposite to the applied magnetic field as shown in Figure 2.18a. At temperature T, statistical mechanics tells us that the probability of finding the magnetic dipole moment m along or opposite to the magnetic field H is given by

$$P_{\pm} = \frac{\exp\left[\pm \frac{mH}{kT}\right]}{\exp\left[+\frac{mH}{kT}\right] + \exp\left[-\frac{mH}{kT}\right]}. \tag{2.239}$$

The average magnetization is then given by

$$M = Nm(P_+ - P_-) = Nm \cdot \frac{\exp\left[+\frac{mH}{kT}\right] - \exp\left[-\frac{mH}{kT}\right]}{\exp\left[+\frac{mB}{kT}\right] + \exp\left[-\frac{mH}{kT}\right]} = Nm \tanh\left(+\frac{mH}{kT}\right). \tag{2.240}$$

If the interaction energy between the dipole and the field is small compared with thermal energies, or $mH \ll kT$, the hyperbolic tangent can be expanded in a series of the form $\tanh x = x - \frac{1}{3}x^3 + \dots$. Applying the magnetic analog of Equation 1.33 to Equation 2.240 then yields

$$\chi_m^{(1)} = \left.\frac{\partial M}{\partial H}\right|_{H=0} = \frac{Nm^2}{kT}, \qquad (2.241)$$

$$\chi_m^{(2)} = \left.\frac{1}{2}\frac{\partial^2 M}{\partial H^2}\right|_{H=0} = 0, \qquad (2.242)$$

and

$$\chi_m^{(3)} = \left.\frac{1}{6}\frac{\partial^3 M}{\partial H^3}\right|_{H=0} = -\frac{Nm}{3}\left(\frac{m}{kT}\right)^3. \qquad (2.243)$$

Note that as the temperature is increased, the linear and nonlinear response vanishes. This behaviour is expected because thermal jiggling fights the alignment torque due to the magnetic field. In the limit $T \to 0$, the odd susceptibilities diverge. Recall that Equations 2.241 through 2.243 define susceptibilities only for small enough field strength so that the linear susceptibility is larger than the nonlinear ones, or when $mH \ll kT$.

Linear and Nonlinear Permittivity of Induced Dipoles

The previous example describes the alignment of magnetic dipoles in response to a magnetic field. Since there are no magnetic monopoles, a material without magnetic dipoles will not polarize in response to a static magnetic field. However, a time varying field can induce a dipole.

Consider first a conducting circular loop of radius a and resistance R as shown in Figure 2.18. We can imagine a material made of of tiny loops of this sort. For now, let's assume that the loop's orientation is fixed so that θ, the angle between the applied magnetic field $\mathbf{H} = H\hat{z}$ and the dipole moment, is a constant. A time-varying magnetic flux through the loop induces a current, which is a time-varying magnetic dipole moment.

For a time harmonic field $B = B_0 \cos(\omega t)$, the flux Φ through the loop is given by

$$\Phi = B_0 \cdot \pi a^2 \cdot \cos\theta \cdot \cos(\omega t). \qquad (2.244)$$

Note that we assume that the loop is in a vacuum, so the magnetic field and the flux density are the same. The electromotive force ($\mathscr{EMF}$) $\mathscr{E}$ is given by Faraday's Law, yielding

$$\mathscr{E} = -\frac{1}{c}\frac{\partial \Phi}{\partial t}, \qquad (2.245)$$

Using Equations 2.244 and 2.245, Ohm's law ($\mathscr{E} = iR$) gives the current

$$i = \frac{\mathscr{E}}{R} = \frac{H_0 \pi a^2 \omega \cos\theta}{cR} \sin(\omega t), \qquad (2.246)$$

from the magnitude of the magnetic dipole moment is

$$m(t) = \frac{H_0 \pi^2 a^4 \omega}{c^2 R} \cos\theta \sin(\omega t). \tag{2.247}$$

The z-component of the dipole moment is given by

$$m_z(t) = \frac{H_0 \pi^2 a^4 \omega}{c^2 R} \cos^2\theta \sin(\omega t). \tag{2.248}$$

Equation 2.249 can be expressed in complex form

$$m^\omega \cdot \frac{e^{-i\omega t}}{2} + \text{c.c.} = \frac{\pi^2 a^4 \omega}{c^2 R} \cos^2\theta \cdot H_0^\omega \frac{e^{-i\omega t}}{2i} + \text{c.c.}, \tag{2.249}$$

from which we can get the magnetic polarizability

$$\alpha_{mzz}(-\omega;\omega) = -\frac{i\pi^2 a^4 \omega}{c^2 R} \cos^2\theta. \tag{2.250}$$

The magnetic polarizability behaves as we would expect. Since the induced current is the time derivative of the flux, the induced dipole moment should increase with the frequency of the magnetic flux density. If the loop's axis is perpendicular to the magnetic flux, then the flux vanishes. Finally, the magnetic polarizability scales as the square of area of the loop (the larger the loop, the larger the flux AND the larger the dipole moment) and inversely with the resistance.

If the loop is free to rotate, the the magnetic field will exert a torque on it, inducing oscillations, leading to higher-order nonlinearities. This problem is too complex to merit taking up space in a book; but, it is worthwhile to give it a try.

> **Problem 2.5.3-1(a):** Calculate the first and second hyperpolarizabilities β at 2ω and γ at ω and 3ω of a freely rotating loop in response to a time-harmonic electric field. It may be wise to express the solution as a small oscillation about a fixed angle θ_0. Note that this is not a simple problem unless your bag of tricks is bigger than mine.

2.5.4 Electric Quadrupole

Figure 2.19 shows a model system that can be used to study electric quadrupolar excitations. It is made of two springs of unequal force constants

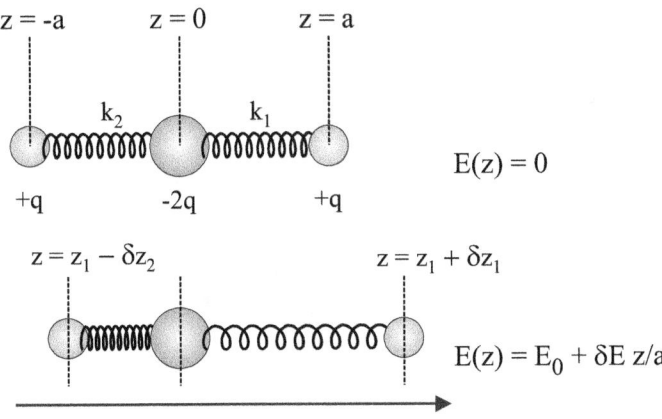

Figure 2.19: (a) An electric quadrupole modeled as charges connected with two unequal springs. The moments change under the application of an electric field with a field gradient of $\delta E/a$.

but of equal equilibrium lengths. The most general form of applied electric field needed to illustrate a wealth of excitations is

$$E(z) = E_0 + \frac{\delta E}{a} z. \tag{2.251}$$

The charge $-2q$ will be assumed massive so that it remains stationary, as do nuclei. We assign the nuclear charge a negative sign to avoid the need for tracking minus signs. In the end, we can set $q = -e$.

The equilibrium position of each charge is determined from the balance between the electric force and the restoring spring force, $qE(z_1) = k_1(z_1 - a)$ and $qE(z_2) = k_2(z_1 + a)$; so, using the electric field from Equation 2.251, the displacements z_1 and z_2 are given by

$$z_1 = \frac{k_1 a + qE_0}{k_1 - q\frac{\delta E}{a}} \quad \text{and} \quad z_2 = \frac{-k_2 a + qE_0}{k_2 - q\frac{\delta E}{a}}. \tag{2.252}$$

Note that we ignore the electric interactions between the charges, which may at first seem like a blatant disregard for reality. However, there are no little springs inside atoms and molecules. Instead, the superposition of all electrical forces leads to an effective potential well, which for small oscillations, gives simple harmonic motion. This is, in effect, what we are being modeled.

Induced Dipole

The electric dipole moment is given by $\sum_i q_i z_i$, yielding

$$p\left(E_0, \frac{\delta E}{a}\right) = qz_1\left(E_0, \frac{\delta E}{a}\right) + qz_2\left(E_0, \frac{\delta E}{a}\right), \tag{2.253}$$

where we must keep in mind that the induced dipole moment depends on both the electric field E_0 and the field gradient $\delta E/a$.

From Equations 2.252 and 2.253, the polarizability is

$$\alpha = \left.\frac{\partial p}{\partial E_0}\right|_{E_0=0, \frac{\delta E}{a}=0} = q^2\left(\frac{1}{k_1} + \frac{1}{k_2}\right). \tag{2.254}$$

and since the springs are linear, the hyperpolarizability and higher order ones vanish. The polarizability is in the form that we would expect.

Can a field gradient induce a dipole moment? Let's check by calculating a new type of polarizability, which we shall call α', and define according to

$$\alpha' = \left.\frac{\partial p}{\partial\left(\frac{\delta E}{a}\right)}\right|_{E_0=0, \frac{\delta E}{a}=0} = aq^2\left(\frac{1}{k_1} - \frac{1}{k_2}\right), \tag{2.255}$$

Note that this term vanishes if the two springs have the same force constants. This makes sense because the gradient force acts on the two charges in opposite directions, so the induced dipole moments cancel when the spring constants are the same.

Interestingly, the second derivative is also nonzero, giving the non-zero nonlinearity

$$\beta' = \frac{1}{2}\left.\frac{\partial^2 p}{\partial\left(\frac{\delta E}{a}\right)^2}\right|_{E_0=0, \frac{\delta E}{a}=0} = aq^3\left(\frac{1}{k_1^2} - \frac{1}{k_2^2}\right). \tag{2.256}$$

We might also imagine a wacky term that comes from taking a first derivative with respect to the electric field and another with respect to the gradient, giving a mixed term. This one we shall call $\xi^{(1,1)}$ where the first argument labels the number of field derivatives and the second one the number of gradient derivatives. Applying this to the quadrupole model yields

$$\xi^{(1,1)} = \left.\frac{\partial^2 p}{\partial E_0 \partial\left(\frac{\delta E}{a}\right)}\right|_{E_0=0, \frac{\delta E}{a}=0} = q^3\left(\frac{1}{k_1^2} + \frac{1}{k_2^2}\right), \tag{2.257}$$

where nonzero $\xi^{(1,1)}$ does not require that the force constants be different.

We can generalize these new terms to tensor form, yielding an induced dipole moment

$$\begin{aligned} p_i &= \alpha_{ij}E_j + \alpha'_{ijk}\partial_j E_k + \beta_{ijk}E_j E_k + \xi^{(1,1)}_{ijkl}\partial_j E_k E_l \\ &+ \beta'_{ijklm}\partial_j \partial_k E_l E_m + \gamma_{ijkl}E_j E_k E_l + \ldots, \end{aligned} \qquad (2.258)$$

where $\partial_i = \frac{\partial}{\partial x_i}$. This illustrates how nonlinear optics, in the way it is normally presented, is just the tip of the iceberg of nonlinear phenomena. One can also include contributions from double, triple and multiple gradients as well as mixtures of these, which are characterised by $\xi^{(a,b,c,\ldots)}_{ijk\ldots}$ where the derivative with respect to the electric field appears a times, the gradient appears b times, the second-order gradient appears c times, and so on.

> **Problem 2.5.4-1:** If a molecule is centrosymmetric, which terms in Equation 2.258 must vanish? Do your results agree with what you would have guessed from Equation 2.254 to 2.257?

Induced Quadrupole

The diagonal component of the Quadrupole moment along the three charges in Figure 2.19 is defined by Equation 2.232, so is of the form

$$\mathcal{Q}\left(E_0, \frac{\delta E}{a}\right) = 2qz_1^2\left(E_0, \frac{\delta E}{a}\right) + 2qz_2^2\left(E_0, \frac{\delta E}{a}\right). \qquad (2.259)$$

The quadrupole moment with zero electric field applied is given by $\mathcal{Q} = 4qa^2$. The quadrupolarizability $\alpha_{\mathcal{Q}}$ can be calculated from Equation 2.259 and the position of the charges give by Equation 2.252, yielding

$$\alpha_{\mathcal{Q}} = \left.\frac{\partial \mathcal{Q}}{\partial E_0}\right|_{E_0=0, \frac{\delta E}{a}=0} = 4aq^2\left(\frac{1}{k_1} - \frac{1}{k_2}\right). \qquad (2.260)$$

Similarly, the effects of the electric field gradient can be characterized by the gradient quadrupolarizability $\alpha'_{\mathcal{Q}}$

$$\alpha'_{\mathcal{Q}} = \left.\frac{\partial \mathcal{Q}}{\partial\left(\frac{\delta E}{a}\right)}\right|_{E_0=0, \frac{\delta E}{a}=0} = 4a^2q^2\left(\frac{1}{k_1} + \frac{1}{k_2}\right). \qquad (2.261)$$

We can continue to get higher order terms, but this exercise has little pedagogical payoff. However, these higher order terms might be important in certain applications or materials classes. In those cases, the patterns shown here can be generalized to both higher-order terms or to higher moments.

As we did for the polarization, the quadrupole expansion can can be generalized to tensor form, yielding

$$\mathcal{Q}_{ij} = \alpha_{\mathcal{Q}ijk}E_k + \alpha'_{\mathcal{Q}ijkl}\partial_j E_l + \beta_{\mathcal{Q}ijkl}E_k E_l + \xi_{\mathcal{Q}ijklm}^{(1,1)}\partial_k E_l E_m + \ldots \tag{2.262}$$

> **Problem 2.5.4-2:** Determine which of the three lowest-order quadrupolarizabilities must vanish if the molecule is centrosymmetric. How does this compare with the electric polarization? Is there a pattern?
>
> **Problem 2.5.4-3:** A simple model system for an induced magnetic quadrupole consists of two coaxial and parallel loops of wire of the same size but made of different materials with different resistance. Use the results from Section 2.5.3 to develop a parallel formalism for the magnetic quadrupole as we did for electric quadrupole in Section 2.5.4.

2.5.5 Mixed Moments

There are other complications that can come into play. So far, we have considered only purely electric or purely magnetic phenomena; an electric field induces an electric moment and a magnetic field indices a magnetic moment. It is also possible for a magnetic field to induce an electric moment, or *vice versa*. It is also possible for a combination of electric and magnetic fields to act together to lead to moments of both types. Rather than present the most general theory, we point out ways that such processes can be viewed.

First, picture an electric dipole glued to a magnetic dipole. Such coupled systems are found in abundance. For example, non-centrosymmetric molecules have electric dipole moments and the electrons have magnetic moments. Thus, an applied electric field will align the electric dipoles; and, the magnetic dipole goes along for the ride. If the dipolar alinement is a nonlinear process, it will lead to mixed susceptibilities, which come about from the actions of both electric and magnetic fields.

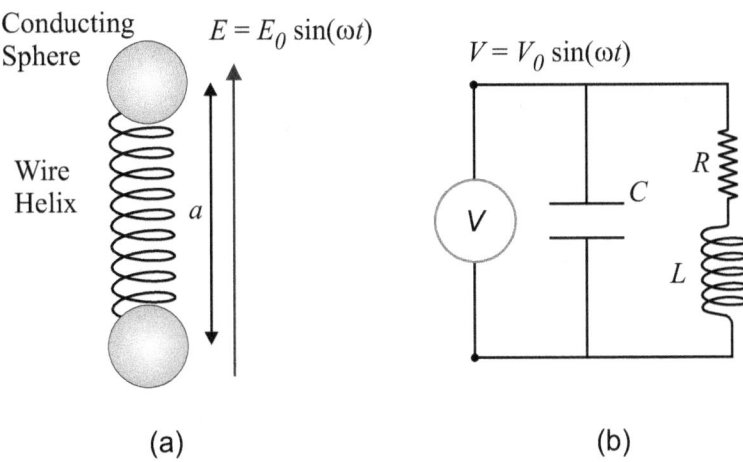

Figure 2.20: (a) A helical molecule modeled as a resistive coil with conducting ends. (b) The equivalent circuit of the molecule, where the coil is modeled as an inductor and resister in series, and the spherical conductors as a capacitor.

For example, we can imaging that an electric quadrupole moment is induced under the action of an electric and magnetic field, which would be quantified by the Qudrupolarization,

$$Q_{ij} = \chi^{eem}_{ijkm} E_j B_m, \tag{2.263}$$

where χ^{eem}_{ijkm} yields an electric moment (first e) under the actions of an electric (second e) and magnetic field (m). Note that there are never enough symbols nor room to include every possible variable, so abbreviations such as χ^{eem}_{ijkm} are used, and the reader needs to determine the other relevant factors from the context of the physical system and the process under consideration.

We finish this section with an example that illustrates mixed susceptibilities. Consider the helical molecule shown in Figure 2.20. The helical part is made of a resistive material with resistance R and the ends are metallic, giving the system a capacitance C. The ends are separated by a distance a and the helix is made of N turns. An oscillating uniform electric field, originating in a beam of light – for example, is applied along the molecules axis.

Figure 2.20b shows the equivalent electrical circuit for the the molecule. The two spherical ends, separated by distance a, can be viewed as a capacitor to which a voltage is applied. The potential difference can be calculated form the electric field, and is given by

$$V(t) = -aE_0 \cos(\omega t) = \mathrm{Re}[aE_0 e^{-i\omega t}]. \tag{2.264}$$

The charge on the capacitor is then given by $q(t) = CV(t)$, so the induced dipole moment in complex form is

$$p(t) = aq(t) = -a^2 CE_0 e^{-i\omega t}. \tag{2.265}$$

Equation 2.265 shows that an electric dipole is excited by an electric field, so this correspond to a purely electric process characterized by the polarizability

$$\alpha_{zz}^{ee}(-\omega;\omega) = -a^2 C. \tag{2.266}$$

Next, we consider the magnetic dipole moment that is induced, which originates in the flow of current in the inductor. Thus, we use Kirchhoff's circuit law on the outer loop – which consists of the driving potential difference, the resistor and the inductor – to determine the current in the inductor, which is given by

$$I(t) = V(t)/(R + i\omega L). \tag{2.267}$$

Note that we are using the complex form of Kirchhoff's law, where the impendence of an inductor and capacitor are $i\omega L$ and $1/i\omega C$.

The magnetic dipole moment, as define by Equation 2.238 is given by

$$m(t) = \frac{NI(t)A}{c}, \tag{2.268}$$

where N is the number of turns in the helix, A is the area of a single loop, and $I(t)A/c$ is the magnetic dipole moment per turn. Applying Equations 2.264 and 2.267 to Equation 2.268 yields

$$m(t) = -\frac{NAa}{c(R + i\omega L)} E_0 e^{-i\omega t}. \tag{2.269}$$

Equation 2.269 defines the complex "magnetic-electric" polarizability, which is given by

$$\alpha_{zz}^{me} = -\frac{NAa}{c(R + i\omega L)}. \tag{2.270}$$

At low frequency, the resistor determines the current, making the polarizability real so that the field and dipole moment are in phase. At high frequency, the voltage drop is across the inductor, so the polarizability becomes purely imaginary, introducing a $\pi/2$ phase shift between the electric field and the induced magnetic dipole moment.

The inductance of a helix in Gaussian units is given by

$$L = \frac{4\pi}{c^2} \cdot \frac{A}{\ell} \cdot N^2, \tag{2.271}$$

where ℓ is the length of the inductor. This allows us to express the magnetic-electric polarizability in terms of geometric factors of the "molecule" using Equation 2.270.

> **Problem 2.5.5-1:** Pick reasonable material parameters for the "molecule" in Figure 2.20 under the action of the electric field (i.e. assume that the helix of N turns is made of an ohmic material such as copper, the capacitance is estimated from the parallel plate case, etc.) and make the molecule's size about of 20Å long and 4Å wide from which your are to get numerical values of the electric-electric and magnetic-electric polarizabilities. Use Gaussian units throughout.
>
> **Problem 2.5.5-2(a):** Instead of an applied electric field, use an applied oscillating magnetic field to the "molecule" in Figure 2.20 to calculate the magnetic-magnetic and electric-magnetic polarizabilities.
>
> **(b)** Use the same material parameters as you did in Problem 2.5.5-1 to get numerical values, and compare the four polarizabilities to each other and order them by rank. Note that in SI units, all these quantizes would have different units.

2.6 Symmetry

Section 2.3.9 describes the use of symmetry to determine which tensor components of the electric susceptibility vanish. Symmetries can also be applied to magnetic systems as well as mixed magentic/electric susceptibilities. However, magnetic fields behave differently.

As an example, consider the magnetic polarization $\mathbf{M}$ to second-order in the magnetic flux density $\mathbf{B}$, which are related to each other through the second-order magnetic susceptibility $\chi^{(2)}$ according to

$$M_x = \chi^{(2)}_{xxx}(\mathbf{r}')B_x B_x. \tag{2.272}$$

As with the electric susceptibilities, the argument $\mathbf{r}'$ represents coordinates within the material used to determine the magnetic susceptibility.

We first consider a centrosymmetric material for which $\chi^{(2)}_{xxx}(-\mathbf{r}') = \chi^{(2)}_{xxx}(\mathbf{r}')$. The complication is that $\mathbf{M}$ and $\mathbf{B}$ are computed from a cross product of two vectors, making them pseudo-vectors. To see that $\mathbf{M}$ is a pseudo-vector, recall that it is proportional to an area vector. For a rectangular area of sides x and y, the area vector is given by $\mathbf{A} = \mathbf{x} \times \mathbf{y}$. The inversion operation changes the sign of **x** and **y**, leaving **A** unchanged. Thus, Applying inversion to Equation 2.272 returns Equation 2.272, thus not requiring the nonlinear susceptibility to vanish.

Next, consider inversion along just one of the coordinate axes. This will change a righthanded coordinate system to a left-hand one, and therefore changes the sign of all pseudo-vectors. Applying this change in handedness to Equation 2.272, and assuming that the nonlinear magnetic susceptibility is invariant to change in handedness, yields

$$-M_x = \chi^{(2)}_{xxx}(\mathbf{r}')(-B_x)(-B_x). \tag{2.273}$$

Adding Equations 2.272 and 2.273 gives $\chi^{(2)}_{xxx}(\mathbf{r}') = 0$ for a material that is invariant upon inverting one coordinate axis. Note that if a system is separately invariant to inversion along each of the three cartesian axes, it is also centrosymmetric.

> **Problem 2.6-1:** Determine which tensor components of all possible mixed third-order susceptibilities are required to vanish if the material is invariant upon inversion about the z-axis. Consider both the induced polarization and induced magnetization in response to all possible combinations of electric and magnetic field components.

2.6.1 Time Reversal Symmetry

The time reversal operator has the action $\Delta t \to -\Delta t$, the exact operation performed when playing a video in reverse. Therefore, quantities that are defined by the first (or odd-order) derivative with respect to time – such as velocity and current density – change sign under time reversal.

The invariance of Maxwell's Equations to time reversal can be used to determine the time symmetries of the fields. Since the charge density in Gauss's law (the first of Equations 3.1) is independent of time derivatives, the electric field must be invariant under time reversal. The third of Maxwell's equations establishes that the magnetic flux density **B** has the opposite symmetry to **E** because of the time derivative, making **B** antisymmetric to time reversal.

Such arguments can be applied to other quantities. For example, Equation 2.230 requires the magnetization **M** to be of the same symmetry as the magnetic flux density, and thus changes sign too. The magnetic contribution to the Lorentz force, on the other hand, is of the form $\mathbf{F}_B = q\mathbf{v} \times \mathbf{B}$. Time reversal changes the sign of both the velocity and magnetic flux density leaving the force invariant. This is as expected since the force is proportional to acceleration, which is a second time derivative and thus invariant. Convince yourselves that playing a video in reverse of a charge orbiting in a uniform magnetic field requires that the velocity be reversed together with the magnetic flux density vector in order for the path to be the same but with an orbit in the opposite direction.

As an example of how these symmetries are applied to the magnetic susceptibility, consider applying time reversal to the second-order magnetization

$$M_x = \chi^{(2)}_{xxx} B_x B_x. \tag{2.274}$$

The signs of both the magnetization and the magnetic flux density change, so

$$-M_x = \mathcal{T}\left[\chi^{(2)}_{xxx}\right](-B_x)(-B_x), \tag{2.275}$$

where $\mathcal{T}\left[\chi^{(2)}_{xxx}\right]$ is the time-reversed second-order magnetic susceptibility. Thus, if the second-order magnetic susceptibility is invariant upon time reversal, $\mathcal{T}\left[\chi^{(2)}_{xxx}\right] = \chi^{(2)}_{xxx}$ and adding equations 2.274 and 2.275 shows that this susceptibility must vanish. The same argument demands that all time-symmetric even-order magnetic susceptibilities vanish.

The magnetic susceptibility is invariant under time reversal when it is independent of properties derived from time derivatives such as currents or magnetic fields. Thus, the second-order magnetic susceptibility will not vanish when a net current flows in the material or in the presence of a spontaneous magnetization as is found in a ferromagnetic. All even-order magnetic susceptibilities can be non-vanishing when time reversal symmetry is broken.

Problem 2.6-2: What is the time symmetry of the scalar potential, vector potential and electric polarization?

Problem 2.6-3: Repeat Problem 2.6-1 if the mixed third-order susceptibility is invariant under time reversal.

Problem 2.6-4: If the sign of the nth-order mixed magnetic/electric susceptibility changes upon time reversal, determine which susceptibilities are required to vanish.

Problem 2.6-5: Comment on the statement: "Time reversal is to magnetic susceptibilities as the inversion operation is to electric susceptibilities."

Chapter 3

Nonlinear Wave Equation

This chapter is concerned with understanding the source of nonlinearity in Maxwell's Equations, which originates in the constitutive equations. These are then used to derive the nonlinear wave equation. The remainder of the chapter describes techniques for solving the nonlinear wave equation, followed by examples of specific nonlinear-optical processes.

3.1 General Technique

This section starts with a derivation of the nonlinear wave equation. Nonlinear differential equations are inherently difficult to solve, so we apply several tricks that work under the conditions of typical experiments. For example, the fact that only a small number of monochromatic fields are present allows the differential equations to be decoupled through Fourier transforms, as we did in Section 2.2, but in a more general way. For low-enough fields, the nonlinear terms are small, so the equations can be further decoupled by collecting terms of the same order. This protocol is spelled out in well-defined steps that can be generally applied to a variety of nonlinear-optical processes.

We start with Maxwell's equations, which are given by

$$\begin{aligned} \nabla \cdot \mathbf{D} &= 4\pi \rho_f, \\ \nabla \cdot \mathbf{B} &= 0, \\ \nabla \times \mathbf{E} &= -\frac{1}{c}\frac{\partial \mathbf{B}}{\partial t}, \\ \nabla \times \mathbf{H} &= \frac{4\pi}{c}\mathbf{J} + \frac{1}{c}\frac{\partial \mathbf{D}}{\partial t}. \end{aligned} \quad (3.1)$$

We assume that there are no free charges, so $\rho_f = 0$, and that the material is described by bound charges that act as the source of the nonlinear polarization. We assume that there are no free or induced currents, so $\mathbf{J} = 0$. Finally, we ignore the magnetic response, which implies that $\mathbf{B} = \mathbf{H}$. This assumption holds for a broad range of materials; however, we must exercise care to identify those cases where the magnetic susceptibilities contribute.

Using the above approximations and assumptions, and taking the curl of the third of Maxwell's equations yields

$$\nabla \times \nabla \times \mathbf{E} + \frac{1}{c}\frac{\partial(\nabla \times \mathbf{B})}{\partial t} = 0 \qquad (3.2)$$

while the fourth of Maxwell's equations is given by

$$\nabla \times \mathbf{B} = \frac{1}{c}\frac{\partial \mathbf{D}}{\partial t}. \qquad (3.3)$$

Substituting Equation 3.3 into Equation 3.2 yields

$$\nabla \times \nabla \times \mathbf{E} + \frac{1}{c^2}\frac{\partial^2 \mathbf{D}}{\partial t^2} = 0. \qquad (3.4)$$

The electric displacement, $\mathbf{D}$, is related to the polarization according to

$$\mathbf{D}(\mathbf{r}, t) = \mathbf{E}(\mathbf{r}, t) + 4\pi \mathbf{P}(\mathbf{r}, t), \qquad (3.5)$$

where $\mathbf{E}(\mathbf{r}, t)$ is the applied electric field and $\mathbf{P}(\mathbf{r}, t)$ is the polarization. Equation 3.5 explicitly shows that all the quantities depend on position and time. More importantly, the displacement is local in space and instantaneous in time, meaning that the electric displacement at point $\mathbf{r}$ is the sum of the electric field at point $\mathbf{r}$ and 4π times the polarization at point $\mathbf{r}$. The same is true for time. The electric displacement cares not about the electric fields and displacements at any other time but the present. However, as we saw in Section 2.3, the response functions can be nonlocal in space and non-instantaneous in time, so the polarization at one point in the material and at a given time depends on the fields at all times in the past and all other positions in the material. Recall that we ignored spatial non-locality, and took into account only temporal non-instantaneousness, which is usually a good approximation.

The most general form of the wave equation where all quantities are evaluated at the same position and time is

$$\nabla \times \nabla \times \mathbf{E} + \frac{1}{c^2}\frac{\partial^2 \mathbf{E}}{\partial t^2} + \frac{4\pi}{c^2}\frac{\partial^2 \mathbf{P}}{\partial t^2} = 0. \qquad (3.6)$$

The polarization in Equation 3.6 is usually separated into two pieces,

$$\mathbf{P} = \mathbf{P}^L + \mathbf{P}^{NL}, \qquad (3.7)$$

where $\mathbf{P}^L$ is the linear polarization that is the fare of traditional E&M courses and $\mathbf{P}^{NL}$ contains the nonlinearity that will be of interest to us.

The non-locality between the polarization and the electric field is in general difficult to treat. However, when only a finite number of monochromatic fields make up the incident and emitted waves, we can use the results in Section 2.3.2 to express the polarizations as

$$P_i^{\omega_\sigma}(\mathbf{r}) = \chi_{ij}^{(1)}(-\omega_\sigma;\omega_\sigma,\mathbf{r})E_j^{\omega_\sigma}(\mathbf{r}), \tag{3.8}$$

$$P_i^{\omega_\sigma}(\mathbf{r}) = K(-\omega_\sigma;\omega',\omega'',\mathbf{r})\chi_{ijk}^{(2)}(-\omega_\sigma;\omega',\omega'',\mathbf{r})E_j^{\omega'}(\mathbf{r})E_k^{\omega''}(\mathbf{r}), \tag{3.9}$$

etc., where the K factor is the one defined by Equation 2.166. The linear and nonlinear susceptibilities explicitly show the local position dependence within the material. Since materials are usually homogeneous, we drop the position dependence from here on – though, it can easily be added when needed. As usual, summation convention applies.

The polarizations and electric fields in Equations 3.8 and 3.9 are Fourier amplitudes, so the time dependence can be recovered by multiplying each equation by $\exp[-i\omega_\sigma t]$, yielding

$$P_i^{\omega_\sigma}(t) = \frac{P_i^{\omega_\sigma}}{2}e^{-i\omega_\sigma t} + \text{c.c.} = \frac{1}{2}\chi_{ij}^{(1)}(-\omega_\sigma;\omega_\sigma) \cdot E_j^{\omega_\sigma} e^{-i\omega_\sigma t} + \text{c.c.}, \tag{3.10}$$

and

$$\begin{aligned} P_i^{\omega_\sigma}(t) &= \frac{P_i^{\omega_\sigma}}{2}e^{-i\omega_\sigma t} + \text{c.c.} \\ &= \frac{1}{2}K(-\omega_\sigma;\omega',\omega'')\chi_{ijk}^{(2)}(-\omega_\sigma;\omega',\omega'') \cdot E_j^{\omega'} e^{-i\omega' t} \cdot E_k^{\omega''} e^{-i\omega'' t} + \text{c.c.}, \end{aligned} \tag{3.11}$$

where the higher-order terms can be treated in the same way. In the case of a finite number of monochromatic fields, then, the time-dependent polarization will be given by the sums of Equations 3.10 and 3.11. The way the resulting equation is solved depends somewhat on the details of the problem, as we discuss later.

For a linear material and a single monochromatic field, substituting Equation 3.10 into Equation 3.6 yields

$$\nabla \times \nabla \times \mathbf{E} + \frac{\epsilon}{c^2} \cdot \frac{\partial^2 \mathbf{E}}{\partial t^2} = 0, \tag{3.12}$$

where $\epsilon_{ij} = \delta_{ij} + 4\pi\chi_{ij}^{(1)}$ and $(\epsilon \cdot \mathbf{E})_i \equiv \epsilon_{ij}E_j$. Equation 3.12 is solved in the usual way as described in the textbooks.[10] We will not solve it here but will return to it later when the need arises.

The general strategy for solving the nonlinear wave equation can be broken down into six steps. We start with listing the steps, then refer to them when considering specific cases:

1. **Assume a finite number of monochromatic fields.** With a finite number of monochromatic fields, the electric field can be written as

$$\mathbf{E} = \sum_n^{N_f} \mathbf{E}_n, \qquad (3.13)$$

which includes all N_f incident and outgoing fields. Each index n represents a field of one Fourier component at frequency ω_n.

2. **Assume that the fields are *approximately* plane waves.** Each monochromatic field can then be expressed as a plane wave with an amplitude that depends on position,[1]

$$\mathbf{E}_n = \left[\frac{1}{2} A_n(z,\rho) e^{i(k_n z - \omega_n t)} + \frac{1}{2} A_n^*(z,\rho) e^{-i(k_n z - \omega_n t)} \right] \hat{\rho}_n. \qquad (3.14)$$

The plane wave propagates along $\hat{z}$ and $\hat{\rho}_n$ is the transverse coordinate along which the electric field $\mathbf{E}_n$ is polarized, making it is a transverse field, where $\hat{\rho} \perp \hat{z}$. $\hat{\rho}_n$ can be complex and therefore can describe any polarization including elliptically polarized light. The slowly varying amplitude approximation, which assumes that the field amplitude varies negligibly when ρ and z varies on the order of a wavelength, is what makes Equation 3.14 approximately a plane wave. These conditions are mathematically expressed as

$$\frac{\partial A_n}{\partial z} \ll A_n k_n, \qquad (3.15)$$

and

$$\frac{\partial A_n}{\partial \rho} \ll A_n k_n. \qquad (3.16)$$

Figure 3.1 illustrates how the electric field oscillates along $\hat{z}$ with an envelope A_n, which obeys Equation 3.15.

The polarization for each frequency can be immediately expressed in the form given by Equation 3.11, with

$$\mathbf{E}^{\omega_n}(z,\rho) = A_n(z,\rho) e^{i k_n z} \hat{\rho}_n. \qquad (3.17)$$

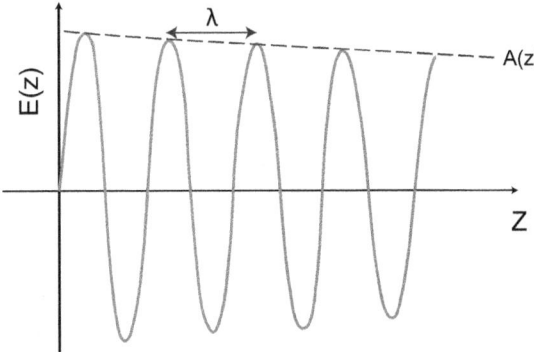

Figure 3.1: Slowly varying envelope approximation: $A(z)$ varies negligibly over a wavelength.

Then, the polarization is a sum over all possible outgoing fields.

3. **Substitute the approximate plane wave fields into Maxwell's equations.** This step is a huge mess. First, one needs to substitute Equation 3.14 into 3.13 and the result into Equations 3.10, 3.11, and higher-order terms to get the time dependence of the polarizations. The total polarization is the sum of the polarizations of various order. Then we must substitute the fields and polarization into Maxwell's wave equation, given by Equation 3.6.

   If you follow these instructions, you'll get a huge differential equation in terms of the fields and the nonlinear susceptibilities. The expression will most likely be too messy to fit onto a sheet of paper. Instead, let's do this step metaphorically at this point, and we will use the steps that follow to simplify the expression as we are calculating it. The specific examples that follow this section will illustrate how this is done.

4. **Separate the wave equation into its Fourier components.** From the messy wave Equation that you got in Item #3, you will get a differential equation for each frequency by taking a Fourier transform with respect to each of the N_f fields. Note that these equations can be determined without writing the whole expression as we are directed to do in Item #3. Instead, just read off all the terms with the coefficient of $\exp[-i\omega_1 t]$ to get the equation for ω_1, and so on, for each frequency.

---

[1]Note that the rules that we are defining here are specifically applicable to problems in which the amplitudes of the waves are changing, as is the process where energy is exchanged between beams of different colors.

Formally, we are implementing the transform

$$\int dt e^{i\omega_n t} \left[\nabla \times \nabla \times \mathbf{E} + \frac{1}{c^2} \frac{\partial^2 \mathbf{D}}{\partial t^2} \right] = 0, \tag{3.18}$$

where the electric displacement $\mathbf{D}$ contains a contribution from the nonlinear polarization, such as $\exp[-i\omega_1 t]\exp[-i\omega_2 t]$, which results in a field at frequency $\omega_1 + \omega_2$.

5. **Separate each equation further by grouping them into terms of the same order of magnitude.** For each differential equation that results from taking a Fourier projection as described in Item #4, further separate each of these equations into equations of the same order. To do so, use the fact that $\partial A_n(\rho,z)/\partial z \ll kA_n$ to assign $\partial A_n(\rho,z)/\partial z$ to be of order $\xi k A_n$, where ξ is a small parameter that keeps track of the order-of-magnitude size of this term.

Since the change of the amplitude is usually due to the exchange of energy between beams due to a nonlinear process, the nonlinear process will also be of order ξ. The examples that follow will make clear why this is so.

Note that the expression

$$\nabla \times \nabla \times \mathbf{E} = -\nabla^2 \mathbf{E} + \nabla(\nabla \cdot \mathbf{E}) \tag{3.19}$$

can often be simplified. For example, if A_n is constant along the direction of polarization, $\hat{\rho}$, then $\nabla \cdot \mathbf{E} = 0$, and we can ignore the second term of the right hand side of Equation 3.19. Note that $\nabla \cdot \mathbf{E}$ will often not vanish, so care must be take to correctly evaluate it.

6. **Solve each equation, starting from the lowest order ones.** Solve each of the differential equations that you get in Item #5, starting from the lowest order equation of the incident fields, then move on to the radiated fields.

Solutions to higher-order terms will usually involve solutions to lower-order ones because the polarization, which is the source of the radiated fields, depends on products of the incident fields.

3.2 Sum Frequency Generation - Non-Depletion Regime

We apply the general technique above to sum frequency generation in the non-depletion regime, where the sum frequency $\omega_3 = \omega_1 + \omega_2$ is generated by

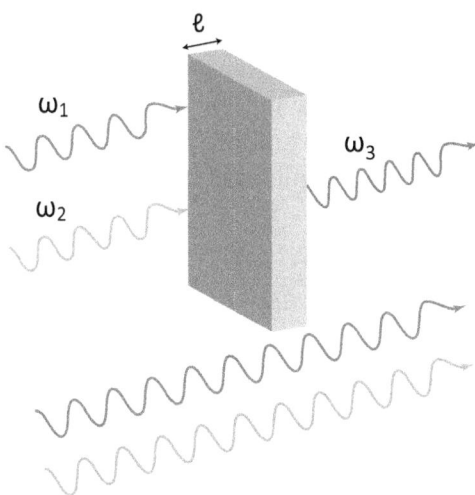

Figure 3.2: Through the second-order susceptibility, two incident light waves of frequencies ω_1 and ω_2 interact to generate light at frequency $\omega_3 = \omega_1 + \omega_2$. Most of the incident light passes through the sample without being converted.

two incident driving waves of frequency ω_1 and ω_2. A schematic representation of the process is shown in Figure 3.2. The above procedure can be applied to the most general case. However, we start with several simplifying assumptions that are often valid under experimental conditions and that make the problem easier to solve.

First, we assume that there is no loss in the linear and nonlinear processes, so the refractive index and the second-order susceptibility are real. Secondly, the intensity of the sum-frequency beam is assumed to be so small that the incident beams are negligibly depleted. Furthermore, the beams are assumed to be much wider than the wavelength, so that all amplitudes are independent of the transverse coordinate ρ. Finally, we assume that all three waves are polarized along the same axis. Thus, the incident field amplitudes $A_1(z) = A_1$ and $A_2(z) = A_2$ are constant and the sum-frequency amplitude $A_3(z)$ depends on z. Then, the three Fourier amplitudes are of the form

$$\mathbf{E}_n = \left[\frac{1}{2} A_n(z) e^{i(k_n z - \omega_n t)} + \frac{1}{2} A_n^*(z) e^{-i(k_n z - \omega_n t)} \right] \hat{\rho}. \qquad (3.20)$$

As per Item #1 on page 114, the electric field is expressed as a sum over the three fields defined by Equation 3.20, which are of the approximately plane wave form specified by Item #2 on page 114. Note that the case discussed here is more restrictive than given by Equation 3.14 in that all beams are polarized along the same axis, there is no ρ dependence of the amplitude, and only the sum-frequency beam's intensity changes.

Next, as Item #3 on page 115 suggests, we metaphorically make the substitution of the sum of three fields and the polarizations into Maxwell's wave Equation, which even in this simple case is a huge mess. However, we can write it in the compact form

$$-\nabla^2 \mathbf{E} + \frac{1}{c^2}\frac{\partial^2(\epsilon \cdot \mathbf{E})}{\partial t^2} = -\frac{4\pi}{c^2}\frac{\partial^2 \mathbf{P}^{NL}}{\partial t^2}, \qquad (3.21)$$

where $\mathbf{E} = \sum_{n=1}^{3} \mathbf{E}_n$, the linear polarizations are buried in the dielectric constant ϵ, the electric displacement $\epsilon \cdot \mathbf{E} = \sum_{n=1}^{3} \epsilon_n \mathbf{E}_n$, and the nonlinear polarization generates light at $\omega_3 = \omega_1 + \omega_2$, so it's amplitude is given by

$$\left(P_\rho^{NL}\right)^{\omega_3} = \chi^{(2)}_{\rho\rho\rho}(-\omega_1-\omega_2;\omega_1,\omega_2)E_\rho^{\omega_1}E_\rho^{\omega_2}, \qquad (3.22)$$

where we have used Equation 3.9 with $K = 1$, its value when both fields vanish. Note that there is no summation implied in Equation 3.22. The electric fields are polarized along ρ, which is parallel to the nonlinear polarization, $\mathbf{P}^{NL}$.[2]

According to Item #4 on page 115, we next determine the differential equations for each of the three Fourier components. First, let's project out the ω_1 and ω_2 Fourier components of Equation 3.21, which yields

$$-\nabla^2\left[\frac{1}{2}A_j e^{i(k_j z - \omega_j t)}\hat{\rho}\right] + \frac{\epsilon_j}{c^2}\frac{\partial^2}{\partial t^2}\left[\frac{1}{2}A_j e^{i(k_j z - \omega_j t)}\hat{\rho}\right] = 0, \qquad (3.23)$$

where $j = 1$ or $j = 2$ and

$$\epsilon_j = 1 + 4\pi\chi^{(1)}_{\rho\rho}(-\omega_j;\omega_j). \qquad (3.24)$$

Item #5 on page 116 instructs us to keep terms of the same order. Since Equation 3.23 represents the incident fields, and the polarization is linear, all terms are of the same order. Don't be concerned if this is confusing at this point. We'll come back to it later with a set of better-defined rules. Finally, Item #6 on page 116 instructs us to solve the differential equations. Since the amplitudes are constant, Equation 3.23 can be easily evaluated through differentiation, yielding

$$k_j^2 = \epsilon(\omega_j)\frac{\omega_j^2}{c^2} = \left(1 + 4\pi\chi^{(1)}_{\rho\rho}(-\omega_j;\omega_j)\right)\frac{\omega_j^2}{c^2}, \qquad (3.25)$$

---

[2] Keep in mind that the word "polarization" here refers to the induced dipole moment per unit volume, $\mathbf{P}$. "Polarization" can also refer to the axis along which the electric field is polarized; i.e. the axis along which the field oscillates. The student should be able to distinguish between the two by context.

Nonlinear Optics: A student's perspective

which is the so-called linear dispersion, and describes how the wave vector depends on the frequency.

Finally, we project out the ω_3 Fourier component of Equation 3.21 as per Item #4 on page 115, and take the dot product with $\hat{\rho}$, which yields

$$-\nabla^2 \left[\frac{1}{2} A_3(z) e^{i(k_3 z - \omega_3 t)} \right] + \frac{\epsilon_3}{c^2} \frac{\partial^2}{\partial t^2} \left[\frac{1}{2} A_3(z) e^{i(k_3 z - \omega_3 t)} \right]$$
$$= -\frac{1}{2} \cdot \frac{4\pi}{c^2} \frac{\partial^2}{\partial t^2} \chi^{(2)}_{\rho\rho\rho}(-\omega_1 - \omega_2; \omega_1, \omega_2) \cdot A_1 e^{i(k_1 z - \omega_1 t)} \cdot A_2 e^{i(k_2 z - \omega_2 t)}, \quad (3.26)$$

Recall that the righthand side of Equation 3.26 is the polarization of the radiated field, which we got from Equation 3.11 with the help of Equation 3.22 and the definition given by Equation 3.17.

As per Item #5 on page 116, we separate Equation 3.26 according to its order of magnitude which is tracked by multiplying each field amplitude by ξ. Secondly, the slowly varying envelope approximation given by Equations 3.15 and 3.16 implies that $dA_3(z)/dz \ll k_3 A_3(Z)$. As such, we multiply each derivative by ξ; so a first derivative gets multiplied by ξ and a second derivative by ξ^2.

Given these simple rules, we can rewrite Equation 3.26 as

$$\left[\left(-\xi^2 \frac{d^2(\xi A_3(z))}{dz^2} - 2ik_3 \xi \frac{d(\xi A_3(z))}{dz} + k_3^2 (\xi A_3(z)) \right) e^{ik_3 z} \right]$$
$$-\epsilon_3 \frac{\omega_3^2}{c^2} \left[\xi A_3(z) e^{ik_3 z} \right] = 4\pi \frac{\omega^2}{c^2} \chi^{(2)}_{\rho\rho\rho}(-\omega_1 - \omega_2; \omega_1, \omega_2) \left[\xi A_1 e^{ik_1 z} \right] \left[\xi A_2 e^{ik_2 z} \right], \quad (3.27)$$

where we have first taken the derivatives, then added the ξ factors, then multiplied the result by $2\exp[i\omega_3 t]$.

Collecting all terms of order ξ in Equation 3.27 yields

$$k_3^2 = \epsilon_3 \frac{\omega_3^2}{c^2} = \left(1 + 4\pi \chi^{(1)}_{\rho\rho}(-\omega_3; \omega_3) \right) \frac{\omega_3^2}{c^2}. \quad (3.28)$$

Equation 3.28 is simply the linear dispersion of the sum-frequency wave, analogous to the expressions for the incident waves given by Equation 3.25 More compactly, Equations 3.25 and 3.28 represent plane waves with

$$k_j = \frac{\omega_j}{c} n_j, \quad (3.29)$$

where $j = 1, 2,$ and 3 and n_j is the refractive index at frequency ω_j.

Next we collect terms of order ξ^2 from Equation 3.27, which yields

$$-ik_3\frac{\partial A_3}{\partial z}e^{ik_3 z} = \frac{2\pi\omega_3^2}{c^2}\chi^{(2)}(-\omega_3;\omega_1,\omega_2)A_1 e^{ik_1 z}A_2 e^{ik_2 z}. \qquad (3.30)$$

The righthand side of Equation 3.30 is a product of the incident fields, which propagate as plane waves, and act as the source that generates a wave at frequency ω_3. Equation 3.30 can be expressed as

$$\frac{\partial A_3}{\partial z} = \frac{2i\pi k_3}{n_3^2}\chi^{(2)}(-\omega_3;\omega_1,\omega_2)A_1 A_2 e^{i\Delta k z}, \qquad (3.31)$$

where we have used Equation 3.29 and where $\Delta k = k_1 + k_2 - k_3$ is called the phase mismatch.

Since A_1 and A_2 are constants, Equation 3.31 can be integrated over the length ℓ of the sample (from $z=0$ to $z=\ell$) to yield the electric field amplitude at $z = \ell$:

$$A_3(\ell) = \frac{2i\pi k_3 \chi^{(2)}(-\omega_3;\omega_1,\omega_2)}{n_3^2 \Delta k}A_1 A_2\left[e^{i\Delta k\ell} - 1\right]. \qquad (3.32)$$

Equation 3.32 represents that field amplitude of the sum-frequency wave at the output plane of the material, but does not determine how much of the light makes it out off the sample – which requires that surface reflections be taken into account. We note that in the small-depletion limit, at the microscopic level, only a small number of photons are converted to the sum frequency. The rest of them pass through the sample, unchanged. Equation 3.32 can be re-expressed in terms of a sine function,

$$A_3(\ell) = -\frac{4\pi k_3 \chi^{(2)} A_1 A_2}{n_3^2 \Delta k}\sin\left(\frac{\Delta k\ell}{2}\right)\exp\left(\frac{i\Delta k\ell}{2}\right), \qquad (3.33)$$

where we drop the frequency arguments of the second-order susceptibility to reduce clutter.

Since we are interested in measurable quantities, we next determine the intensity from the field amplitude using Poynting's theorem, which in Gaussian units is of the form

$$I_j = \frac{c}{8\pi}n_j|A_j|^2, \quad \text{where} \quad j = 1,2,3. \qquad (3.34)$$

Substituting Equation 3.33 into Equation 3.34, we get the sum-frequency intensity,

$$I_3 = \frac{c}{8\pi}n_3|A_3|^2 = \frac{cn_3}{8\pi}\left(\frac{4\pi k_3 \chi^{(2)} A_1 A_2}{n_3^2 \Delta k}\right)^2 \sin^2\left(\frac{\Delta k\ell}{2}\right). \qquad (3.35)$$

Since A_1 and A_2 are real, $A_1^2 = |A_1|^2$ and $A_2^2 = |A_2|^2$, so we can rewrite Equation 3.35 in terms of the intensities using $|A_i|^2 = 8\pi I_i/cn_i$, which yields

$$I_3 = \frac{32\pi^3 \omega_3^2}{c^3} \left(\chi^{(2)}\right)^2 \frac{I_1 I_2}{n_1 n_2 n_3} \cdot L_c^2 \sin^2\left(\frac{l}{L_c}\right), \qquad (3.36)$$

where we have made the substitutions $k_i = n_i \frac{\omega_i}{c}$ and

$$L_c = \frac{2}{\Delta k} = \frac{2c}{n_1 \omega_1 + n_2 \omega_2 - n_3 \omega_3} \qquad (3.37)$$

is called the coherence length.

According to Equation 3.36, the output intensity at the sum frequency oscillates as a function of the sample thickness with a period given by πL_c. The oscillation in the intensity is produced by the phase mismatch between the three waves that are traveling at different speeds, $v_i = c/n_i$, which causes interference. The conversion efficiency is maximum when the waves at ω_1 and ω_2 produce light at frequency ω_3, which is in phase with the light at ω_3 that had been generated upstream. This condition for maximum conversion efficiency is called phase matching.

The intensity of the sum-frequency wave is proportional to the intensity of the incident waves, and on $\chi^{(2)}$. For fixed sample length, a higher incident intensity and larger $\chi^{(2)}$ leads to a higher conversion efficiency. However, for a fixed second-order susceptibility and input intensity, the coherence length strongly influences the conversion efficiency. Since the coherence length depends on the refractive index at the three wavelengths, as shown by Equation 3.37, it is often possible to orient the material in a way that takes advantage of its birefringence. We will discuss this approach later when the tensor nature of both the refractive index and the nonlinear susceptibilities will be taken into account. For now, we investigate how the sum-frequency intensity depends on the coherence length through the linear dispersion of the refractive index.

The phase-matching condition is defined by $\Delta k \to 0$, so Equation 3.37 demands that $L_c \to \infty$. The rightmost factor in Equation 3.36 contains the coherence length, and under the phase-matching condition becomes

$$\lim_{L_c \to \infty} L_c^2 \sin^2(l/L_c) = \ell^2. \qquad (3.38)$$

The sum-frequency intensity grows quadratically with sample length, as shown by the thick curve in Figure 3.3. Clearly, the intensity can not increase quadratically for arbitrarily long lengths, which would result in more

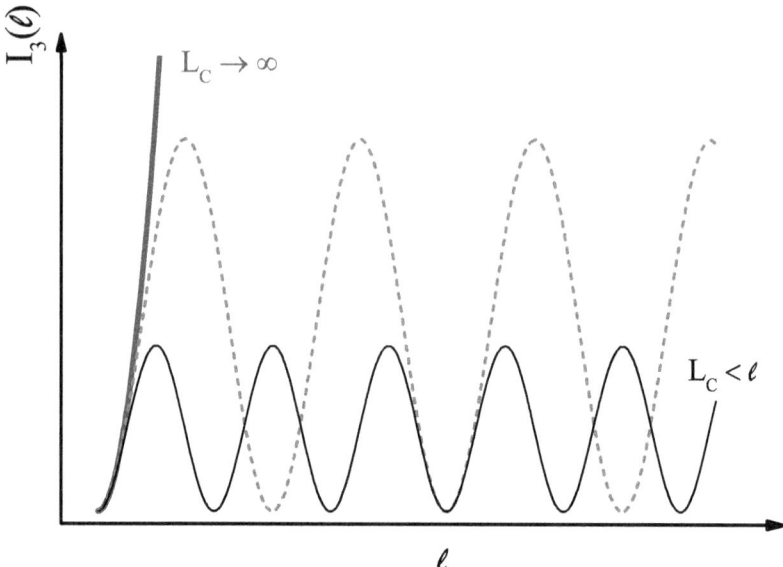

Figure 3.3: The intensity of the sum frequency light as a function of sample length ℓ under the phase-matching condition when $L_c \to \infty$ (thick solid curve), and when $L_c < \ell$ (thin solid curve and dashed curve). The period and amplitude increases with the coherence length, as seen in the sequence of three curves.

energy being converted than is available in the two input beams. This seeming paradox is resolved when we recall that the non-depletion limit requires that the intensity of light generated is small compared to the intensity of the incident waves.

As the coherence length is made smaller than the length of the sample, the sum-frequency intensity oscillates with a period of the coherence length. As the coherence length gets shorter, the amplitude falls and the period decreases, as shown by the dashed and thin curves in Figure 3.3.

Problem 3.2-1: The calculation leading to Equation 3.36 assumes that there is no loss due to linear and nonlinear absorption. We can add a smidgen of loss by adding small imaginary parts to both the refractive index and second-order susceptibility to make them complex.

(a) Re-derive the sum frequency generation calculation in this section, but this time make the refractive index and second-order susceptibility complex. Don't forget that the k-vector is complex since it is related to the refractive index.

(b) Do the solutions that you get in Part (a) show gain? If so, under what conditions does this happen.

3.3 Spatially Varying Second-Order Susceptibility

Next we consider a material in which the second-order susceptibility depends on position under the non-depletion approximation. This will be the case when the material's composition varies with position. We will consider only the special case where variations are along the propagation direction, so the second-order susceptibility is of the form $\chi^{(2)}(z)$ and is constant on planes of constant z.

There are many examples of materials that are artificially fabricated in planar layers. Examples include Langmuir-Blodgett films, which are transferred one molecular monolayer at a time from a water surface to a substrate, or molecular beam epitaxy, where atoms or molecules are essentially sprayed onto a surface one layer at a time.

Figure 3.4 shows an example of such a layered structure made of a variety of materials. The solid plot below shows the second-order susceptibility as a function of position and the dashed curve shows an approximation to a smooth function. Since each layer can have a thickness on the order of a molecular size, the position dependence is essentially a continuous curve since the layers are much smaller than a wavelength of light.

We continue to use the nondepleted approximation as was used to arrive at Equation 3.31. The point of departure now is that the second-order susceptibility depends on the position z, so Equation 3.31 can be integrated to yield the field amplitude

$$A_3(\ell) = \frac{2i\pi k_3}{n_3^2} A_1 A_2 \int_0^\ell dz\, \chi^{(2)}(-\omega_3;\omega_1,\omega_2;z) e^{i\Delta k z}. \quad (3.39)$$

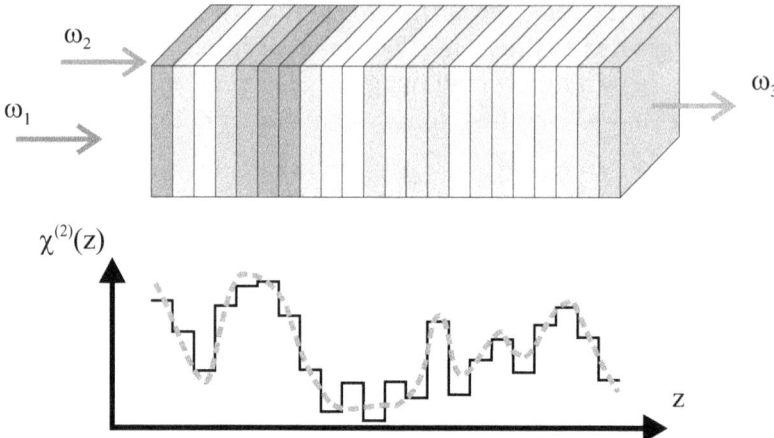

Figure 3.4: A layered material in which the second-order susceptibility depends on position as shown in the plot below. If the layers are much thinner than all wavelengths involved in the process, the smooth curve (dashed) is well approximated by the discrete function (solid).

The sum frequency generated for any arbitrary second-order susceptibility profile comes down to integrating Equation 3.39 .

The following section gives examples of $\chi^{(2)}$ profiles to illustrate the technique and to fortify our intuitive grasp of the meaning of phase matching .

> **Problem 3.3-1:** In this problem, you are to investigate if it is possible to tailor the second order susceptibility profile to get any arbitrary function $A_3(\ell)$. Use this result to determine the profile that gives $A(\ell)$=constant, $A(\ell),= \ell^2$, $A(\ell),= e^{-\ell}$, etc.

3.3.1 Applications

Two Thin Sheets of a $\chi^{(2)}$ Material

Consider two sheets of a second-order material as shown in Figure 3.5. Light at frequencies ω_1 and ω_2 are incident from the left. Note that the beams are on top of each other but are separated in the figure for clarity.

The two light beams generate the sum frequency beam at ω_3 at the first sheet, slightly depleting each incident beam. The three beams travel together in the passive medium until encountering the second nonlinear-optical sheet.

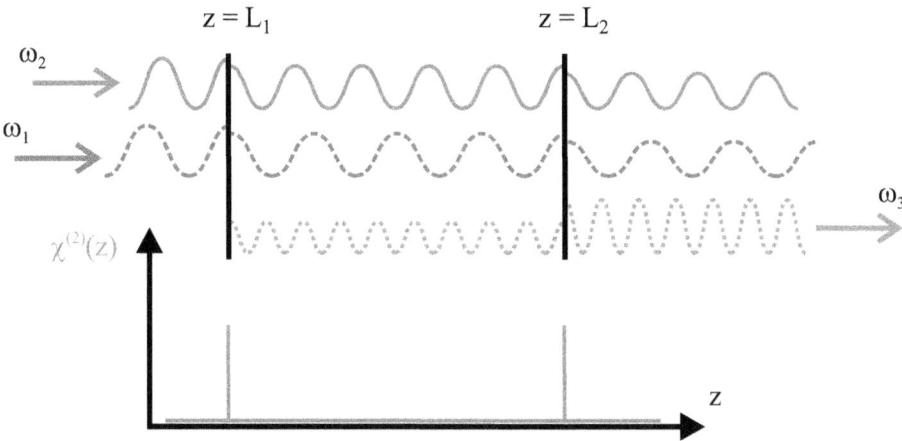

Figure 3.5: Two sheets of a material that together yield a second-order susceptibility that depends on position as shown in the bottom plot.

For illustration, the separation between the two sheets in the figure has been chosen such that all the waves are in phase at the second sheet, which will lead to constructive interference, hence a maximal amount of sum frequency light. The second sheet, if displaced by the right amount, can lead to destructive interference, cancelling the ω_3 wave. The calculation that follows is the general case for any two arbitrary positions, $z = L_1$ and $z = L_2$.

We model the second-order susceptibility profile as delta functions, or

$$\chi^{(2)}(z) = a\chi_0^{(2)}[\delta(z-L_1) + \delta(z-L_2)], \tag{3.40}$$

where $\chi_0^{(2)}$ is a constant and a represents the strength of the delta functions. Note that the delta functions here have units of inverse length, so the strength a has units of length. Then, $\chi_0^{(2)}$ and $\chi^{(2)}(z)$ have the same units.

Substituting Equation 3.40 into Equation 3.39 yields

$$A_3(\ell > L_1, L_2) = \frac{2i\pi k_3}{n_3^2} A_1 A_2 a \chi_0^{(2)} \left(e^{i\Delta k L_1} + e^{i\Delta k L_2}\right), \tag{3.41}$$

where the interval between $z = 0$ and $z = \ell$ must include the two delta functions. The terms in parentheses are the ones of interest because they depend on Δk. These terms can be rewritten as

$$() = 2e^{i\Delta k(L_1+L_2)/2} \cos\left(\Delta k \cdot \frac{L_1 - L_2}{2}\right), \tag{3.42}$$

which gives the intensity,

$$I_3 \propto \cos^2\left(\Delta k \cdot \frac{L_1-L_2}{2}\right) = \cos^2\left(\frac{L_1-L_2}{L_c}\right). \tag{3.43}$$

Equation 3.41 shows that constructive interference requires that the spacing between the two sheets obeys the condition

$$\Delta k \cdot \frac{L_1-L_2}{2} = m\pi, \tag{3.44}$$

where m is an integer. Since $\Delta k = 2/L_c$, this yields

$$\frac{L_1-L_2}{L_c} = m\pi, \tag{3.45}$$

We note that the coherence length is defined by the refractive index of the passive material between the two sheets, giving the technologist an additional degree of flexibility in making an efficient sum frequency generator. Unlike a monolithic material, in which both the second-order susceptibility and refractive index must both be ideal to fully take advantage of the material, the user can mix and match the components of a composite structure to get the best properties of each part. In addition, the user can control the spacing.

Periodic Thin Sheets of $\chi^{(2)}$

In this section, we calculate the sum-frequency intensity for a periodic series of sheets made of a second-order nonlinear-optical material as shown in Figure 3.6 with a spacing of $\delta\ell$.

If the first sheet is located at $z = 0$, and the spacing is $\delta\ell$, then the mth sheet will be located at $z = m \cdot \delta\ell$. Thus, for $N+1$ sheets, we can generalize Equation 3.41 by noting that sheet m contributes $\exp[i\Delta k \cdot m \cdot \delta\ell]$, yielding

$$A_3(\ell > L_N) = \frac{2i\pi k_3}{n_3^2} A_1 A_2 a \chi_0^{(2)} \sum_{m=0}^{N} e^{2im(\delta\ell/L_c)}. \tag{3.46}$$

Equation 3.46 is a geometric series that sums to

$$\sum_{m=0}^{N} e^{2im(\delta\ell/L_c)} = \frac{e^{2i(N+1)(\delta\ell/L_c)} - 1}{e^{2i\delta\ell/L_c} - 1}. \tag{3.47}$$

The sum-frequency intensity is proportional to the square of the magnitude of Equation 3.47, or

$$I_3 \propto \left|\sum_{m=1}^{N} e^{2im\delta\ell/L_c}\right|^2 = \frac{\sin^2\left((N+1)\frac{\delta\ell}{L_c}\right)}{\sin^2\left(\frac{\delta\ell}{L_c}\right)}. \tag{3.48}$$

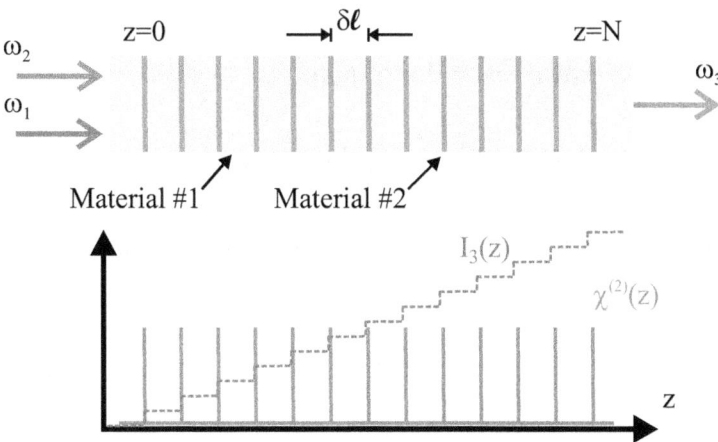

Figure 3.6: A layered material in which the second-order susceptibility depends on position as shown in the plot below under optimum phase matching
.

Equation 3.48 appears to diverge for the phase-matching condition $\delta\ell/L_c = m\pi$; but, a careful analysis shows that it does not. Furthermore, under phase matching, the sum-frequency intensity increases quadratically with the number of layers.

Phase matching in a material requires the refractive indices at the desired wavelengths meet the very special condition $\omega_3 n(\omega_3) = \omega_1 n(\omega_1) + \omega_2 n(\omega_2)$. In a periodic structure, a material of any coherence length can be used. The only condition for phase matching is that the spacing $\delta\ell$ meet the condition,

$$\delta\ell = m\pi L_c, \qquad (3.49)$$

a condition that is easily met by making a layered structure. Any periodic structure that results in a sum-frequency intensity that grows quadratically with propagation distance is called quasi phase matching.

Problem 3.3.1-1(a): Verify that Equation 3.48 with $N = 2$ reduces to Equation 3.41.

(b) Show that the sum-frequency intensity increases quadratically with the number of layers.

(c) When the spacing meets the phase-matching condition $\delta\ell/L_c \approx m\pi$ so that $\delta\ell/L_c = m\pi + \epsilon$ – where ϵ is a small parameter, determine how the intensity depends on ϵ to lowest non-trivial order in ϵ.

(d) Sketch, based on intuition (not using a plotting program, only your brains) how I depends on ϵ for $N = 1$, $N = 2$ and $N \to \infty$.

(e) Calculate the case where the sheets are made of layers with alternating sign of $\chi^{(2)}$.

(f) Could you imagine how this effect could be used to make a temperature or force sensor? What would be the ideal geometry for making the sensor highly sensitive to small changes in temperature or force?

Sinusoidal Modulation of $\chi^{(2)}$

Finally, we consider a material in which the second-order susceptibility varies sinusoidally – an example of a finite continuous function. For a second-order susceptibility profile

$$\chi^{(2)}(z) = \chi_0^{(2)} \sin(Kz), \tag{3.50}$$

Equation 3.39 yields

$$A_3(\ell) = \frac{2i\pi k_3}{n_3^2} A_1 A_2 \chi_0^{(2)} \int_0^\ell dz \, \frac{e^{iKz} - e^{-iKz}}{2i} e^{i\Delta kz}, \tag{3.51}$$

where we have used the fact that $\sin(Kz) = (e^{iKz} - e^{-iKz})/2i$

The integral in Equation 3.51 is given by

$$\int_0^\ell dz \, \sin(Kz) e^{i\Delta kz} = \frac{\sin\left[\frac{K-\Delta k}{2}\ell\right]}{(K-\Delta k)} \cdot e^{i\frac{K-\Delta k}{2}\ell} + \frac{\sin\left[\frac{K+\Delta k}{2}\ell\right]}{(K+\Delta k)} \cdot e^{i\frac{K+\Delta k}{2}\ell}. \tag{3.52}$$

At this point, we can substitute Equation 3.52 into Equation 3.51, take the square of the absolute magnitude, and get an expression for the sum frequency intensity as a function of the incident intensities, coherence length, etc. However, the complicated equation that we would get, though filled with useful information, would not help in the development of our intuition.

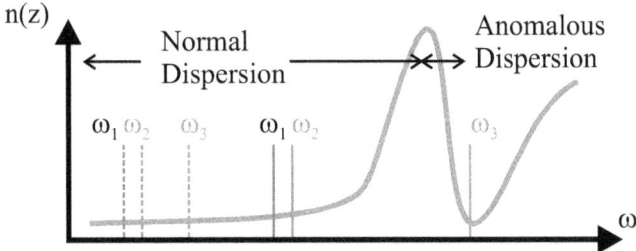

Figure 3.7: In the normal dispersion part of the spectrum, the refractive index increases with frequency, so if all three waves are in the normal dispersion regime (dashed vertical lines), $\Delta k < 0$. If the ω_3 wave is in the anomalous regime, where the refractive index of the sum-frequency wave is lower than for the incident waves (solid vertical lines), $\Delta k > 0$

Before proceeding, we digress briefly to discuss the meaning of the sign of Δk. Recall that,

$$\Delta k = n_1 \omega_1 + n_2 \omega_2 - n_3 \omega_3, \tag{3.53}$$

so the sign depends on the value of $n_1\omega_1 + n_2\omega_2$ relative to $n_3\omega_3$. In the spectral regime often referred to as the normal dispersion regime, the refractive index increases at shorter wavelength. The normal dispersion regime spans the spectral range where each of the photon energies have much lower energy than the energy levels of the material. This is usually true since the energy range of normal dispersion defines the transparency window, where light passes through a material without being absorbed. Since the sum-frequency photon energy is closer to a material resonance than for the incident photons, both its frequency and refractive index are larger, resulting in a negative value of δk. The dashed vertical lines in Figure 3.7 illustrate this property.

Just above a material resonance is a spectral region where the refractive index decreases as the frequency increases. This frequency dependence is called anomalous dispersion. When the sum-frequency photon is in the anomalous dispersion regime, the case illustrated by solid vertical lines in Figure 3.7, one can get $\delta k > 0$. As such, the sign of δk depends on the frequencies used. Because applications often require specific input and output frequencies, the phase matching condition (i.e. $\delta k = 0$) is difficult to meet using material dispersion alone.

Thus, depending on the spectral region of operation, δk can be positive or negative. For the sake of argument, we assume that the wavelengths are in the normal dispersion regime, so $\delta k < 0$. In this case, the sum-frequency efficiency is the highest when $K + \delta k = 0$. We thus evaluate Equation 3.52 in the vicinity of $K + \delta k = 0$, where only the second term is large, yielding

$$\int_0^\ell dz\, \sin(Kz)e^{i\Delta kz} \approx \frac{\sin\left[\frac{K+\Delta k}{2}\ell\right]}{(K+\Delta k)} \cdot e^{i\frac{K+\Delta k}{2}\ell}, \qquad (3.54)$$

so

$$\left|\int_0^\ell dz\, \sin(Kz)e^{i\Delta kz}\right|^2 = \frac{\sin^2\left[\frac{K+\Delta k}{2}\ell\right]}{(K+\Delta k)^2} \approx \ell^2/4. \qquad (3.55)$$

According to Equation 3.51 with the help of Equation 3.55, the sum-frequency intensity grows in proportion to the square of the sample length – the same dependence as in phase matching. Thus, this is an example of quasi phase matching, with an effective coherence length of

$$L_c^{\text{eff}} = \frac{2}{K+\Delta k} \qquad (3.56)$$

that diverges when the denominator vanishes (recall that for normal dispersion, $\delta k < 0$). Thus, even if the refractive index has an unfavorable dispersion for phase matching, sinusoidally modulating the second-order susceptibility profile with a period that matches the coherence length (πL_c) yields quasi phase marching with an efficiency that is one fourth the value obtained using true phase matching. This is one example of how materials can be coaxed to meet the demands of an application. Note that the nonlinear profile need not be sinusoidal. However, other profiles will yield a lower conversion efficiency.

There are many ways of making a material that is quasi phase matched. One method uses an electric field to break the symmetry in an isotropic material to induce a second-order susceptibility at the required periodicity. This idea was demonstrated in a dye-doped polymer.[11]

Phase matching and anomalous dispersion are discussed in more detail later in this chapter.

Problem 3.3.1-2(a): Calculate the sum frequency intensity as a function of the incident intensities in the quasi phase matching case (i.e. $K + \Delta k \to 0$) and show that the intensity is one quarter the value given by true phase matching (Equation 3.36).

Problem 3.3.1-3(a): Determine how the sum frequency intensity depends on the sample length if the nonlinearity has the continuous profile

$$\chi^{(2)}(z) = \chi_0^{(2)} + \chi_1^{(2)} \exp[-\alpha z], \tag{3.57}$$

where $\chi_0^{(2)}$, $\chi_1^{(2)}$, and α are constants.

(b) Sketch $I_3(z)$, making sure to label $\chi_0^{(2)}$, $\chi_1^{(2)}$, and α on your plot.

Problem 3.3.1-4(a): Write python code that will output the intensity $I_3(z)$ for any second-order susceptibility profile.

(b) Check your code by comparing the results you get with the analytical expressions in Section 3.3.1, both in the text and in the problems.

(c) Use your code to determine the optimum second-order susceptibility profile under the constraint that $\chi^{(2)}(z) \leq \chi^{(2)}_{max}$ that gives the largest conversion efficiency when the material is much longer than the coherence length.

3.4 Sum Frequency Generation - Small Depletion Regime

In Section 3.2, we carefully showed how the rules are applied to the non-depletion limit to solve the nonlinear wave equation. Here, and in the remainder of this chapter, we proceed without explicitly referring to the steps. However, it may be helpful to the student if (s)he practices by independently doing the derivations using the rules, or at least mentally tracks which step has used which rule. Note that the rules are sometime applied in another order if it simplifies the calculation.

In this section, we consider sum frequency generation in the small depletion regime, relaxing the condition that the incident beams are not depleted. As shown in Figure 3.8, light is generated at the sum frequency, which interacts with the incident waves to be converted back to ω_1 and ω_2. From the quantum perspective, two incident photons of frequencies ω_1 and

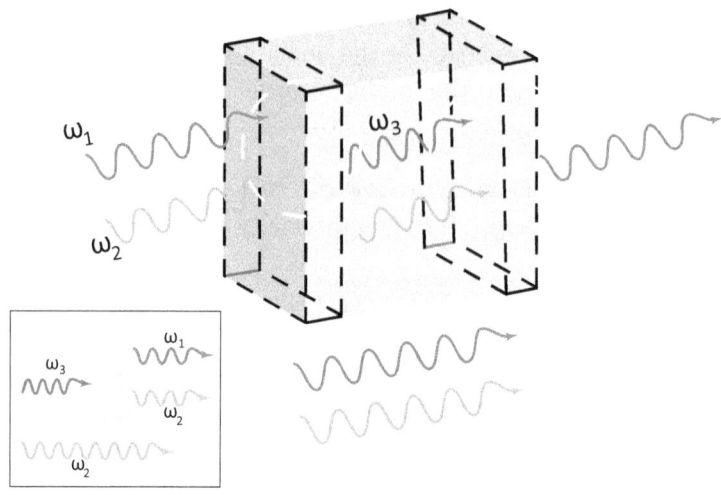

Figure 3.8: Sum frequency generation in the small depletion regime is represented by the nonlinear interaction of two incident waves of frequencies ω_1 and ω_2 within the material, to generate a photon of frequency ω_3. Downstream, this generated photon and a photon of frequency ω_2 interact and create a photon of frequency ω_1. The inset shows how the energy is conserved in the process of the destruction of a photon of frequency ω_3 and creation of two photons of frequencies ω_1 and ω_2.

ω_2, and amplitudes A_1 and A_2 interact to produce a photon at frequency ω_3. Downstream, the photon at frequency ω_3 interacts with another photon of frequency ω_2 to produce a photon of frequency ω_1. Similarly, a photon at ω_2 can result. In the no depletion approximation, not enough light is generated at frequency ω_3 to result in back conversion; but, in the small depletion regime, ω_3 can convert back to ω_1 by mixing with ω_2.

It is instructive to pause here to consider how the classical wave equation can have contributions to the nonlinear polarization in which light at frequency ω_3 and ω_2 interact to form a wave at ω_1, which appears to violate energy conservation when viewed from the microscopic perspective of photons. As we already saw in Section 1.7, this issue can be resolved by considering the actual process from the microscopic perspective, as shown in the inset of Figure 3.8. Here, the photon at frequency ω_2 stimulates the emission of another photon with frequency ω_2, thus also producing a photon at ω_1 as the photon of frequency ω_3 is annihilated. For the purposes of the macroscopic calculations presented here, it is comforting that these contributions yield the correct mixing efficiency. The root cause is the delicate symmetry between the emission and absorption of a photon, as we will describe in detail in Section 4.2.

The derivation begins by rewriting Equation 3.31 with amplitudes that all explicitly depend on z and are complex:

$$\frac{\partial A_3(z)}{\partial z} = \frac{2i\pi k_3}{\epsilon_3} \chi^{(2)}(-\omega_3;\omega_1,\omega_2) A_1(z) A_2(z) e^{i\Delta k z}, \tag{3.58}$$

and its complex conjugate

$$\frac{\partial A_3^*(z)}{\partial z} = -\frac{2i\pi k_3}{\epsilon_3} \chi^{(2)}(\omega_3;-\omega_1,-\omega_2) A_1^*(z) A_2^*(z) e^{-i\Delta k z}. \tag{3.59}$$

Here we have used the fact that the complex conjugate of the response function is equivalent to a sign change to all its arguments. This notation elucidates the underlying process. Recall that incident waves are represented with positive frequencies and outgoing waves with negative ones. Thus, Equation 3.58 describes photons of frequency ω_1 and ω_2 that combine to give a photon at frequency ω_3, and the complex conjugate given by Equation 3.59 describes a photon of frequency ω_3 that splits into two photons. These processes are shown in Figure 3.8 and its inset.

We can equivalently write $\chi^{(2)}(\omega_3;-\omega_1,-\omega_2)$ as $\chi^{(2)}(-\omega_1;\omega_3,-\omega_2)$ to rewrite Equation 3.59 as

$$\frac{\partial A_3^*(z)}{\partial z} = -\frac{2i\pi k_3}{\epsilon_3} \chi^{(2)}(-\omega_1;\omega_3,-\omega_2) A_1^*(z) A_2^*(z) e^{-i\Delta k z}. \tag{3.60}$$

This conjugate equation represents the process where ω_3 is being depleted and focuses on the ω_1 wave that is generated. Since we are interested in the wave at ω_1, we would like to place the field amplitude A_1 on the left-hand side of Equation 3.60 and the amplitude A_3 on the right, so we switch them and in the process must take their complex conjugates, which yields

$$\frac{\partial A_1(z)}{\partial z} = \frac{2i\pi k_1}{\epsilon_1} \chi^{(2)}(-\omega_1;\omega_3,-\omega_2) A_2^*(z) A_3(z) e^{-i\Delta k z}. \tag{3.61}$$

Note that we also relabelled all quantities associated with the wave vectors and dielectric constants.

We could have arrived at Equation 3.61 by solving the nonlinear wave equation for the field amplitude A_1; but, our intent here is to illustrate how we can use patterns to quickly get all the required equations. This becomes particularly useful when solving higher-order wave equations. We can get the differential equation for A_2 by interchanging indices 1 and 2, yielding

$$\frac{\partial A_2(z)}{\partial z} = \frac{2i\pi k_2}{\epsilon_2}\chi^{(2)}(-\omega_2;\omega_3,-\omega_1)A_1^*(z)A_3(z)e^{-i\Delta kz}, \quad (3.62)$$

where we have taken the complex conjugate of the the amplitudes of waves that are emitted.

It is, of course, most prudent to derive all equations from Maxwell's wave equation. However, with some experience, equations of the form given by Equations 3.61 and 3.62 can be written without much thought. The reader should stare at these equations, and similar ones that follow, to observe the pattern and determine it's origin from Maxwell's wave equation. Then apply these patterns in later calculations to see if they work in other cases.

Equations 3.60, 3.61, and 3.62 can be decoupled by converting electric field amplitudes to intensities. The relationship that we need for this task is obtained by differentiating Equation 3.34 with respect to z, which yields

$$\frac{\partial I_i}{\partial z} = \frac{cn_i}{8\pi}\frac{\partial |A_i|^2}{\partial z} = \frac{cn_i}{8\pi}\left(\frac{\partial A_i}{\partial z}A_i^* + \frac{\partial A_i^*}{\partial z}A_i\right), \quad \text{where} \quad i=1,2,3. \quad (3.63)$$

Before proceeding we impose several simplifying but reasonable approximations. First, we assume that all light frequencies are far from the Bohr frequencies, so $\chi^{(2)}$ is real. Secondly, we assume full permutation symmetry, which implies that $\chi^{(2)}$ is independent of the wavelength. Finally, we assume the refractive indices are real.

The procedure, then, is as follows. We multiply Equation 3.60 by A_1^*, multiply the complex conjugate of Equation 3.60 by A_1, then add the two together, which we substitute into the righthand side of Equation 3.63, yielding

$$\frac{\partial I_1}{\partial z} = \frac{cn_1}{8\pi}\left(\frac{2\pi ik_1\chi^{(2)}}{\epsilon_1}A_2^*A_3A_1^*e^{-i\Delta kz} - \frac{2\pi ik_1\chi^{(2)}}{\epsilon_1}A_2A_3^*A_1 e^{i\Delta kz}\right). \quad (3.64)$$

Doing the same for Equations 3.62 and 3.60 yields

$$\frac{\partial I_2}{\partial z} = \frac{cn_2}{4}\frac{ik_2\chi^{(2)}}{\epsilon_2}\left(A_1^*A_3 A_2^* e^{-i\Delta kz} - A_1 A_3^* A_2 e^{i\Delta kz}\right), \quad (3.65)$$

and

$$\frac{\partial I_3}{\partial z} = \frac{cn_3}{4}\frac{ik_3\chi^{(2)}}{\epsilon_3}\left(A_1 A_2 A_3^* e^{i\Delta kz} - A_1^* A_2^* A_3 e^{-i\Delta kz}\right). \quad (3.66)$$

Using the dispersion relationships $k_j = n_j \omega_j / c$ and $\epsilon_j = n_j^2$, Equations 3.64 - 3.66 become,

$$\frac{\partial I_1}{\partial z} = \frac{i\omega_1 \chi^{(2)}}{4} \left(A_2^* A_3 A_1^* e^{-i\Delta kz} - A_2 A_3^* A_1 e^{i\Delta kz} \right), \tag{3.67}$$

$$\frac{\partial I_2}{\partial z} = \frac{i\omega_2 \chi^{(2)}}{4} \left(A_1^* A_3 A_2^* e^{-i\Delta kz} - A_1 A_3^* A_2 e^{i\Delta kz} \right), \tag{3.68}$$

and

$$\frac{\partial I_3}{\partial z} = \frac{i\omega_3 \chi^{(2)}}{4} \left(A_1 A_2 A_3^* e^{i\Delta kz} - A_1^* A_2^* A_3 e^{-i\Delta kz} \right). \tag{3.69}$$

The sum of Equations 3.67-3.69 is

$$\begin{aligned}
\frac{\partial I_1}{\partial z} + \frac{\partial I_2}{\partial z} + \frac{\partial I_3}{\partial z} &= -\frac{i\chi^{(2)}}{4} \Big[e^{i\Delta kz} \\
&\quad \times \left(\omega_1 A_2 A_3^* A_1 + \omega_2 A_1 A_3^* A_2 - \omega_3 A_1 A_2 A_3^* \right) \\
&\quad + e^{-i\Delta kz} \left(-\omega_1 A_2^* A_3 A_1^* - \omega_2 A_1^* A_3 A_2^* + \omega_3 A_1^* A_2^* A_3 \right) \Big] \\
&= \frac{i\chi^{(2)}}{4} \left(e^{i\Delta kz} - e^{-i\Delta kz} \right) (\omega_3 - \omega_1 - \omega_2) \\
&\quad \times \left[A_1^* A_2^* A_3 + A_1 A_2 A_3^* \right].
\end{aligned} \tag{3.70}$$

The fact that $\omega_3 = \omega_1 + \omega_2$ implies

$$\frac{\partial I_1}{\partial z} + \frac{\partial I_2}{\partial z} + \frac{\partial I_3}{\partial z} = 0, \tag{3.71}$$

which in turn implies that

$$I_1 + I_2 + I_3 = \text{constant}. \tag{3.72}$$

This is a simple restatement of the conservation of energy. In the case of sum frequency generation, the sum total of the intensities remains unchanged and is given by the initial intensity of the two input beams, $I_1(z=0) + I_2(z=0)$.

Note that Equation 3.72 is an approximate solution to the order of nonlinearity treated. The linear Maxwell's equations obey superposition, so the total electric field is a sum over all Fourier components. The nonlinear polarization, on the other hand, makes superposition no longer valid. However, for a weak nonlinearity, superposition holds approximately as does energy conservation. At higher intensities, energy can flow to other Fourier components due to higher order terms that we have not taking into account. If the bookkeeping is done right, energy must always be conserved.

When the light is far from material resonances, as we assume, all the energy is stored in the electric fields. On resonance, energy can be transferred between the field and the material, in which case the total energy includes the fields and the material. Thus, we must be cautious when assuming that the energy of the electromagnetic field is constant, especially at higher intensities or near resonance, when energy is transferred to the material.

> **Problem 3.4-1:** A fourth-order nonlinear process is characterized by the nonlinear susceptibility $\chi^{(4)}(-\omega_5;\omega_1,-\omega_2,-\omega_3,\omega_4)$, where $\omega_5 = \omega_1 - \omega_2 - \omega_3 + \omega_4$.
>
> **(a)** Determine the coherence length for this process. Could you have done so without doing the calculation, but rather by using the observed pattern in the derivations of this section?
>
> **(b)** Using the patterns that you observed that lead to Equations 3.60, 3.61 and 3.62, write down without calculation the differential equations for the five fields. Note that you still need to determine one of the equations from the nonlinear wave equation, but you cat get the rest through trickery.

3.4.1 Physical Interpretation of the Manley-Rowe Equation

Let's revisit Equations 3.67 through 3.69. If we divide each equation by the frequency associated with the left-hand side of the equation, the righthand side will be the same for all frequencies, but only differs in the sign. Consequently, we get

$$\frac{1}{\omega_1}\frac{\partial I_1}{\partial z} = \frac{1}{\omega_2}\frac{\partial I_2}{\partial z} = -\frac{1}{\omega_3}\frac{\partial I_3}{\partial z}. \tag{3.73}$$

Equations 3.73 are called the Manley-Rowe equations.

To understand the meaning of Manley-Rowe, consider the process shown in Figure 3.9. If we integrate the Manley-Rowe equations over an infinite plane that is perpendicular to the beam's wave vector in a region of space where all three beams, I_1, I_2, and I_3 are present, we get

$$\frac{\partial}{\partial z}\left[\frac{1}{\omega_1}\int I_1 ds = \frac{1}{\omega_2}\int I_2 ds = -\frac{1}{\omega_3}\int I_3 ds\right]. \tag{3.74}$$

Nonlinear Optics: A student's perspective

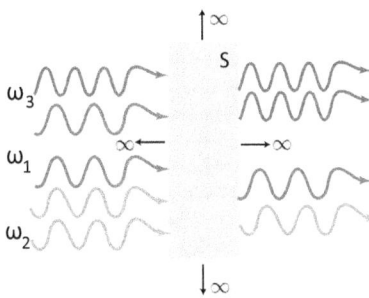

Figure 3.9: The Manley-Rowe equation expresses the fact that the absolute change in the number of photons at each frequency is the same. For example, the destruction of one photon at ω_1 and one photon at ω_2 is accompanied by the creation of one photon at frequency ω_3.

Recall that intensity, in units of $\left[\frac{\text{Power}}{\text{Area}}\right]$, when integrated over the plane yields the power, in units of $\left[\frac{\text{Energy}}{\text{Time}}\right]$, which passes through the plane. If we divide the power by $\hbar\omega$, the energy per photon, we get the number of photons that pass through the plane per unit time. Equation 3.74 thus indicates that the *magnitude* of the change in the number of photons in each beam is the same, and for sum frequency generation yields

$$\Delta N_1 = \Delta N_2 = -\Delta N_3. \tag{3.75}$$

This is exactly the picture we have been using: two photons are destroyed at frequencies ω_1 and ω_2, and a photon is created at frequency ω_3. These results in the small depletion regime are illustrated in Figure 3.10. I_1 and I_2 have high intensity and can be depleted, and wherever these two are minimum, I_3 is peaked.

Problem 3.4.1-3(a): Show that the Manley-Rowe relations are a consequence of $I_1(z) + I_2(z) + I_3(z) = I_1(0) + I_2(0) + I_3(0)$.

(b) If all three field amplitudes A_1, A_2, and A_3 in a second-order phase-matched frequency mixing process are real, calculate the total intensity $I_1 + I_2 + I_3$ as a function of z if a tiny bit of energy is dissipated from the ω_3 beam by adding to the second-order susceptibility – corresponding to the generation of the ω_3 wave – a small imaginary part. Approximate your solution to first order in the imaginary part of the second-order susceptibility that produces the ω_3 light.

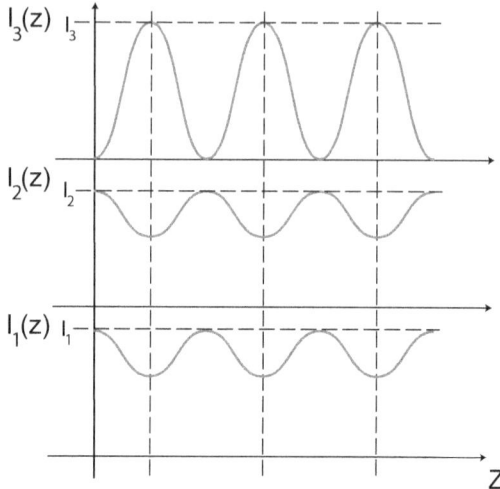

Figure 3.10: Comparison of the intensities of the three beams of frequency ω_1, ω_2, and ω_3 in sum frequency generation. I_1 and I_2 have their minima when I_3 peaks.

3.5 Sum Frequency Generation with Depletion of One Input Beam

Previously, we considered the process of sum frequency generation for two undepleted beams. We now turn to the case where one of the two beams can be depleted. We call the undepleted beam, at frequency ω_2, the pump and the depleted beam, ω_1, the idler.

In this calculation, we assume:

1. a strong undepleted pump at frequency ω_2

2. a weak idler at frequency ω_1, which can be fully depleted

3. the sum frequency and probe intensity are much smaller than the pump, so $I_1 + I_3 \ll I_2$ where subscripts denote the frequencies ω_1, ω_2, and ω_3

4. The phase matching condition holds, therefore, $\Delta k = 0$.

5. The frequencies are far of resonance so that the second-order susceptibilities are all the same

The above conditions are easy to obtain experimentally.

We start with Equations 3.61 and 3.58 with $\Delta k = 0$ and A_2 constant, which yields

$$\frac{\partial A_1(z)}{\partial z} = \frac{2i\pi k_1}{\epsilon_1}\chi^{(2)} A_2^* A_3(z), \tag{3.76}$$

and

$$\frac{\partial A_3(z)}{\partial z} = \frac{2i\pi k_3}{\epsilon_3}\chi^{(2)} A_1(z) A_2. \tag{3.77}$$

The student is reminded to translate such equations into the physical processes that they describe. For example, Equation 3.76 describes the output wave at frequency ω_1 for an incident wave at frequency ω_3. Equation 3.77 describes the output sum frequency wave at ω_3 for two input waves at frequencies ω_1 and ω_2. Since A_2 is constant, there is no light generated at ω_2 and there is no equation that has as its output an ω_2 wave. As such, these two differential equations describe all the processes that exchange energy, and are all that is needed to solve the problem.

Because there is no depletion of the pump, so A_2 is a constant, Equations 3.76 and 3.77 can be written as

$$\boxed{\frac{dA_1}{dz} = \kappa_1 A_3} \tag{3.78}$$

$$\boxed{\frac{dA_3}{dz} = \kappa_3 A_1,} \tag{3.79}$$

where κ_1 and κ_2 are constants, given by

$$\kappa_1 = \frac{2i\pi k_1}{\epsilon_1}\chi^{(2)} A_2^*, \tag{3.80}$$

and

$$\kappa_3 = \frac{2i\pi k_3}{\epsilon_3}\chi^{(2)} A_2. \tag{3.81}$$

Substituting Equation 3.79 in the z derivative of Equation 3.78 gives

$$\frac{d}{dz}\left(\frac{1}{\kappa_3}\frac{dA_3}{dz}\right) = \kappa_1 A_3, \tag{3.82}$$

and therefore

$$\frac{d^2 A_3}{dz^2} = \kappa_1 \kappa_3 A_3. \tag{3.83}$$

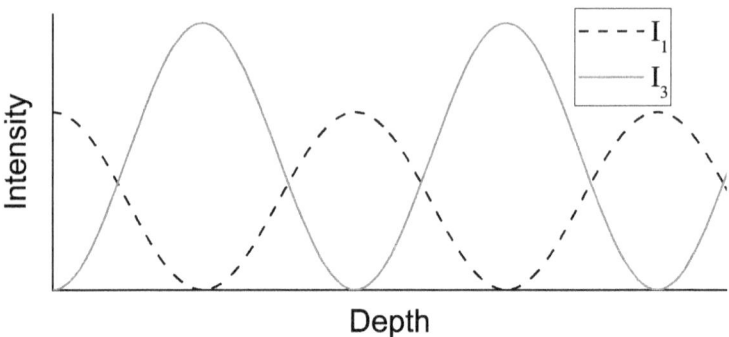

Figure 3.11: The intensities at frequencies ω_1 and ω_3 plotted as a function of propagation distance.

Off resonance, where $\chi^{(2)}$ is real, $\kappa_1 \kappa_2 = -K^2$ and the resulting one-dimensional Helmholtz equation

$$\frac{d^2 A_3}{dz^2} = -K^2 A_3, \tag{3.84}$$

which has the solution

$$A_3(z) = B\cos(Kz) + C\sin(Kz). \tag{3.85}$$

The solution to A_1 is found by substituting Equation 3.85 into Equation 3.79, which yields

$$A_1(z) = \frac{-BK}{\kappa_3}\sin(Kz) + \frac{CK}{\kappa_3}\cos(Kz). \tag{3.86}$$

According to our initial assumptions, the weak idler beam get's depleted, so it is maximum at the input surface of the material. Therefore, $A_1(0)$ is maximum, implying that $B = 0$ and

$$A_1(z) = A_1(0)\cos(Kz). \tag{3.87}$$

To evaluate the integration constant for A_3, we recognize that $\frac{\kappa_3}{K} = \frac{\sqrt{\kappa_3}}{\sqrt{\kappa_1}}$, which using Equations 3.80 and 3.81 becomes

$$\frac{\kappa_3}{K} = \sqrt{\frac{n_1}{n_3}\frac{\omega_3}{\omega_1}\frac{A_2}{A_2^*}}. \tag{3.88}$$

Equation 3.88 can be re-written as

$$\frac{\kappa_3}{K} = -ie^{i\phi_2}\sqrt{\frac{n_1\,\omega_3}{n_3\,\omega_1}}, \tag{3.89}$$

where we have expressed the complex constant A_2 as

$$A_2 = A_{20}e^{i\phi_2}, \tag{3.90}$$

where A_{20} is the amplitude and ϕ_2 the phase.

Using the boundary condition that there is no input wave at frequency ω_3, $A_3(0) = 0$ so $B = 0$ in Equation 3.85. Substituting Equation 3.85 with $B = 0$ into Equation 3.81 allows C to be determined, yielding

$$C = \frac{\kappa_3 A_1(0)}{K}. \tag{3.91}$$

Finally, with the help of Equation 3.89 and 3.91, Equation 3.85 becomes

$$A_3(z) = -ie^{i\phi_2}\sqrt{\frac{n_1\,\omega_3}{n_3\,\omega_1}}A_1(0)\sin(Kz). \tag{3.92}$$

The intensity of each field is calculated using Poynting's theorem as expressed in Equation 3.34, which gives:

$$\boxed{I_1 = \frac{cn_1}{8\pi}|A_1(0)|^2\cos^2(Kz).} \tag{3.93}$$

$$\boxed{I_3 = \frac{cn_3}{8\pi}\left[|A_1(0)|^2\frac{n_1\,\omega_3}{n_3\,\omega_1}\sin^2(Kz)\right].} \tag{3.94}$$

Figure 3.11 plots the intensities as a function of z.

3.6 Difference Frequency Generation and Parametric Amplification

Next we consider difference frequency generation. For $\omega_1 > \omega_2$, the difference frequency beam at frequency $\omega_3 = \omega_1 - \omega_2$ is generated in response to beams at frequencies ω_1 and and ω_2. The process is shown by Figure 3.12.

As was previously discussed in Section 1.7, violation of energy conservation implied by Figure 3.12 is resolved when viewed from the quantum perspective, as shown in Figure 1.11. In reality, the photon at frequency ω_1 splits into two; one at frequency ω_2, which is stimulated by the ω_2 beam, and

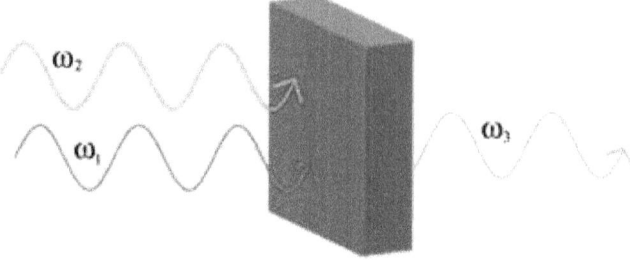

Figure 3.12: In a second-order nonlinear-optical material, the pump beam at frequency ω_1 and the amplified beam at frequency ω_2 results in the generation of a difference frequency beam at frequency $\omega_3 = \omega_1 - \omega_2$.

another one at frequency $\omega_1 - \omega_2$, which balances the energy account. The one photon added to the ω_2 beam for every ω_1 annihilated makes this an amplifier as well as a difference frequency generator; and, the ω_1 beam is called a pump since it adds energy to the amplified beam as well as producing light at the difference frequency.

Since this process is slightly different than the others that we have described in this chapter, we will use difference frequency generation as practice in determining the appropriate differential equations from intuition.

First, let's start by listing the simplifying assumptions used in the derivation that follows. They are:

1. Strong undepleted pump at frequency ω_1.

2. The beam frequencies are far from material resonances so there is full permutation symmetry.

3. $\chi^{(2)}$ is independent of frequency.

4. The phase matching condition holds, therefore, $\Delta k = 0$.

Our goal is to determine the appropriate differential equations in analogy to the derivations of Equations 3.58, 3.61 and 3.62. Rather than do the calculation, we will illustrate the intuitive approach.

We start with the equation for the amplified beam. A_2 appears on the left hand side since it is the beam if interest. The beam is generated by an incident photon at ω_1, which is anihilated to for a photon at ω_2 and $\omega_3 = \omega_1 - \omega_2$. Thus, the amplitude of the ω_3 wave being on the right appears as a complex conjugate, yielding

$$\frac{dA_2(z)}{dz} = \frac{2i\pi k_2}{\epsilon_2} \chi^{(2)} A_1 A_3^*(z), \qquad (3.95)$$

where we have used $\Delta k = 0$. The beam at ω_1 is assumed to be more intense than all others since it is the power source for amplification and difference frequency generation. As a consequence, a negligible amount will be depleted, so A_1 can be assumed constant. Since the field being solved for is A_2, the coefficient on the righthand side of Equation 3.95 has subscripts for the ω_2 field.

Equation 3.95 can then be expressed as

$$\frac{dA_2}{dz} = K_2 A_3^*, \qquad (3.96)$$

where the constant K_2 is given by

$$K_2 = \frac{2i\pi k_2}{\epsilon_2} \chi^{(2)} A_1. \qquad (3.97)$$

The propagation of the ω_3 wave is governed by an equation that can be written from intuition as was Equation 3.95. The reader should convince themselves that interchanging the indices 2 and 3 will do the trick, leading to

$$\frac{dA_3(z)}{dz} = \frac{2i\pi k_3}{\epsilon_3} \chi^{(2)} A_1 A_2^*(z), \qquad (3.98)$$

which can be expressed as

$$\frac{dA_3}{dz} = K_3 A_2^*, \qquad (3.99)$$

where

$$K_3 = \frac{2i\pi k_3}{\epsilon_3} \chi^{(2)} A_1. \qquad (3.100)$$

Using the standard decoupling strategy, we differentiate Equation 3.96 with respect to z, yielding

$$\frac{d^2 A_2}{dz^2} = K_2 \frac{dA_3^*}{dz}, \qquad (3.101)$$

which can be rewritten using the complex conjugate of Equation 3.99, which gives

$$\frac{d^2 A_2}{dz^2} = K_2 K_3^* A_2. \qquad (3.102)$$

Since $K_2 K_3^* > 0$, it can be written as $K^2 = K_2 K_3^*$, so the solution to Equation 3.102 is

$$A_2(z) = C \sinh(Kz) + D \cosh(Kz). \tag{3.103}$$

Substituting Equation 3.103 into Equation 3.99 and integrating, $A_3(z)$ is found to be of the form

$$A_3(z) = \frac{-K}{K_2^*} [C \cosh(Kz) + D \sinh(Kz)]. \tag{3.104}$$

There is no difference frequency light at the input interface, defined to be at $z = 0$, so $A_3(z=0) = 0$ and $A_2(z=0) \neq 0$. These boundary demand that $C = 0$ and $D = A_2(0)$, so

$$A_2(z) = A_2(0) \cosh(Kz). \tag{3.105}$$

Substituting Equation 3.105 into Equation 3.96 we get

$$A_3(z) = -i \frac{A_1}{|A_1|} A_2(0) \sqrt{\frac{k_3 \, \epsilon_2}{k_2 \, \epsilon_3}} \sinh(Kz), \tag{3.106}$$

where the constant A_1 is the electric field amplitude of the pump beam.

Converting the field amplitudes to intensities using Equation 3.34 yields

$$\boxed{I_2(z) = I_2(0) \cosh^2(Kz)} \tag{3.107}$$

and

$$\boxed{I_3(z) = I_2(0) \frac{\omega_3 \, n_2}{\omega_2 \, n_3} \sinh^2(Kz).} \tag{3.108}$$

Note that we made the substitution $k_j = n_j \omega_j / c$ and $\epsilon_j = n_j^2$ in arriving at Equation 3.108.

The intensities $I_2(z)$ and $I_3(z)$ are plotted in Figure 3.13. Equations 3.107 and 3.108 show that the intensity I_2 is amplified – called parametric amplification – and light at ω_3 is generated – called difference frequency generation. For each photon generated at the sum frequency, a photon is added to the ω_2 beam. Since Equations 3.107 and 3.108 diverge for large z, the sample length must be short enough so that the pump is not depleted.

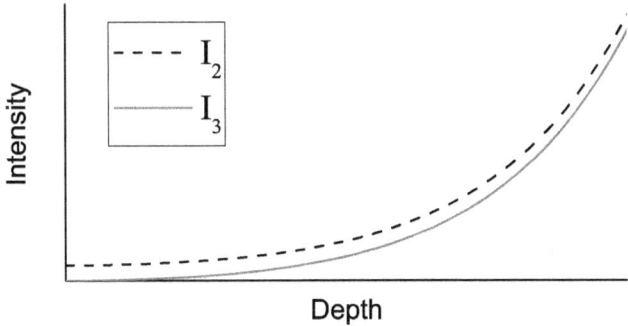

Figure 3.13: The intensities at frequencies ω_2 and ω_3 as a function of depth, z, into the material.

Problem 3.6-1(a): We made the approximation that the pump beam, A_1 is not depleted. Use the intuitive approach to determine the differential equation for the pump amplitude $A_1(z)$.

(b) Substitute Equation 3.105 for $A_2(z)$ and Equation 3.106 for $A_3(z)$ into your intuitively-derived differential Equation to determine $A_1(z)$. In Equation 3.106, assume that $A_1 = A_1(0)$.

(c) Determine a condition that must hold for the undepleted pump approximation to hold by demanding that the change in the intensity I_1 is small.

(d) Solve the three differential equations using Python to determine the intensity of the three waves as a function of z. Assume the initial condition that $A_1(0)$ and $A_2(0)$ are nonzero with $A_1(0) > A_2(0)$; and, $A_3(0) = 0$. You will start from these boundary conditions and use an iterative method, starting at $z = 0$, to solve the equations with step size Δz as you cycle through the three differential equations in the appropriate order. Plot the numerical results together with the analytical ones that we determined in the undepleted pump approximation, and investigate when they differ.

Figure 3.14: (left) In second harmonic generation, two incident photons of frequency ω interact with a nonlinear medium, generating an output photon of frequency 2ω. (right) Down conversion – where the second harmonic light interacts with the fundamental – is difference frequency generation, as described in the previous section.

3.7 Second Harmonic Generation

We now consider second harmonic generation, a special case of sum frequency generation with two incident photons at the same frequency $\omega \equiv \omega_1 \equiv \omega_2$, which generate an output photon of frequency $2\omega \equiv \omega_3$, as shown in Figure 3.14(left). In a lossless dielectric material (which is the case we are considering throughout our discursion), energy will be transferred between the ω and 2ω wave, but the total intensity remains constant.

The two relevant nonlinear polarizations are

$$\left(P^{NL}\right)^{2\omega} = \frac{1}{2}\chi^{(2)}\left(E^{\omega}\right)^2, \qquad (3.109)$$

for the second harmonic process, and

$$\left(P^{NL}\right)^{\omega} = \chi^{(2)} E^{-\omega} E^{2\omega}, \qquad (3.110)$$

for the down-conversion process as shown in Figure 3.14(right). E^{ω} and $E^{2\omega}$ are the usual complex amplitudes. To understand the factor of 1/2 in Equation 3.109, recall that the nonlinear polarization is calculated from the square of the total electric field, which contains the fundamental at ω and the second harmonic at 2ω, or $\left(E^{\omega} + E^{2\omega}\right)^2 = (E^{\omega})^2 + 2E^{\omega}E^{2\omega} + (E^{2\omega})^2$. The factor comes from the first term being half of the second one. This is an example of the K-factor, which is defined by Equation 2.166 and described in detail in Section 2.3.5.

Applying our intuitive approach, we need to account for the factor of two difference between the two terms and we need to track the wave vector in the exponents. To make the procedure clear, we write out the exponent explicitly. For the second harmonic generation process, with two incident photons and

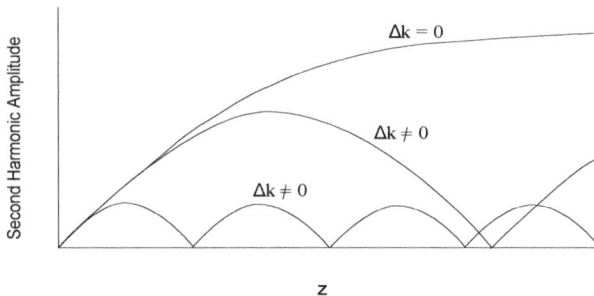

Figure 3.15: Phase-mismatching reduces the efficiency of generating second harmonic wave significantly.

an emitted second harmonic photon, we get

$$\frac{dA_2(z)}{dz} = \frac{i\pi k_2}{\epsilon_2}\chi^{(2)} A_1(z)A_1(z)e^{i(k_1+k_1-k_2)z}. \quad (3.111)$$

The wave vectors in the exponent are associated with incident photon #1 (k_1), incident photon #2 (k_1), and emitted photon ($-k_2$). The negative sign is accounts for the photon being emitted. Note that there are no complex conjugated amplitudes because all photons on the righthand side of the equation are incident and all photons on the left-hand side are emitted.

In the process where a photon at ω is generated through down conversion, the true process is an incident photon at 2ω splitting into two photons at frequency ω due to a stimulating photon at ω. Then, we get

$$\frac{dA_1(z)}{dz} = \frac{2i\pi k_1}{\epsilon_1}\chi^{(2)} A_2(z)A_1^*(z)e^{i(k_2-k_1-k_1)z}. \quad (3.112)$$

The field on the left-hand side of the equation is the one of interest, but the other emitted photon appears on the righthand side, hence the complex conjugation. The incident photon contributes k_2 to the exponent while the two outgoing ones contribute $-k_1$ and $-k_1$.

3.7.1 General Case

With the definition $\Delta k = 2k_1 - k_2$, the exponents in Equations 3.111 and 3.112 contain $\pm \Delta k$. The solutions of Equations 3.111 and 3.112 give intensities in the form of Jacobi elliptic functions. Figure 3.15 shows the amplitude as a function of z for the exact solutions.

Without solving any Equations, we can sketch the behavior of the intensities. For example, under the phase-matching condition, we know that the

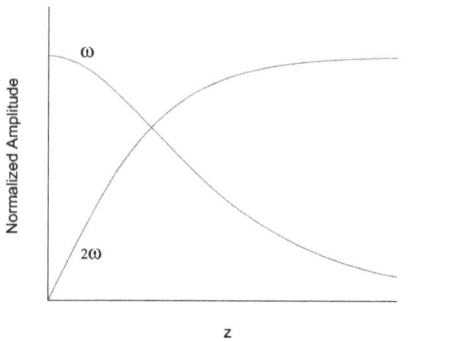
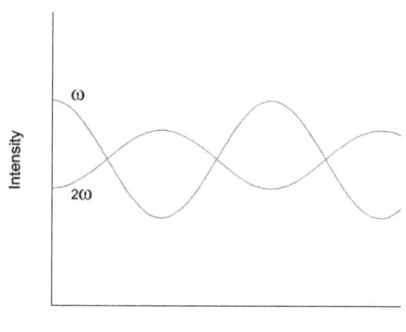

Figure 3.16: (left) The fundamental wave is depleted as it is converted to the second harmonic wave. At $z = 0$ the slope of the amplitude of the fundamental wave vanishes and the second harmonic amplitude's slope is nonzero. (right) When both fundamental and second harmonic intensities are incident on the material, they oscillate in the absence of phase-matching.

intensity increases with distance of propagation, then levels off when all the light is converted to the second harmonic. Furthermore, when only the fundamental wave is incident on the nonlinear medium, so that $I_1(z = 0) \neq 0$ and $I_2(z = 0) = 0$, then $A_1(0) \neq 0$ and $A_2(0) = 0$ and Equations 3.111 and 3.112 imply that $\left(\frac{dA_2}{dz}\right)_{z=0} \neq 0$ and $\left(\frac{dA_1}{dz}\right)_{z=0} = 0$. Figure 3.16 shows that the exact results agree with our expectations.

If the incident intensities of the fundamental and second harmonic waves are given by $I_1(z = 0)$ and $I_2(z = 0)$, where both are nonzero, $I_1(z)$ and $I_2(z)$ will oscillate as a function of z; but the total intensity will be constant as required by energy conservation. Figure 3.16(right), which plots the analytical results, shows this behavior.

Problem 3.6-1(a): Solve Equations 3.111 and 3.112 numerically using Python for the case where $A_1(0) \neq 0$ and $A_2(0) = 0$ and reproduce the results shown in Figure 3.15.

(b) For the phase-matched case where $A_1(0) \neq 0$ and $A_2(0) = 0$, use your code to reproduce Figure 3.16(left).

(c) Under the phase-matched condition, when $I_1(0) \neq I_2(0) \neq 0$, reproduce Figure 3.16(right) with your code.

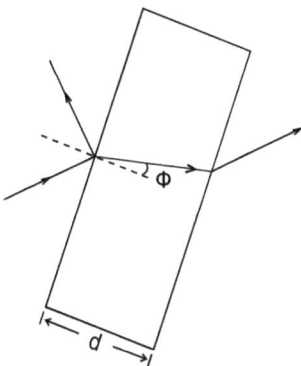

Figure 3.17: Rotating a quartz sample changes the path length ℓ that the light travels in the material. Consequently, the intensity of the second harmonic varies as the orientation of the quartz sample changes.

3.7.2 No Depletion Approximation

Next, we solve for the second harmonic intensity as a function of propagation distance in the undepleted limit. Rather than solve this problem from scratch, we instead use Equation 3.36, which was obtained for sum frequency generation. We can use it after making the substitution $\omega_1 = \omega_2 = \omega$ and $\omega_3 = 2\omega$. The only other modification is to add the factor of 1/2 (from the K-factor) in the amplitude of the second harmonic wave, which will add a factor of 1/4 to the intensity. This factor of 1/4 will cancel the factor of 4 that comes from $(2\omega)^2$. Do the math to make sure you agree. Thus, we get

$$I_2 = \frac{32\pi^3 \omega^2}{c^3} \left(\chi^{(2)}\right)^2 \frac{I_1^2}{n_1^2 n_2} \cdot L_c^2 \sin^2\left(\frac{l}{L_c}\right). \tag{3.113}$$

When the intensity of the second harmonic light that is generated is low compared with the incident intensity, the no-depletion approximation is a good one. The simplest experimental technique to characterize a material is to measure the second harmonic intensity as a function of the sample length. If the sample's length could be changed, the resulting second harmonic intensity would appear as shown in Figure 3.17. Since the length is fixed, the distance through which the beams travel can be changed by rotating the sample by an angle ϕ, as shown in Figure 3.18.

If the sample thickness is d, then the length of propagation of the beams is given by

$$\ell = d/\cos\phi. \tag{3.114}$$

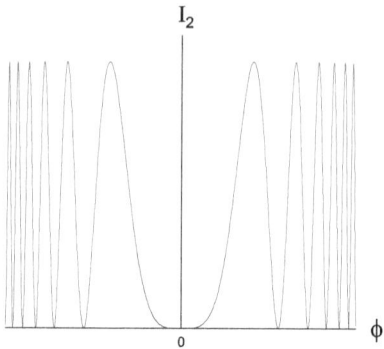

Figure 3.18: The intensity of the second harmonic wave as a function of the angle of refraction, ϕ, of the fundamental wave. Here we neglect the birefringence of the sample, the tensor nature of $\chi^{(2)}$ and the Fresnel reflections at the sample's surfaces.

Substituting this length into Equation 3.113 gives an intensity dependence as shown in Figure 3.18. The figure does not take into account the angular dependence of the reflection coefficient, which yields the maximum transmittance at normal incidence. Also not taken into account are the tensor properties of second-order susceptibility and the birefringence, which complicates the observed pattern due to the fact that different tensor components contribute as the sample is rotated. As a consequence, the intensity of the second harmonic signal will depend on the orientation of the quartz sample in a more complex way. If these effects are properly taken into account, one will observe an intensity as a function of angle as plotted in Figure 3.19, which was experimentally observed by Maker et al. in 1962.[12] The observed pattern of fringes are called Maker Fringes.

3.7.3 Phase Matching

The second harmonic intensity is maximal under the phase-matching condition

$$\Delta k = 2k_1 - k_2 = 2 \cdot \frac{\omega}{c} \cdot n_1 - \frac{2\omega}{c} \cdot n_2 = 0 \qquad (3.115)$$

or

$$n(\omega) = n(2\omega). \qquad (3.116)$$

Equation 3.116 suggests that phase-matching is achievable using the dispersion of the refractive index of a material. This section describes methods for

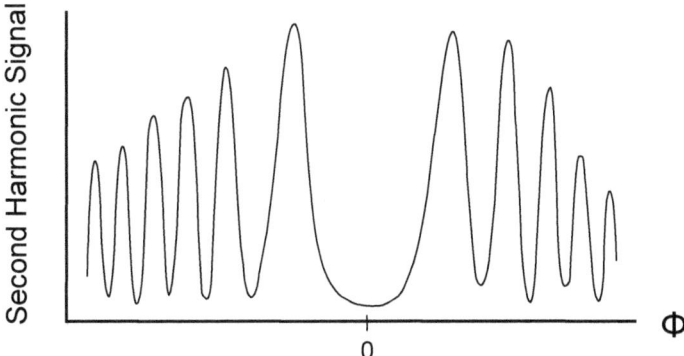

Figure 3.19: The observed angular dependence of the second harmonic intensity in quartz.

attaining phase machining materials.

anomalous dispersion phase matching

As we saw in Figure 3.7, anomalous dispersion – which is found on the high-energy side of the first excited state of a material – makes it possible for the the refractive index at the fundamental in the normal dispersion regime to be the same as the refractive index of the second harmonic wave in the anomalous dispersion regime. It is thus possible for Equation 3.116 to hold. However, the phase condition will hold only for a specific fundamental wavelength, making anomalous dispersion impractical due to its lack of flexibility.

Birefringence Phase Matching

The best way to achieve the phase-matching condition is by using a material's birefringence because it offers a way to tune the refractive index difference between any two wavelengths simply by rotating the crystal. Birefringence is pictorially quantified by an index ellipsoid. Our focus here is on uniaxial birefringence characterized by the index ellipsoid shown in Figure 3.20a. The components of the refractive index in the xy-plane are independent of the polarization in that plane; so, $n_x = n_y = n_o$, and $n_z = n_e \neq n_o$, where n_o and n_e are called ordinary and extraordinary refractive indices, respectively.

Consider a plane wave with its k-vector in the yz-plane and inclined by an angle θ from the z axis. So, when $\theta = 0$, the wave propagates along the z-axis and when $\theta = \pi/2$, the wave propagates along the y-axis. When θ is between these extremes, imagine a plane perpendicular to **k** that includes the center

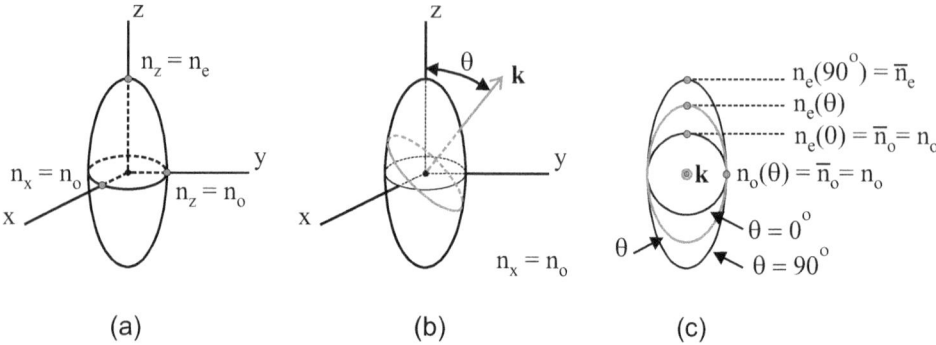

Figure 3.20: (a) Uniaxial birefringence is characterized by an ellipsoid having the ordinary refractive index n_o for any electric field polarization in the xy plane and the extraordinary refractive index n_e along z. (b) A wave incident in the yz plane and a cross-section of the ellipse in the plane perpendicular to $\mathbf{k}$. (c) The cross-section of the ellipsoid shown with $\mathbf{k}$ pointing out of the page for $\theta = 0$, $0 < \theta < 90°$ and $\theta = 90°$.

of the ellipsoid. The intersection between the ellipsoid and the plane gives an ellipse as shown in Figure 3.20b. Figure 3.20c shows a front-on view (along $\mathbf{k}$) of this ellipse for three angles. When $\theta = 0$, the ellipse is a circle of radius n_0 and when $\theta = \pi/2$, the ellipse has semi-major and semi-minor axes of n_e and n_0. In between these two extremes, we call the two axes $\bar{n}_e$ and $\bar{n}_o$ as shown in the figure. The student should study Figure 3.20 until (s)he understands the spatial relationships between the ellipsoid, the coordinate axes, $\mathbf{k}$, and the ellipse. The refractive index is given by the distance from the ellipse center to the ellipse in the direction of the polarization.

Now that we have fixed the direction of $\mathbf{k}$, all that remains is to describe the polarization vector of the electric field. Note that in a transverse wave $\mathbf{k}$ is always perpendicular to $\mathbf{E}$. Define the perpendicular polarization $E_\perp$ to be perpendicular to the z axis, so it has a refractive index of $\bar{n}_o$. We call this wave the ordinary ray. The polarization orthogonal to $E_\perp$, call it $E_\parallel$, has the refractive index $\bar{n}_e$. We call this the extraordinary ray.

Finally, imagine that these two orthogonal polarizations remain fixed according to the above definitions while the k-vector is varied by adjusting θ, which recall is the angle between the z-axis and the k-vector. The extraordinary refractive index will depend on the angle θ through the ellipse's eccentricity according to

$$\frac{1}{n_e^2(\theta)} = \frac{\cos^2\theta}{\bar{n}_o^2} + \frac{\sin^2\theta}{\bar{n}_e^2}. \tag{3.117}$$

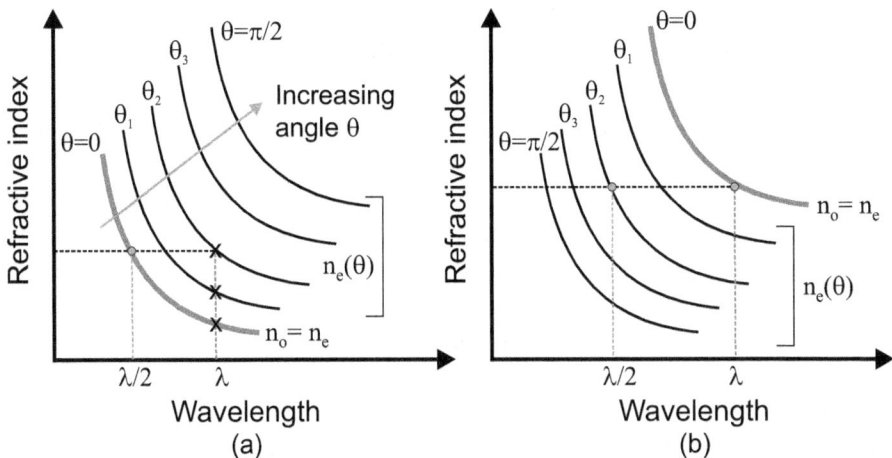

Figure 3.21: The dispersion of the refractive index for several angles θ for (a) a positive uniaxial crystal and (b) a negative uniaxial crystal. Phase matching demands that the fundamental and second harmonic are of opposite types. If the fundamental is an extraordinary ray, the second harmonic is an ordinary ray, or vice versa, as shown at the indicated points.

Under this construction, n_o does not depend on the angle, so the ordinary ray's refractive index is given by

$$n_o(\theta) = \bar{n}_o = n_0. \tag{3.118}$$

To summarize, $\bar{n}_o$ and $\bar{n}_e$ are the principle refractive indices of a plane wave polarized along the semi-major axis and semi-minor axis, respectively. Evidently, $n_o(\theta) = \bar{n}_o$, $n_e(0) = \bar{n}_o = \bar{n}_o = n_o(\theta)$ and $n_e(\pi/2) = \bar{n}_e$.

The principal refractive indices depend on wavelength, so adding dispersion to Equation 3.117 is straightforward, yielding

$$\frac{1}{n_e^2(\theta,\lambda)} = \frac{\cos^2\theta}{\bar{n}_o^2(\lambda)} + \frac{\sin^2\theta}{\bar{n}_e^2(\lambda)}. \tag{3.119}$$

Figure 3.21a shows a plot of the refractive index as a function of wavelength for a positive uniaxial crystal ($n_o < n_e$) for several values of θ. Since n_o is independent of θ, there is only one dispersion-curve for the ordinary ray, which is represented by the thicker curve. Figure 3.21b shows the dispersion curves for a negative uniaxial crystal with $n_o > n_e$. When $\theta = 0$, the ordinary and extraordinary dispersion curves are the same.

The phase matching condition in a positive uniaxial crystal for a fundamental wave at wavelength λ and second harmonic at $\lambda/2$ is determined as follows. Since in the normal dispersion regime – as shown in Figure 3.21 – the refractive index at the fundamental wavelength is lower than for the second harmonic, the fundamental wave must be an extraordinary ray and the second harmonic wave must be an ordinary ray, as explained below.

To get the correct phase-matching angle, draw a horizontal line at the refractive index of the second harmonic on the ordinary ray curve labelled $\theta = 0$ (shown as a point in Figure 3.21a) and draw a vertical line at the fundamental wavelength. The intersection between the two is shown as an "x" in Figure 3.21a on the curve labelled θ_2. Because the intersection point is on the curve corresponding to $\theta = \theta_2$, that is the phase matching angle.

This approach works because the refractive index of the second harmonic, which is polarized along the x-axis in Figure 3.20, remains constant as the second harmonic crystal is rotated; but, the refractive index of the fundamental, being an extraordinary ray, changes with angle. As the crystal is rotated from $\theta = 0$, the extraordinary ray's refractive index increases with angle as shown by the series of "x"s in Figure 3.21a. At angle θ_2, the fundamental wave has the same refractive index as the second harmonic wave, shown by the horizontal dashed line that connects the dot and the "x." For a negative uniaxial crystal, the refractive index of the fundamental beam defines the required refractive index of the second harmonic wave.

Observing phase matching in the lab is quite a treat, especially when the fundamental is invisible, as is the case in high power near infrared Nd:YAG lasers. The invisible light at a wavelength of 1064nm enters the crystal, and nothing is observed to exit. The researcher turns the knob that controls the angle θ, and a bright green flash at 532nm is observed as the angle sweeps past the phase matching angle. The process is sensitive to the angle, so one must make the adjustment slowly to find the peak output. When a second harmonic crystal is designed for a particular wavelength, the crystal is cut in a way so that the fundamental beam enters at normal incidence, thus optimizing the amount of light that gets into the crystal, among other properties.

Phase matching is a property of the linear refractive index. It might be the case that a particular ray of light is phase matched but the second-order susceptibility tensor vanishes for that ray. Fortunately, there are many possible phase matching directions in a crystal, so the trick is to find a ray trajectory that is both phase matched and selects the largest nonlinear-optical tensor component. All properties are not often simultaneously optimized, so it becomes a matter of tradeoffs.

There are different tensor components that can be used in a second harmonic crystal. For example, an extraordinary ray of frequency 2ω that is generated from two incident ordinary rays of frequency ω is called type *I* phase-matching, which corresponds to $\chi^{(2)}_{ijj}$ with i and j referring to the extraordinary and two ordinary rays, respectively. Type II phase-matching corresponds to an extraordinary ray of frequency 2ω generated from one ordinary ray and one extraordinary ray at frequency ω. Type II phase-matching requires $\chi^{(2)}_{iij}$, where the first i refers to the outgoing extraordinary ray and the second i and j refer to the incident extraordinary ray and ordinary ray, respectively.

Quasi Phase Patching

As we saw in Section 3.3, a spatially varying second-order susceptibility profile can lead to quasi phase matching even in materials where the refractive index dispersion makes phase matching impossible. According to Equation 3.39, the amplitude of the generated wave is proportional to the integral,

$$A_3(\ell) \propto \int_0^\ell dz\, \chi^{(2)}(z) e^{i\Delta k z}. \qquad (3.120)$$

Equation 3.120 is the Fourier transform of the spatial variations of the second-order susceptibility, so gives the Fourier amplitude at the period corresponding to the coherence length (or more presciently, at the period πL_c). This will be maximum when the second order susceptibility has a period equal to πL_c.

> **Problem 3.7.3-1:** Evaluate the integral in Equation 3.120 for a periodic square wave and compare your results with the sinusoidal waveform and periodic Dirac delta functions that were discussed as two examples in Section 3.3.1.

Mode Index Phase Matching in a Waveguide

When light is confined to dimensions comparable to a wavelength of light, the ray description no longer applies. For example, in a small-diameter optical fiber, there are certain transverse intensity profiles that remain unchanged as they travel down the length of a fiber. These so-called stationary solutions are called eigenstates or modes.

Figure 3.22 shows the intensity profile of the two lowest-order modes in an optical fiber. The important point here is that the group velocity of each mode is different. Thus, even when the refractive index of the material is uniform, the modes will travel according to a quantity analogous to the refractive

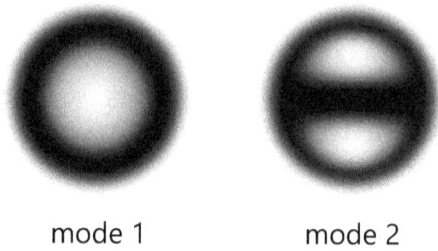

mode 1　　　mode 2

Figure 3.22: The transverse intensity profile of the two lowest modes of an optical fiber.

index but is called the mode index , which depends on the refractive index, wavelength, and fiber diameter. As such, phase matching is made possible by matching the mode indices to offset material dispersion.

This offers yet another avenue to trick nature into phase matching . Optical fibers have the additional benefit of confining light tightly, so the intensity is higher over longer distances than what is possible in free space. The down side is that the conversion efficiency depends also on the spatial overlap between the two modes carrying the fundamental and second harmonic waves. So if the phase matching condition is accompanied by poor spatial overlap, the second harmonic efficiency will be suppressed.

3.8 Applications of Frequency Mixing

This chapter has discussed a variety of nonlinear-optical processes. Why are they important. There is of course the challenge of understanding nature, but there are lots of applications.

Second harmonic generation has been used from the day lasers were invented to double the frequency of lasers . Light of shorter wavelength contains photons of higher energy, which can excite materials or push them to their limits of stability. Interestingly, all the original green laser pointers used tiny second harmonic crystals to convert near IR lasers to green because green lasers were difficult to make. This is also true in large industrial-scale lasers, in which crystals are used to convert light to higher frequency. Even the Blue Ray players needed nonlinear crystals to produce blue light, which by virtue of the shorter wavelength can cram more information onto the surface of a DVD.

Second harmonic generation is also used in medical imaging to bring out feature that are not normally visible with regular light. It is also used in

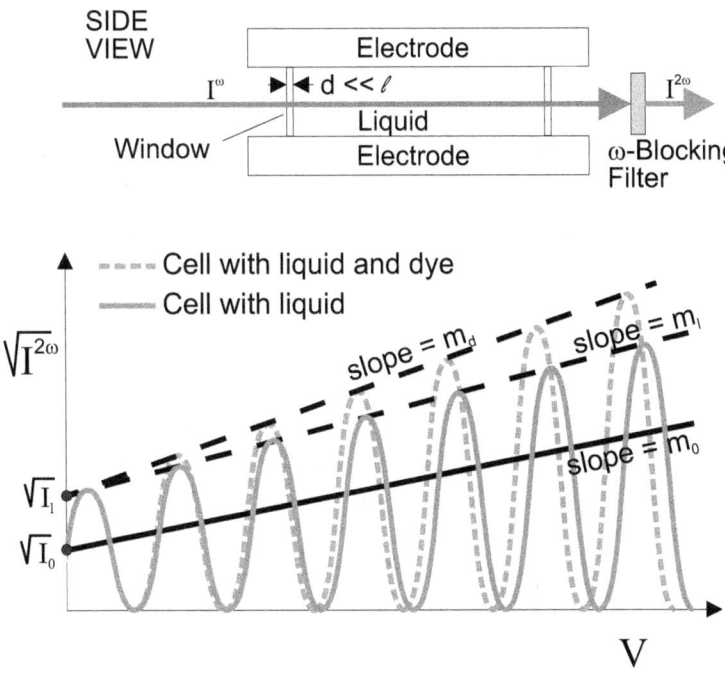

Figure 3.23: Figure for practical problem.

microscopy to show regions of polar alignment is cells.

Sum frequency generation is used to make tunable lasers that span incredible wavelength ranges. The same process is used to amplify weak light pulses.

3.9 Practical Problems

We end this chapter with practical problems. These are the sorts of problems that researchers are faced with daily. They require one to draw on a wide body of knowledge, and to apply the right principles under reasonable approximations. Note that these problems are intended to be open ended and that bits and pieces of information may be missing. In addition to applying the material we have learned, these problems are designed to train you to think like a researcher.

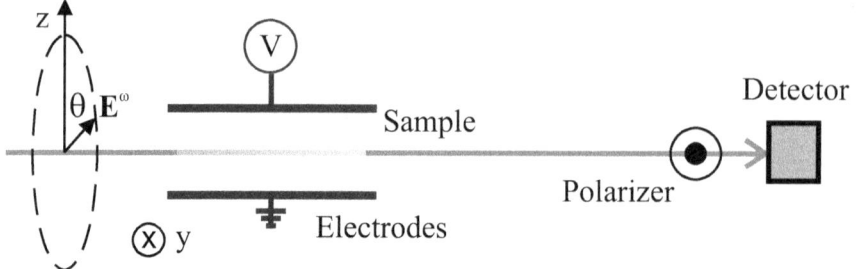

Figure 3.24: Figure for practical problem.

Problem 3.9-1: A liquid cell is constructed with transparent windows and electrodes as shown in the cut-away drawing in Figure 3.23. The windows are much thinner than the wavelength of light and the separation between the windows is large ($\approx$ 1cm). An experimenter intends to use this apparatus to determine the nonlinear response of a dye solution by measuring the intensity of second harmonic light in the presence of a static field. The three experimental steps are as follows. First the empty cell is placed in a beam and the second harmonic is measured as a function of applied voltage to the electrodes. A plot of $\sqrt{I^{2\omega}}$ gives a straight line of slope m_0. Secondly, when the cell is filled with the liquid solvent, the data oscillates and grows with applied voltage as shown by the solid curve in the figure . Finally, when the dye molecules are dissolved into the solvent, the solution shows the same oscillatory behavior with decreased fringe spacing and greater amplitude (dotted line). In all cases, the intensity of the second harmonic is much smaller than the input intensity.

(a) Propose as many hypotheses as you can imagine that might explain the source of the fringes that would be consistent with the data.

(b) What experiment would you perform to test each hypotheses?

(c) The liquid cell is removed and replaced with known reference material whose thickness is much smaller than the wavelength of light. With no field applied, the observed second harmonic intensity is I_q and the voltage dependence is found to be linear with slope m_q. If the refractive index function is approximately equal for all the materials (i.e. reference material, liquid, windows, and dye solution), determine all of the bulk susceptibilities that the data has to offer.

NOTE: The fringes are not necessarily equally spaced.

Problem 3.9-2: A voltage $V(t) = V_0 \sin(\Omega t)$ is applied to a sample (in the form of a rectangular solid) along the z-axis in coincidence with a beam of light at frequency ω (with $\Omega \ll \omega$) whose polarization makes an angle θ with respect to the 'static field' (which you can adjust), as shown in Figure 3.24. The sample's linear susceptibility is isotropic. The light passes through a polarizer (whose angle you can also adjust), and the intensity at the detector is found to have a static component as well as components at frequency Ω and 2Ω. As such, you use a lockin amplifier to measure the amplitude of the intensity oscillation at zero, Ω and 2Ω, or I_0, I_Ω and $I_{2\Omega}$, respectively. Determine the real and imaginary parts of the $\chi^{(1)}$, $\chi^{(2)}$, and $\chi^{(3)}$ tensors for the sample from these measurements. Note that you can measure the tensor components in the $y-z$ plane, i.e. components of the form $\chi^{(2)}_{zzz}$, $\chi^{(2)}_{zyy}$, etc. Note that for simplicity, you may assume Kleinman symmetry.

Be sure to discuss any other measurements that you may need to do, keeping in mind that the detector cannot measure an absolute intensity and that its reading can change if it is slightly moved. Note that you should use the approach of solving the nonlinear wave equation in the material.

Chapter 4

Quantum Theory of Nonlinear Optics

The origin of nonlinear-optical effects is in the interaction of photons with electrons in atoms, molecules, crystals, etc. Electrons are most precisely described within a quantum-mechanical framework. A majority of students are familiar with the Schrödinger Equation and how the potential describes the interactions of the quantum system with the outside world. In such treatments, the electric field is treated classically.

Solutions to the Schrödinger Equation give wavefunctions, from which one can calculate expectations of operators – the stuff we measure. While this approach gives the right answer, it hides the beauty of what is happening in the quantum realm. The shame of the matter is that the quantization of photons leads to a deeper understanding and appreciation of the physics of the problem, which feeds the intuition.

Learning about the quantization of photons takes a little bit of extra effort, but it's worth it. In the end, we will learn how to apply Feynman-like diagrams with aplomb to any nonlinear process, we will be amazed at how stimulated emission arises naturally without having to shoe-horn it into the theory, and the intuition and tools gained will allow us to surmount more complex problems with ease.

4.1 Dirac Notation

Dirac notation is simple and beautiful; but, gives some students difficulties because it represents abstract objects, which are unfamiliar to those who are inexperienced with linear algebra. This section will provide a non-rigourous introduction that is meant to get novices up to speed quickly. In the long run, everyone should become an expert with Dirac notation .

For students who are unfamiliar with Dirac notation , it is recommended that they read through this section once without fully mastering the topic. Then, if there is confusion in the sections that follow, the student can come back here to review definitions and concepts. The best way of gaining a familiarity with Dirac notation is to use it.

It is always useful to use a crutch at first, and vectors in three-dimensional space will the familiar analogy that we will call upon. Consider two vectors **A** and **B**, which represents *things* that can be pictured as arrows in three-dimensional space. We define a dot product between the two vectors as

$$\mathbf{A} \cdot \mathbf{B} = \text{complex number}. \quad (4.1)$$

The number that results is independent of the coordinate system.

The state of a quantum system can also be represented by a vector, but this vector can live in an infinite-dimensional space – called a Hilbert space . Consider two state vectors, $|\psi\rangle$ and $|\phi\rangle$. We can also define a dot product between them, written as

$$\langle\psi| \cdot |\phi\rangle \equiv \langle\psi|\phi\rangle \text{ complex number}, \quad (4.2)$$

where the righthand side of Equation 4.2 is the standard shorthand notation. One can define rotations in this infinitely-dimensional space in a way analogous to rotations of a vector in our familiar physical space, and this inner product , as it is called, is invariant under such rotations. The only departure in this analogy is that for vectors, $\mathbf{A} \cdot \mathbf{B} = \mathbf{B} \cdot \mathbf{A}$ but for state vectors, we have

$$\langle\psi|\phi\rangle = \langle\phi|\psi\rangle^*, \quad (4.3)$$

where the ' * ' is the complex conjugate. The inner product $\langle\psi|\phi\rangle$ is an expression straddled by brackets; so, Dirac coined the phrase *bra* for the left part of the bracket, $\langle\psi|$, and *ket* for the right part, $|\psi\rangle$.

In our familiar world, we can define a set of three special vectors that each have unit length and are aligned along the three Cartesian axes, which we call $\hat{\mathbf{x}}$, $\hat{\mathbf{y}}$, and $\hat{\mathbf{z}}$. By definition, $\hat{\mathbf{x}} \cdot \hat{\mathbf{x}} = 1$, $\hat{\mathbf{y}} \cdot \hat{\mathbf{y}} = 1$, and $\hat{\mathbf{z}} \cdot \hat{\mathbf{z}} = 1$. We can use these basis vectors to determine the shadow casted by the vector onto any axis. For example, the projection of the vector **A** onto the *x*-axis is computed from the dot product,

$$A_x = \hat{\mathbf{x}} \cdot \mathbf{A}. \quad (4.4)$$

Similarly, we can choose a special set of unit vectors in Hilbert space that define a basis, which we call $|\phi_n\rangle$, where $\langle\phi_n|\phi_m\rangle = \delta_{n,m}$. There are complications here because there are two types of spaces to worry about; one is discrete – where n takes on integer values, and the other continuous, where n can be an irrational number. To make matters worse, the Hilbert space can be a mixture of both. Independent of these complications, we can still determine the projection of any state vector $|\psi\rangle$ onto a basis vector using the inner product

$$a_n = \langle\phi_n|\psi\rangle. \tag{4.5}$$

If the space is continuous, the convention is to use $a(n)$ rather than a_n to remind ourselves that the projection is a continuous function of the parameter n. Since n so often denotes an integer, many physicists will label continuous indices with a different symbol, such as ξ so the projection is given by $a(\xi)$. Note that Equation 4.3 informs us that $\langle\psi|\phi\rangle \neq \langle\phi|\psi\rangle$, so by convention, the state vector is represented by the ket $|\psi\rangle$, and the projection is calculated by taking the inner product with the bra $\langle\phi_n|$.

In regular space, physics behaves the same relative to any coordinate system made of orthogonal unit vectors. A coordinate system can be made by rotating another one, and/or moving the origin to a different place. Non-cartesian unit vectors also do the job. Similarly, there are many possible basis vectors in Hilbert space . One very special basis set is the one in which the basis state vectors are the energy eigenstates . In this case, we label them $|n\rangle$. We will return to this topic later.

Next, we discuss operators, which transform one vector into another. The rotation operator is an example of one that changes the components of a vector, but keep the length fixed. The projection operator is another example, which returns one component. Let's begin by considering the operator that projects a vector onto the x axis. We write it as

$$P_x = \hat{\mathbf{x}}(\hat{\mathbf{x}} \cdot. \tag{4.6}$$

This strange-looking construction is evaluated by starting with the operation on the far right and working left. Thus, first take the dot product of $\hat{\mathbf{x}}$ of whatever is to its right, then place $\hat{\mathbf{x}}$ adjacent to the result. Thus,

$$P_x \mathbf{A} = \hat{\mathbf{x}}(\hat{\mathbf{x}} \cdot \mathbf{A}) = \hat{\mathbf{x}} A_x. \tag{4.7}$$

Now let's define the projection operator ,

$$P = P_x + P_y + P_z = \hat{\mathbf{x}}\hat{\mathbf{x}} \cdot + \hat{\mathbf{y}}\hat{\mathbf{y}} \cdot + \hat{\mathbf{z}}\hat{\mathbf{z}} \cdot. \tag{4.8}$$

Operating on the vector **A**, we get

$$P\mathbf{A} = \hat{\mathbf{x}}(\hat{\mathbf{x}} \cdot \mathbf{A}) + \hat{\mathbf{y}}(\hat{\mathbf{y}} \cdot \mathbf{A}) + \hat{\mathbf{z}}(\hat{\mathbf{z}} \cdot \mathbf{A}) = \hat{\mathbf{x}}A_x + \hat{\mathbf{y}}A_y + \hat{\mathbf{z}}A_z = \mathbf{A}. \tag{4.9}$$

Thus, each projection operator returns a single vector component, but the three added together returns the original vector. Thus, P is the identity operator, or $P = \mathbb{1}$.

State vectors have the same properties in Hilbert space. We can define the projection operator onto state $|\phi_n\rangle$ as $P_{\phi_n} = |\phi_n\rangle\langle\phi_n|$, so

$$P_{\phi_n}|\psi\rangle = |\phi_n\rangle\langle\phi_n|\psi\rangle = |\phi_n\rangle a_n. \tag{4.10}$$

where the last equality comes from Equation 4.5. In analogy to the three-dimensional case, we can show that

$$\sum_n |\phi_n\rangle\langle\phi_n| = \mathbb{1}, \tag{4.11}$$

where the sum is over all states of the system. Note that if the states are continuous, the sum in Equation 4.11 becomes an integral, or

$$\int d\xi |\xi\rangle\langle\xi| = \mathbb{1}. \tag{4.12}$$

Finally, if the system has a set of discrete states and a set of continuous states, such as is the case for the bound and unbound states of an electron in a hydrogen atom,

$$\boxed{\sum_n^{\text{discrete}} |\phi_n\rangle\langle\phi_n| + \int_{\text{continuous}} d\xi |\xi\rangle\langle\xi| = \mathbb{1},} \tag{4.13}$$

where the discrete and continuous parts are separately evaluated and added together. Equation 4.13 is called closure, and is one of the most powerful identities in quantum mechanics. Many quantum problems can be solved by inserting closure in the right place. Thus, **closure is the first thing to try when you are stuck.** Note that for simplicity, we will often express closure as given by Equation 4.11, but, it is understood that the sum represents sums and/or integrals, where appropriate.

At this point, we discard our crutch and move forward briefly reviewing the important properties of Dirac notation.

Using closure, the state vector can be written as

$$|\psi\rangle = \mathbb{1}|\psi\rangle = \sum_n |\phi_n\rangle\langle\phi_n| \cdot |\psi\rangle = \sum_n a_n |\phi_n\rangle, \tag{4.14}$$

where we have moved a_n to the left of the kets to put the final result in standard notation. Since a_n is a complex number, it can be moved to any place the heart desires. Equation 4.14 is a statement that any state vector can be expanded as a sum over the basis vectors.

The basis vectors is one example of an operator. Any operator can be massaged into the ket-bra form. For example, consider the Hamiltonian operator H, which gives energy eigenvalues when it operates on the energy eigenstate vector $|n\rangle$, yielding $H|n\rangle = E_n|n\rangle$. The Hamiltonian can be molded in the ket-bra form by using closure,

$$H = \mathbb{1}H\mathbb{1} = \sum_n |n\rangle\langle n| H \sum_m |m\rangle\langle m| = \sum_{n,m} |n\rangle\langle n|E_m|m\rangle\langle m| = \sum_{n,m} |n\rangle E_m \delta_{m,n} \langle m|, \quad (4.15)$$

where we have used that fact the $\langle n|E_m|m\rangle = E_m \langle n|m\rangle = E_m \delta_{m,n}$. Finally, this leads to

$$H = \sum_n E_n |n\rangle\langle n|. \quad (4.16)$$

Any operator can be converted to the ket-bra representation using closure. For an arbitrary operator O, we get

$$O = \sum_{n,m} O_{n,m} |n\rangle\langle m|, \quad (4.17)$$

where

$$O_{n,m} = \langle n|O|m\rangle. \quad (4.18)$$

Note that $\langle n|O|m\rangle$ is the operator in matrix form just as a_m is a component of the state vector $|\psi\rangle$ in the form of a column vector. Similarly, $\langle \psi|$ can be represented as a row vector with elements a_n^*.

Operators can also be converted to Dirac notation in an intuitive manner. For example, consider a spin 1/2 particle that can be in state spin up, $|+\rangle$, and spin down, $|-\rangle$. If we need an operator F that flips the spin, we can immediately write it down as

$$F = |+\rangle\langle -| + |-\rangle\langle +|, \quad (4.19)$$

where it is easy to verify that $F|\pm\rangle = |\mp\rangle$. Also, we can easily calculate the matrix elements of 4.19, yielding the matrix

$$F \rightarrow \begin{pmatrix} 0 & 1 \\ 1 & 0 \end{pmatrix}. \quad (4.20)$$

There are many types of state vectors that can be used as a basis. The energy eigenstates and spin states are two examples that we have discussed.

Two other useful bases are $|\mathbf{r}\rangle$, the position basis and $|\mathbf{p}\rangle$, the momentum basis. $|\mathbf{r}\rangle$ is a state vector that represents a particle at position $\mathbf{r}$ and $|\mathbf{p}\rangle$ is a state vector for a particle with momentum $\mathbf{p}$. There exists such a state vector corresponding to every point in space and one corresponding to every possible momentum so both are continuous bases. Note that $|\mathbf{r}\rangle$ should not be confused with $\mathbf{r}$. $|\mathbf{r}\rangle$ can be pictured as a column vector that has a zero in every entry for every single position in space (a huge vector!) except for the one position corresponding to the point $\mathbf{r}$. $\mathbf{r}$, on the other hand, is a vector with three components.

Now that we have covered the mathematics, we can use it as a tool for interpreting physics. To do so requires that physical properties be assigned the to mathematical operations. The inner product and expectation values are the two central operations that connect the mathematics to how we interact with the physics world. We discuss each separately, below, as well as the time evolution operator .

Inner Products and Probabilities

The inner product between two state vectors, such as $\langle \phi | \psi \rangle$, gives a complex number that correspond to a probability amplitude. The square of the absolute magnitude, $|\langle \phi | \psi \rangle|^2$, is a probability. If the state vectors are discrete, $|\langle \phi | \psi \rangle|^2$ is the probability of a measurement finding the system in state $|\phi\rangle$ given that the system is prepared to be in state $|\psi\rangle$. If at least one of the state vectors is continuous, $|\langle \phi | \psi \rangle|^2$ is the probability density of finding the system in state $|\phi\rangle$ given that the system is prepared to be in state $|\psi\rangle$. To see how this works, we consider two examples – the position state vectors and wavefunctions.

Recall that the state vector $|\mathbf{r}\rangle$ describes the state of a particle when it is prepared to be located at position $\mathbf{r}$ and $|\mathbf{r}'\rangle$ is the state vector of a particle at postilion $\mathbf{r}'$. Thus, the inner product $\langle \mathbf{r}' | \mathbf{r} \rangle$ gives the probability amplitude that a particle at position $\mathbf{r}$ will be found at position $\mathbf{r}'$. Since the state vectors are normalized, for discrete eigenstates $|\phi_n\rangle$ and $|\phi_m\rangle$ we get

$$\langle \phi_m | \phi_n \rangle = \delta_{nm} \quad (4.21)$$

and for continuous eigenstates $|\xi\rangle$ and $|\xi'\rangle$ we get

$$\langle \xi' | \xi \rangle = \delta(\xi - \xi') \quad (4.22)$$

so $\langle \mathbf{r}' | \mathbf{r} \rangle = \delta(\mathbf{r} - \mathbf{r}')$. This is just the trivial statement that a particle at position $\mathbf{r}$ is not at position $\mathbf{r}'$ unless $\mathbf{r} = \mathbf{r}'$.

Nonlinear Optics: A student's perspective

Next we discuss the connection between state vectors and wavefunctions. Recall that all information about a quantum system is contained in its state vector $|\psi\rangle$. Therefore, according to the meaning of an inner product, the probability density of finding a particle at position $\mathbf{r}$ that is prepared to be in state $|\psi\rangle$ is $|\langle \mathbf{r}|\psi\rangle|^2$. Thus,

$$\psi(\mathbf{r}) = \langle \mathbf{r}|\psi\rangle \text{ and } |\psi(\mathbf{r})|^2 = |\langle \mathbf{r}|\psi\rangle|^2. \quad (4.23)$$

The probability of finding the particle within a small volume $\delta V(\mathbf{r})$ centered at position $\mathbf{r}$ is given by

$$\mathcal{P}(\mathbf{r}, \delta V(\mathbf{r})) = |\psi(\mathbf{r})|^2 \delta V(\mathbf{r}) = |\langle \mathbf{r}|\psi\rangle|^2 \delta V(\mathbf{r}). \quad (4.24)$$

Equation 4.23 is the connection that converts the abstract object that we call a state vector into a wavefunction.

Finally, a useful inner product that is often used is the one between the position state vector and the momentum state vector,

$$\langle \mathbf{r}|\mathbf{p}\rangle = \frac{1}{(2\pi\hbar)^{3/2}} \exp\left[i\frac{\mathbf{r}\cdot\mathbf{p}}{\hbar}\right]. \quad (4.25)$$

which we state here without proof.

Matrix Elements of an Operator and the Hermitian Adjoint

Equation 4.18 hinted at the fact that operators can be expressed as matrices. In this section, we briefly discuss matrix elements of operators and the Hermitian Adjoint.

An operator A acts on a state vector $|\phi\rangle$ to produce another state vector $|\psi\rangle$, or

$$|\psi\rangle = A|\phi\rangle. \quad (4.26)$$

But, the inner product between two state vector gives a scaler, so the matrix element of an operator is a scalar, as is made obvious from the fact that

$$\langle \chi|\psi\rangle = \langle \chi|A|\phi\rangle. \quad (4.27)$$

We call the quantity $\langle \chi|A|\phi\rangle$ the $\chi\ \phi$ matrix elements of the operator A because any pair of state vectors returns a number just like the row and column of a matrix identifies a number. Using a set of eigenstate $|\phi_n\rangle$, we can construct a matrix with elements A_{nm}

$$A_{nm} = \langle \phi_n|A|\phi_n\rangle. \quad (4.28)$$

Next, we define the Hermitian Adjoint of a matrix A, $A^\dagger$, by

$$\langle \chi | A | \phi \rangle^* = \langle \phi | A^\dagger | \chi \rangle. \tag{4.29}$$

If the matrix element on the left hand side of Equation 4.29 is real, then using Equations 4.3 and 4.27, we get.

$$\langle \chi | A | \phi \rangle^* = \langle \phi | A | \chi \rangle. \tag{4.30}$$

Comparing Equation 4.29 with Equation 4.30, and noting that this is true for any two state functions, this implies that $A^\dagger = A$. When this is so, the operator A is said to be Hermitian.

Since observables are described by expectation values of the form $\langle \phi | A | \phi \rangle$, as described in the next section, and what we observe must be real, Equation 4.29 demands that an operator associated with an observable – such as a dipole moment – must be Hermitian.

Expectation Values and Transitions

A measurement of a quantity can only result in an eigenvalue. For example, a measurement of the energy will yield an energy eigenvalue. If many particles are prepared to be in the same quantum state, an energy measurement can give a different energy for each particle if the state vector is a linear combination of energy eigenvectors. The expectation value of the Hamiltonian then gives the average energy.

For example, if the state vector is given by $|\psi\rangle = \frac{1}{\sqrt{2}}(|0\rangle + |1\rangle)$, then the probability of finding a particle in state $|0\rangle$ is $|\langle 0 | \psi \rangle|^2 = \frac{1}{2}$ and the probability of finding a particle in state $|1\rangle$ is $|\langle 1 | \psi \rangle|^2 = \frac{1}{2}$. Thus, the average energy is $\frac{1}{2}(E_0 + E_1)$. This same result comes from the expectation value,

$$\begin{aligned} \bar{E} = \langle \psi | H | \psi \rangle &= \frac{1}{\sqrt{2}}(\langle 0 | + \langle 1 |) H \frac{1}{\sqrt{2}}(|0\rangle + |1\rangle) \\ &= \frac{1}{\sqrt{2}}(\langle 0 | + \langle 1 |) \frac{1}{\sqrt{2}}(E_0 |0\rangle + E_1 |1\rangle) \\ &= \frac{1}{2}(E_0 + E_1). \end{aligned} \tag{4.31}$$

Finally, we consider a transition $\langle \phi | A(\mathbf{r}, \mathbf{p}) | \psi \rangle$, which is like an expectation value, but, the two states are different. We consider the case where the operator A is a function of the operators $\mathbf{r}$ and $\mathbf{p}$. Using closure, we get

$$\begin{aligned}\langle\phi|A(\mathbf{r},\mathbf{p})|\psi\rangle &= \langle\phi|\left[\int d^3\mathbf{r}'\,|\mathbf{r}'\rangle\langle\mathbf{r}'|\right]A(\mathbf{r},\mathbf{p})\left[\int d^3\mathbf{r}''\,|\mathbf{r}''\rangle\langle\mathbf{r}''|\right]|\psi\rangle \\ &= \int d^3\mathbf{r}'\int d^3\mathbf{r}''\,\phi^*(\mathbf{r}')\,\langle\mathbf{r}'|A(\mathbf{r},\mathbf{p})|\mathbf{r}''\rangle\,\psi(\mathbf{r}'') \\ &= \int d^3\mathbf{r}'\,\phi^*(\mathbf{r}')A\left(\mathbf{r}',-\frac{\hbar}{i}\nabla'\right)\psi(\mathbf{r}'),\end{aligned} \quad (4.32)$$

where we have used the fact that

$$\langle\mathbf{r}'|A(\mathbf{r},\mathbf{p})|\mathbf{r}''\rangle = \delta(\mathbf{r}-\mathbf{r}')A\left(\mathbf{r}',-\frac{\hbar}{i}\nabla'\right) \quad (4.33)$$

The last expression in Equation 4.32 is the transition in terms of an integral over two wavefunctions, the form most familiar to students.

The Time Evolution Operator

Recall that an operator acts on a state vector to give a different state vector. There is a special operator called the time evolution operator $U(t-t_0)$, which has the action,

$$|\psi(t)\rangle = U(t,t_0)|\psi(t_0)\rangle, \quad (4.34)$$

where for a time-independent Hamiltonian H,

$$U(t,t_0) = \exp\left[-\frac{i(t-t_0)H}{\hbar}\right]. \quad (4.35)$$

Thus, the state vector can be determined at any time given the state vector at a fixed time, and the Hamiltonian operator H.

Operators Versus Functions

In Dirac notation, the state vectors can be a function of time, but not a function of position, momentum, etc. Only when the state vector is projected onto the position state vector will we get a wavefunction that can depend on position or momentum.

Operators and functions are distinguished by context. For example, H is the Hamiltonian, so is an operator, but $\mathbf{r}$ can be the position eigenvalue or the position operator. In the context of the eigenvalue equation $\mathbf{r}|\mathbf{r}'\rangle = \mathbf{r}'|\mathbf{r}'\rangle$, it is obvious that $\mathbf{r}$ is an operator and $\mathbf{r}'$ the eigenvalue. We used this notation in Equation 4.32 when inserting closure.

In Dirac notation, operators too are abstract objects that can be converted to regular space by taking matrix elements as shown in Equation 4.33 for the operator A.

Equation 4.25 is another example of using context in determining what is or is not an operator. Being an inner product, the left hand side of Equation 4.25 is a complex number, so the righthand side must also be a complex number. Then, **r** and **p** are not operators, but variables that represent numbers.

Problem 4.1-1: Show that the operator $P_x + P_z$ projects a vector onto the xz-plane.

Problem 4.1-2: The lowering operator a for the harmonic oscillator when acting on the energy eigenstate $|n\rangle$ has the action

$$a|n\rangle = \sqrt{n}|n-1\rangle.$$

Define a general lowering operator in the ket-bra representation that has this result when it acts on any energy eigenstate vector of *any Hamiltonian*. Closure will serve you well.

Problem 4.1-3: Express the position operator in ket-bra notation. Hint: Use closure.

Problem 4.1-4: The most general state vector for two spin-1/2 particles is

$$|\psi\rangle = \alpha|++\rangle + \beta|+-\rangle + \gamma|-+\rangle + \delta|--\rangle.$$

If the operator Ξ has the action that it interchanges Particle #1 and Particle #2, express Ξ in the ket-bra representation and show that $\Xi^2 = \mathbb{1}$, as we would expect.

Problem 4.1-5: A free particle is prepared to be in state $|\mathbf{r}\rangle$ at $t = 0$. Apply the time evolution operator to determine the probability that the particle will be found in state $|\mathbf{r}'\rangle$ at time t.

4.2 A Hand-Waving Introduction to Quantum Field Theory

In this section, we motivate second quantization of the electromagnetic fields in free space using historic arguments that were forwarded by perhaps the greatest minds ever. Section 4.2.1 starts with the plane wave form of the vector potential, from which in the absence of charges and currents the electric and magnetic fields can be determined. We then calculate the energy of the plane wave and determine the amplitude of the fields per photon.

Section 4.2.2 argues that the plane wave solutions with amplitudes corresponding to single photons is in a form reminiscent of a harmonic oscillator, so the complex coefficients are promoted to be raising and lowering operators, which are assumed to commute in the normal way. This leads to the development of the Hamiltonian of the fields in terms of photon number operators. The method for sweeping infinities under the rug is also described.

Section 4.2.3 describes how the interaction potential between light and molecules can be generalized to the second-quantized form. This approach is applied to calculating transition amplitudes and Section 4.2.4 applies these results to show how stimulated emission naturally shows itself when an excited molecule is bathed in a sea of photons.

4.2.1 Continuous Theory

The vector potential $\mathbf{A}$ and the scalar potential ϕ are the fundamental quantities that enter into the Schrödinger Equation. In this section, we start with the classical fields and then get down to the business of doing quantum mechanics in Section 4.2.2. In the absence of free charges and currents, as is the case in free space, we have the flexibility to choose $\nabla \cdot \mathbf{A} = 0$ and $\phi = 0$. Note that though the molecules that we seek to probe are a collection of charges, we are considering the electromagnetic fields in the space between them. In the Coulomb gauge, as this is called, Maxwell's wave equation has plane wave solutions of the form

$$\mathbf{A}(\mathbf{x}, t) = \sum_{r=1}^{2} \sum_{\mathbf{k}} \mathbf{A}_{0r}(\mathbf{k}) e^{i(\mathbf{k} \cdot \mathbf{x} - \omega_k t)}. \tag{4.36}$$

Equation 4.36 is summed over all possible wave vectors $\mathbf{k}$ and their associated frequencies ω_k; and, the two orthogonal polarizations, r (these can represent linear, circular or other polarization types). Also, the transverse field condition demands that

$$\mathbf{k} \cdot \mathbf{A}_{0r}(k) = 0 \tag{4.37}$$

for $r=1,2$ is satisfied.

The electric and magnetic fields can be calculated from $\mathbf{A}$ according to,

$$\mathbf{B} = \nabla \times \mathbf{A}, \tag{4.38}$$

and

$$\mathbf{E} = -\frac{1}{c} \frac{\partial \mathbf{A}}{\partial t}. \tag{4.39}$$

By substituting **A** from Equation 4.36 into Equations 4.38 and 4.39 we get

$$\mathbf{B}(\mathbf{x},t) = \sum_{r=1}^{2}\sum_{\mathbf{k}}[i\mathbf{k}\times\mathbf{A}_{0r}(\mathbf{k})]e^{i(\mathbf{k}\cdot\mathbf{x}-\omega_k t)} \equiv \sum_{r,\mathbf{k}}\mathbf{B}_r(\mathbf{k})e^{i(\mathbf{k}\cdot\mathbf{x}-\omega_k t)}, \quad (4.40)$$

$$\mathbf{E}(\mathbf{x},t) = \sum_{r=1}^{2}\sum_{\mathbf{k}}\frac{i\omega_k}{c}\mathbf{A}_{0r}(\mathbf{k})e^{i(\mathbf{k}\cdot\mathbf{x}-\omega_k t)} \equiv \sum_{r,\mathbf{k}}\mathbf{E}_r(\mathbf{k})e^{i(\mathbf{k}\cdot\mathbf{x}-\omega_k t)}. \quad (4.41)$$

Keep in mind that the fields in Equations 4.40 and 4.41 are in complex form; so, to get the real fields, we will need to get the real parts Re[**E**] and Re[**B**]. Using $k = \omega_k/c$, the magnitude of the fields can be written as

$$|\mathbf{B}_r(\mathbf{k})| = |\mathbf{E}_r(\mathbf{k})| = k|\mathbf{A}_{0r}(\mathbf{k})| \equiv k|A_{0r}(\mathbf{k})|. \quad (4.42)$$

To avoid the infinite energy of a plane wave, we consider a cell of dimensions $L \times L \times L$ and apply periodic boundary conditions to the vector potential, yielding

$$\mathbf{A}(x,y,z,t) = \mathbf{A}(x+L,y,z,t). \quad (4.43)$$

We do the same for each Cartesian component. When such a cell has finite energy but its size is infinite – as is approximately the case for the universe, the energy density vanishes and it is not possible to calculate the energy from the energy density . Making the energy density small but finite in an infinite universe leads to infinite energy. Treating space as a finite cell, then taking the cell size to be arbitrarily large, is a computational convenience that avoids these issues. The periodic boundary conditions , on the other hand, ensure that the field is represented by a travelling wave. If we would have chosen other boundary conditions, for example demanding that the fields vanish on the surface of the cell, this would have resulted in a standing wave . In an optical resonator, fields are standing waves , so these boundary conditions would be the appropriate ones. Periodic boundary conditions avoid reflections from the cell's walls so are appropriate for modeling empty space.

By applying periodic boundary conditions given by Equation 4.43 to Equation 4.36, we get

$$1 = e^{ik_x L}, \quad (4.44)$$
$$\Rightarrow k_x L = 2\pi n_x, \quad (4.45)$$

where n_x is an integer. Doing the same for y and z yields

$$\mathbf{k} = \frac{2\pi}{L}\left(n_x\hat{x} + n_y\hat{y} + n_z\hat{z}\right). \quad (4.46)$$

Nonlinear Optics: A student's perspective

So, **k** is discrete and approaches a continuum in the large-L limit. Subsequent to implementing a calculation, the $L \to \infty$ is applied, allowing for arbitrary **k**.

The energy of a single cell from Poynting's theorem for *real fields* yields

$$H_{rad} = \frac{1}{8\pi} \int_{L^3} d^3\mathbf{x} (\mathbf{E}_R \cdot \mathbf{E}_R + \mathbf{B}_R \cdot \mathbf{B}_R), \tag{4.47}$$

where $\mathbf{E}_R = \mathbf{E}/2 + \text{c.c.}$

Substituting the expressions for the magnetic and electric fields from Equations 4.40 and 4.41 into Equation 4.47 leads to a summation over all the modes characterized by **k** and r. The cross terms vanish by orthonormality of the modes. For example, consider the contribution from the real part of the electric field for mode **k** and one of the polarizations r, which contributes an energy

$$\begin{aligned}
H_{mode} &= \frac{1}{8\pi} \int_0^L E_{0r}^2(\mathbf{k}) \cos^2(kx - \omega t + \phi) dx \tag{4.48} \\
&= \frac{1}{8\pi} \int_0^L E_{0r}^2(\mathbf{k}) \left[\frac{1 + \cos(2kx - 2\omega t + 2\phi)}{2}\right] dx \tag{4.49} \\
&= \frac{1}{8\pi} E_{0r}^2(\mathbf{k}) \left[\frac{1}{2} + \frac{\sin(2kx - 2\omega t + 2\phi)}{2k}\right]_0^L \tag{4.50} \\
&= \frac{1}{8\pi} E_{0r}^2(\mathbf{k}) \cdot \frac{L}{2} = \frac{L}{16\pi} E_{0r}^2(\mathbf{k}), \tag{4.51}
\end{aligned}$$

where we have defined $\mathbf{E} = E_{0r}(\mathbf{k}) \exp[i\phi]$ in terms of the two real quantities $E_{0r}(\mathbf{k})$ and ϕ. So including the x, y and z directions, and, both the electric and magnetic fields, Equation 4.47 – the Hamiltonian for a single mode – can be written as

$$H_{rad}^r(\mathbf{k}) = \frac{L^3}{16\pi} \left(B_{0r}^2(\mathbf{k}) + E_{0r}^2(\mathbf{k})\right). \tag{4.52}$$

Using Equation 4.42, we can re-express Equation 4.52 in terms of the vector potential, yielding

$$H_{rad}^r(\mathbf{k}) = \left\{\frac{L^3}{8\pi} k^2 |\mathbf{A}_{0r}(\mathbf{k})|^2\right\}. \tag{4.53}$$

We can express the vector potential as $\mathbf{A}_{0r} = \boldsymbol{\epsilon}_r(\mathbf{k}) A_{0r}$ where A_{0r} is its magnitude and r its polarization. Let's start by placing one "photon" into the cell of volume L^3. This is the point at which we are making a connection with the quantum properties of light. Then, according to the observation that photons have energy $\hbar \omega_k$, we assign one photon per mode, so,

$$[\text{Energy/mode}] = E_k = \hbar \omega_k. \tag{4.54}$$

The energy is independent of polarization, so there is no need to label it here. Combining Equations 4.53 and 4.54, we get

$$\hbar\omega_k = \frac{L^3}{8\pi}k^2|\mathbf{A}_{0r}(\mathbf{k})|^2 \qquad (4.55)$$

$$\Rightarrow A_{0r}(\mathbf{k}) = |\mathbf{A}_{0r}(\mathbf{k})| = \sqrt{\frac{8\pi\hbar\omega_k}{L^3 k^2}}. \qquad (4.56)$$

The significance of Equation 4.56 is that it determines the field amplitude of a single quantum of energy, or the field amplitude per photon.

Since only real quantities are measured, we take the real part of the fields or potentials, which for **A** is given by

$$\text{Re}[\mathbf{A}(\mathbf{x},t)] = \frac{\mathbf{A}+\mathbf{A}^*}{2}. \qquad (4.57)$$

Combining Equations 4.36, 4.56 and 4.57 yields

$$\mathbf{A}_R(\mathbf{x},t) = \text{Re}[\mathbf{A}(\mathbf{x},t)] = \sum_{\mathbf{k},r}\sqrt{\frac{2\pi\hbar c^2}{L^3\omega_k}}\boldsymbol{\epsilon}_r(\mathbf{k})\left\{a_r(\mathbf{k})e^{i(\mathbf{k}\cdot\mathbf{x}-\omega_k t)} + a_r^*(\mathbf{k})e^{-i(\mathbf{k}\cdot\mathbf{x}-\omega_k t)}\right\}. \qquad (4.58)$$

Here, $a_r(\mathbf{k})$ and $a_r^*(\mathbf{k})$ are dimensionless complex quantities. Since we placed one quantum of energy in each mode, the coefficients have magnitude unity. However, if we placed multiple quanta of energy into a mode, these quantities would have a magnitude in proportion to the square root of the number of photons with wave vector $\mathbf{k}$ and polarization r.

Equation 4.58 can be used to determine the electric field from Equation 4.41, which yields

$$\mathbf{E}_R(\mathbf{x},t) = \text{Re}[\mathbf{E}(\mathbf{x},t)] = i\sum_{\mathbf{k},r}\sqrt{\frac{2\pi\hbar\omega_k}{L^3}}\boldsymbol{\epsilon}_r(\mathbf{k})\left\{a_r(\mathbf{k})e^{i(\mathbf{k}\cdot\mathbf{x}-\omega_k t)} - a_r^*(\mathbf{k})e^{-i(\mathbf{k}\cdot\mathbf{x}-\omega_k t)}\right\}. \qquad (4.59)$$

A similar equation for $\mathbf{B}_R(\mathbf{x},t)$ can be obtained. Thus, we have expressed the field amplitudes in terms of the number of photons. As such, the theory is classical in that the wave behaves as expected from Maxwell's equations; but, the field amplitude is quantized. As a consequence, the energy of mode characterized by $\mathbf{k}$ will be quantized in units of $\hbar\omega_k$, in accord with Einstein's proposition that the energy of an electromagnetic field comes in units of a fundamental quantum.

4.2.2 Second Quantization

This is the point at which we make the big inductive leap that forms the heart of the quantum theory of the photon. The terms in braces in Equation 4.58 are reminiscent of the time evolution of the raising and lowering operators for a harmonic oscillator. This is not at all surprising given that the fields oscillate harmonically.

Recall that the position of the harmonic oscillator can be expressed in terms of the raising and lowering operators as

$$x = \sqrt{\frac{\hbar}{2m\omega}}(a + a^\dagger) \quad (4.60)$$

and the momentum as

$$p = -i\sqrt{\frac{\hbar m\omega}{2}}(a - a^\dagger) \quad (4.61)$$

with $[a, a^\dagger] = 1$ and $[x, p] = i\hbar$. Note the similarity between Equations 4.60 and 4.58; and, between Equations 4.61 and 4.59 by keeping in mind that the time evolution of the raising and lowering operators is of the form $a \to a e^{-i\omega t}$ and $a^\dagger \to a e^{i\omega t}$. This similarity associates the vector potential with the position of a harmonic oscillator via

$$\mathbf{A}_R(\mathbf{x}, t) \to \sqrt{\frac{4\pi m c^2}{L^3}} \cdot x; \quad (4.62)$$

and, the electric field with the momentum according to

$$\mathbf{E}_R(\mathbf{x}, t) \to -\sqrt{\frac{4\pi}{mL^3}} \cdot p, \quad (4.63)$$

where the prefactors are used to make the units the fields match that of p and x.

Using the analogy with a simple harmonic oscillator, we will promote the status of the complex coefficients in Equations 4.58 and 4.59 to operators, so $a_r(\mathbf{k})$ is the *annihilation operator* and $a_r^*(\mathbf{k}) \to a_r^\dagger(\mathbf{k})$ is the *creation operator*, annihilating and creating one quantum of energy with wave vector $\mathbf{k}$ and polarization r. Note that all the other parameters in Equations 4.58 and 4.59 – such as x – are still just parameters in the Hilbert space of the photons, and are not operators. To complete the analogy, we posit that the commutator is given by

$$\left[a_r(\mathbf{k}), a_{r'}^\dagger(\mathbf{k}')\right] = \delta_{k,k'}\delta_{r,r'}, \quad (4.64)$$

where there is a distinct creation and annihilation operator for each wave vector and each polarization.

Next, we calculate the commutator between the electric field and vector potential given by Equations 4.58 and 4.59 with the complex amplitudes promoted to raising and lowering operators, which with the help of Equation 4.64 yields

$$\left[\left(\frac{L^3}{-4\pi c}\right)\mathbf{A}_R(\mathbf{x},t), \mathbf{E}_R(\mathbf{x},t)\right] = i\hbar \sum_{\vec{k},r} \mathbb{1} \cdot \boldsymbol{\epsilon}_r(\mathbf{k})\boldsymbol{\epsilon}_r(\mathbf{k}). \tag{4.65}$$

The term in square brackets is the product of the two prefractors in Equations 4.62 and 4.63, showing that up to a constant multiplicative factor $\mathbf{A}(\mathbf{x},t)$ and $\mathbf{E}(\mathbf{x},t)$ are the two canonical conjugate quantities. Note that the righthand side of Equation 4.65 includes a sum of the identity operator over all the modes of the system, and diverges because there are an infinite number of modes, irrespective of whether or not they are populated. The meaning of the divergence will be described shortly in the context of the Hamiltonian.

Our next step is to calculate the Hamiltonian using these properties. The Hamiltonian for a harmonic oscillator is calculated using Equations 4.60 and 4.61, yielding

$$H = \frac{p^2}{2m} + \frac{1}{2}m\omega^2 x^2 = \hbar\omega\left(a^\dagger a + \frac{1}{2}\right). \tag{4.66}$$

Similarly, we substitute Equation 4.58 into the Hamiltonian of the fields given by Equation 4.53, but taking care to account for the commutation relationships between the raising and lower operators. As before, terms involving two different modes don't contribute to the integrals because

$$\int_0^L e^{ik_x x} e^{ik'_x x} dx = \delta_{k_x, -k'_x} L, \tag{4.67}$$

but the cross term including the Hermitian adjoints of the same mode does contribute, yielding the Hamiltonian

$$H_{rad} = \sum_{\mathbf{k},r} \frac{1}{8\pi}\left\{2\pi\hbar\omega_k 2\left[a_r(\mathbf{k})a_r(\mathbf{k})^\dagger + a_r(\mathbf{k})^\dagger a_r(\mathbf{k})\right]\right\}, \tag{4.68}$$

where we have used $|\boldsymbol{\epsilon}_r(\mathbf{k})| = 1$. With the commutator $[a_r(\mathbf{k}), a_r^\dagger(\mathbf{k})] = 1$ we get

$$H_{rad} = \sum_{\mathbf{k},r} \hbar\omega_k \left[a_r^\dagger(\mathbf{k})a_r(\mathbf{k}) + \frac{1}{2}\right]. \tag{4.69}$$

Finally, $a_r^\dagger(\mathbf{k})a_r(\mathbf{k})$ is the number operator $N_r(\mathbf{k})$ so

$$H_{rad} = \sum_{\mathbf{k},r} \hbar\omega_k \left(N_r(\mathbf{k}) + \frac{1}{2}\right). \tag{4.70}$$

The second term on the right hand side of Equation 4.70 diverges, independent of the number of photons as there are an infinite number of possible modes. Even with no photons present, we have

$$\sum_{\mathbf{k},r} \hbar\omega_k \frac{1}{2} = \frac{\hbar c}{2} \sum_{\mathbf{k},r} |\mathbf{k}| = \infty. \qquad (4.71)$$

Since molecular properties depend only on energy differences, and the creation and annihilation of photons leads to differences of photon energies, this infinite term cancels in all meaningful calculations. We therefore ignore it, using the Hamiltonian given by

$$H_{rad} = \sum_{r,\mathbf{k}} \hbar\omega_k N_r(\mathbf{k}). \qquad (4.72)$$

It is worthwhile to reflect upon the meaning of the raising and lowering operators . For the harmonic oscillator, these operators excite or de-excite the system by one quantum number. There is one system and the state of the system changes. In the case of the photon field, the raising operator adds a photon and the lowering operator removes one. The photon that is added itself has a wavefunction that, for example, can describe the probability of finding a photon at a particular position in space. In this sense, the raising operator creates a wavefunction. To differentiate between this and the mundane quantization that is embodied by quantum numbers of let's say an electron in an atom, the quantum numbers that refer to the number of photons is called second quantization .

We are thus left with two views of the electromagnetic field; it is characterized by states of the vacuum that we perceive as photons or it can be viewed as photons that pop in and out of existence when interacting with matter.

In classical quantum mechanics, particles are immutable objects that are described by a state vector. When energies of a particle become comparable to or larger than twice the rest mass of existing particles, and selection rules allow it, pair production will lead to the generation of particles. Since the photon is massless, it can pop in and out of existence at any energy scale, allowing what we call relativistic effects to be observable by humans' bare eyes. In this regard, the photon holds a special place.

4.2.3 Photon-Molecule Interactions

We now consider the Hamiltonian of an electron in an atom or molecule, given by

$$H_e = \frac{\mathbf{p}^2}{2m} + V. \tag{4.73}$$

Here V includes all internal forces on the electron due to the nucleus and other electrons. In an external electric field, the Hamiltonian becomes

$$H' = \frac{1}{2m}\left(\mathbf{p} - \frac{q\mathbf{A}}{c}\right)^2 + V + \phi_m \tag{4.74}$$

$$= \frac{1}{2m}\left(\mathbf{p}^2 + \frac{e^2|\mathbf{A}|^2}{c} + \frac{2e\mathbf{A}\cdot\mathbf{p}}{c}\right) + V, \tag{4.75}$$

where the charge of an electron is $q = -e$; and, in the Coulomb gauge $\phi_m = 0$ and $[\mathbf{A}, \mathbf{p}] = 0$. Equation 4.74 also neglects higher order terms such as $|\mathbf{A}|^2$.

The total Hamiltonian of the electron is thus given by

$$H' = \frac{1}{2m}\left(\mathbf{p}^2 + \frac{2e\mathbf{A}\cdot\mathbf{p}}{c}\right) + V, \tag{4.76}$$

or

$$H' = H_e + H_I, \tag{4.77}$$

where the interaction Hamiltonian H_I is given by

$$H_I = \frac{e\mathbf{A}\cdot\mathbf{p}}{mc}. \tag{4.78}$$

The Hamiltonian of the whole system including the fields is given by

$$H = H_e + H_{rad} + H_I = H_0 + H_I, \tag{4.79}$$

where eigenfunctions for H_e and H_{rad} are assumed to be known and we treat the the interaction between the two, embodied in H_I, using perturbation theory.

Since H_e and H_{rad} belong to different spaces,

$$[H_e, H_{rad}] = 0. \tag{4.80}$$

The state vector is constructed as a direct-product of the state vectors $|\psi_e\rangle$ and $|\psi_{rad}\rangle$, which are eigenfunctions of H_e and H_{rad}, respectively, yielding

$$|\psi\rangle = |\psi_e\rangle |\psi_{rad}\rangle. \tag{4.81}$$

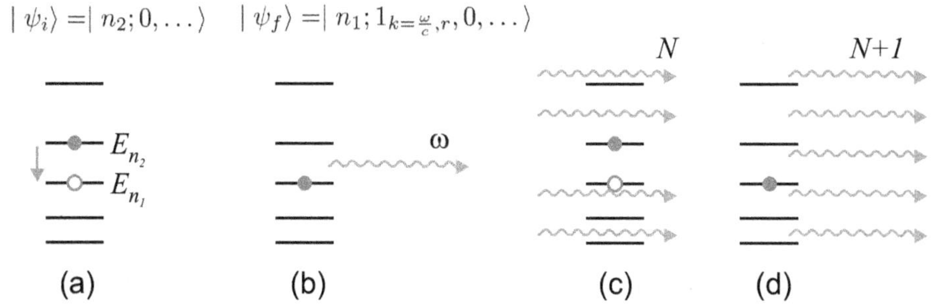

Nonlinear Optics: A student's perspective

$|\psi_i\rangle = | n_2; 0, \ldots \rangle \quad |\psi_f\rangle = | n_1; 1_{k=\frac{\omega}{c}, r}, 0, \ldots \rangle$

(a) (b) (c) (d)

Figure 4.1: (a) An excited molecule (b) emits a photon in a process called spontaneous emission. (c) An excited molecule in a sea of photons (d) emits a photon of the same frequency by stimulated emission.

Then,

$$H_0 |\psi_e\rangle |\psi_{rad}\rangle = (H_e |\psi_e\rangle) |\psi_{rad}\rangle + |\psi_e\rangle H_{rad} |\psi_{rad}\rangle \quad (4.82)$$
$$= (E_{rad} + E_e) |\psi\rangle, \quad (4.83)$$

where E_{rad} and E_e are the energy eigenvalues of the corresponding state vectors.

The state vector of the atom/molecule field can be expressed as

$$|\psi\rangle = | k, \ell, m \ldots; \ldots N_{\mathbf{k}_1, r_1}, N_{\mathbf{k}_2, r_2}, N_{\mathbf{k}_3, r_3} \ldots \rangle, \quad (4.84)$$

where k, ℓ, m represent the quantum numbers of the electrons in the atom/molecule and $N_{\mathbf{k}_1, r_1}$ represents the number of photons with wave vector $\mathbf{k} = \mathbf{k}_1$ and polarization r_1, and so on. This notation is unwieldily because there are an infinite number of photon entries corresponding to all possible modes. The notation can be simplified by only listing the entries that contain at least one photon.

To illustrate how the probability of a transition is calculated, consider the two energy levels n_1 and n_2. The the probability of a transition is proportional to the matrix elements of the interaction Hamiltonian

$$\mathscr{P}_{fi} = \kappa |\langle \psi_f | H_I | \psi_i \rangle|^2 = \kappa |\langle \psi_f | \frac{e\mathbf{A} \cdot \mathbf{p}}{mc} | \psi_i, \rangle|^2, \quad (4.85)$$

where κ is a proportionality constant whose value is of no concern for now. Assuming that the electron is initially in the higher-energy state E_{n_2} as shown in Figure 4.1, energy conservation demands that upon de-excitation, a photon is emitted and $E_{n_2} - E_{n_1} = \hbar\omega$. Since initially there are no photons, and one is emitted when the atom de-excites, the initial and final state vectors are given by

$$|\psi_i\rangle = |n_2;0,0,0\ldots\rangle, \quad (4.86)$$
$$|\psi_f\rangle = |n_1;0,0,1_{k=\frac{\omega}{c},r},0,0\ldots\rangle. \quad (4.87)$$

Before proceeding further, we will show that $\mathbf{A}\cdot\mathbf{p}$ in Equation 4.85 is equivalent to the electric dipole energy

$$U = -\boldsymbol{\mu}\cdot\mathbf{E} = e\mathbf{x}\cdot\mathbf{E}, \quad (4.88)$$

where e is the magnitude of the electron's charge. Here we are ignoring higher order moments and higher-order terms in $\mathbf{E}$. Equation 4.88 is the potential typically used when including the classical electric field in the Schrödinger Equation.

Equation 4.85 can be separated into the product of a photon part and molecule part since the state vectors are a direct product of the two parts as are the operators $\mathbf{A}$ and $\mathbf{p}$, yielding

$$\mathscr{P}_{fi} = \kappa \frac{e^2}{m^2 c^2} |\langle 1_{k=\frac{\omega}{c},r}|\mathbf{A}|0\rangle \cdot \langle n_1|\mathbf{p}|n_2\rangle|^2. \quad (4.89)$$

We can express $\langle n_1|\mathbf{p}|n_2\rangle$ in terms of the position operator by calling upon the commutator

$$[\mathbf{x}, H_e] = i\hbar \frac{\mathbf{p}}{m}, \quad (4.90)$$

and calculating the matrix elements of Equation 4.90, yielding

$$\langle n_1|\mathbf{x}H_e - H_e\mathbf{x}|n_2\rangle = \frac{i\hbar}{m}\langle n_1|\mathbf{p}|n_2\rangle, \quad (4.91)$$
$$(E_l - E_g)\langle n_1|\mathbf{x}|n_2\rangle = \frac{i\hbar}{m}\langle n_1|\mathbf{p}|n_2\rangle, \quad (4.92)$$
$$\Rightarrow \langle n_1|\mathbf{p}|n_2\rangle = \frac{m\omega}{i}\langle n_1|\mathbf{x}|n_2\rangle, \quad (4.93)$$

where we have used the fact that $E_{n_2} - E_{n_1} = \hbar\omega$.

Substituting Equation 4.93 into Equation 4.89 yields

$$\mathscr{P}_{fi} = \kappa \frac{e^2}{m^2 c^2} |\langle 1_{k=\frac{\omega}{c},r}|\mathbf{A}|0\rangle \cdot \left(\frac{m\omega}{i}\right)\langle n_1|\mathbf{x}|n_2\rangle|^2$$
$$= \kappa \frac{\omega^2}{c^2} |\langle 1_{k=\frac{\omega}{c},r}|\mathbf{A}|0\rangle \cdot \langle n_1|e\mathbf{x}|n_2\rangle|^2. \quad (4.94)$$

Since the photon created is at one frequency, we can write the vector potential as
$$\mathbf{A} = a^\dagger_{\omega,r} A_0 \boldsymbol{\epsilon}_r(\mathbf{k}) e^{-i\mathbf{k}\cdot\mathbf{x}}. \tag{4.95}$$

For a molecule that is small compared with the light's wavelength,
$$e^{-i(\mathbf{k}\cdot\mathbf{x})} = 1 - i(\mathbf{k}\cdot\mathbf{x}) + \ldots, \tag{4.96}$$

where $\mathbf{x}$ is the position coordinate of an electron relative to the atom or molecule's center of mass and is on the order of few angstroms. For visible light, the wavelength is 1×10^{-4}cm, so
$$\mathbf{k}\cdot\mathbf{x} \sim \frac{2\pi x}{\lambda} \sim 2\pi \times 10^{-4}. \tag{4.97}$$

Since $\mathbf{k}\cdot\mathbf{x}$ is so small, we can assume that
$$e^{i(\mathbf{k}\cdot\mathbf{x})} \sim 1 = \text{constant}. \tag{4.98}$$

Equation 4.98 is often called the *dipole approximation*. Some materials may not have a dipole moment $\boldsymbol{\mu}$ but have an octuple moment or, in some cases, the size of the system may be comparable to the wavelength of the light. In those cases, the dipole approximation is not valid.

Using the dipole approximation, the transition probability given by Equation 4.94 for the vector potential given by Equation 4.95 becomes
$$\mathscr{P}_{fi} = \kappa A_0^2 \frac{\omega^2}{c^2} |\langle 1_{k=\frac{\omega}{c},r} | \mathbf{a}^\dagger | 0 \rangle \langle n_1 | e\mathbf{x}\cdot\boldsymbol{\epsilon}_r(\mathbf{k}) | n_2 \rangle|^2. \tag{4.99}$$

Equation 4.41 shows that $\frac{\omega^2}{c^2} A_0^2 = E_0^2$, so we can express Equation 4.99 as
$$\begin{aligned}
\mathscr{P}_{fi} &= \kappa |\langle n_1 | e\mathbf{x}\cdot\mathbf{E} | n_2 \rangle|^2 & (4.100) \\
&= \kappa |\langle n_1 | \boldsymbol{\mu}\cdot\mathbf{E} | n_2 \rangle|^2. & (4.101)
\end{aligned}$$

Thus, the interaction Hamiltonian can be equally well expressed as $-\boldsymbol{\mu}\cdot\mathbf{E}$ in the dipole approximation and Equation 4.101 is equivalent to Equation 4.85. We can use whatever form suits our fancy.

4.2.4 Stimulated Emission

Consider once again a transition between two levels when the molecule is initially in an excited state. This time, we bath the molecule in N photons of wave vector $\mathbf{k}$ and polarization r as shown in Figure 4.1c,d. When the system de-excites, a photon will be emitted, which joins its like-minded companions. This process is called stimulated emission because the presence of the photons increases the transition probability, as we describe below.

Initially,

$$N_{\mathbf{k},r} = N \tag{4.102}$$

and the final and initial wavefunctions are

$$|\psi_i\rangle = |n_2; N\rangle, \tag{4.103}$$
$$|\psi_f\rangle = |n_1; N+1\rangle. \tag{4.104}$$

Here we assume all other occupation numbers are 0. Then the transition probability is given by

$$\mathcal{P}_{fi} = \kappa |\langle \psi_f | H_I | \psi_i \rangle|^2 \tag{4.105}$$
$$= \kappa |\langle n_1 | e\mathbf{x} | n_2 \rangle \cdot \langle N+1 | \mathbf{E} | N \rangle|^2 \tag{4.106}$$
$$= \kappa |\langle n_1 | \boldsymbol{\mu} | n_2 \rangle|^2 |\langle N+1 | a_r^\dagger(\mathbf{k}) | N \rangle|^2 \tag{4.107}$$
$$= \kappa |\boldsymbol{\mu}_{n_1 n_2}|^2 \left(\sqrt{N+1}\right)^2 \tag{4.108}$$
$$= \kappa |\boldsymbol{\mu}_{n_1 n_2}|^2 (N+1). \tag{4.109}$$

With $N = 0$, Equation 4.105 gives the spontaneous emission rate. When N is large, the stimulated emission rate is N times larger than the spontaneous rate. Note how this naturally emerges from the matrix element.

> **Problem 4.2-1:** A student discovers a process in which light at frequency ω_1 and ω_2 is converted by a sample to light at frequency $2\omega_1 + \omega_2$ only when all the light is polarized along $\hat{x}$. The intensity of the output peaks at frequency ω_0 and has the form $1/\left([\{(2\omega_1 + \omega_2) - \omega_0\}/\omega_0]^2 + \Gamma^2\right)$. Express the *simplest* interaction Hamiltonian H_I in terms of photon creation and annihilation operators that describes this process. Discuss the process at the level of photons interacting with molecules. The constant of proportionality is unimportant.

4.3 Coherent States

The energy eignestate vectors of the electromagnetic field are eigenstates of the number operator. However, the expectation value of the electric field given by Equation 4.59 vanishes for any state of fixed number of photons, that is,

$$\langle N|\mathbf{E}|N\rangle \propto \langle N|\pm N\rangle = 0 \tag{4.110}$$

because the field is a sum of creation and annihilation operators which each have zero expectation value. As such, the number of photons in a field and the electric field are not simultaneously measurable observables. In this section, we develop a special kind of state called a coherent state, which give an expectation value that is the classical field.

A coherent state is described by the state vector $|\alpha\rangle$, which is an eigenstate of the annihilation operator a, or

$$a|\alpha\rangle = \alpha|\alpha\rangle \tag{4.111}$$

with complex eigenvalue α. Note that coherent states are applicable to systems beyond photons, and were in fact originally studied by Schrödinger for the harmonic oscillator. The approach below is general, and will be applied to photons at the end of this section. We expand $|\alpha\rangle$ as a superposition of the number states $|n\rangle$, given by

$$|\alpha\rangle = \sum_{n=0}^{\infty} c_n |n\rangle, \tag{4.112}$$

where the coefficients c_n are determined from the condition given by Equation 4.111.

Applying the annihilation operator to Equation 4.112 yields

$$a|\alpha\rangle = \sum_{n=0}^{\infty} c_n \sqrt{n-1} |n-1\rangle, \tag{4.113}$$

and setting Equation 4.113 equal to Equation 4.111 with the help of Equation 4.112 gives

$$\alpha \sum_{n=0}^{\infty} c_n |n\rangle = \sum_{n=0}^{\infty} c_n \sqrt{n-1} |n-1\rangle. \tag{4.114}$$

Finally, taking the inner product of Equation 4.114 with $\langle q|$ given the recursion relationship,

$$\alpha c_q = c_{q+1} \sqrt{q}, \tag{4.115}$$

which allows all the coefficients to be expressed in terms of c_0, or

$$c_q = \frac{\alpha^q}{\sqrt{q!}} c_0. \tag{4.116}$$

c_0 can be determined by substituting Equation 4.116 into Equation 4.112 and demanding normalization, which yields

$$|c_0|^2 \sum_{n=0}^{\infty} \frac{|\alpha|^{2n}}{n!} = |c_0|^2 e^{|\alpha|^2} = 1, \qquad (4.117)$$

which gives

$$c_0 = e^{i\delta} e^{-|\alpha|^2/2}, \qquad (4.118)$$

where $e^{i\delta}$ is an arbitrary phase. Putting it all together using Equations 4.112, 4.116 and 4.118, we get the coherent state

$$|\alpha\rangle = e^{i\delta} e^{-|\alpha|^2/2} \sum_{n=0}^{\infty} \frac{\alpha^n}{\sqrt{n!}} |n\rangle = e^{i\delta} e^{-|\alpha|^2/2} \sum_{n=0}^{\infty} \frac{\alpha^n (a^\dagger)^n}{n!} |0\rangle = e^{i\delta - |\alpha|^2/2 + \alpha a^\dagger} |0\rangle. \qquad (4.119)$$

The coherent state has been constructed so that the annihilation operator has the eigenvalue α; so, the expectation value of the annihilation operator is obviously given by

$$\langle \alpha | a | \alpha \rangle = \alpha. \qquad (4.120)$$

Using Equation 4.29, it is trivial to show that the expectation value of the creation operator is given by

$$\langle \alpha | a^\dagger | \alpha \rangle = \langle \alpha | a | \alpha \rangle^* = \alpha^*. \qquad (4.121)$$

Now we apply these results to the second-quantized theory of the electric field by expressing Equation 4.59 in terms of creation and annihilation operators, which gives

$$\mathbf{E}(\mathbf{x}, t) = i \sum_{\mathbf{k}, r} \sqrt{\frac{2\pi \hbar \omega_k}{L^3}} \boldsymbol{\epsilon}_r(\mathbf{k}) \left\{ a_r(\mathbf{k}) e^{i(\mathbf{k} \cdot \mathbf{x} - \omega_k t)} - a_r^\dagger(\mathbf{k}) e^{-i(\mathbf{k} \cdot \mathbf{x} - \omega_k t)} \right\}. \qquad (4.122)$$

The relevant expectation value of the electric field is taken with respect to the coherent state $|\alpha_r(\mathbf{k}), \alpha_{r'}'(\mathbf{k}'), \ldots\rangle$, which includes all possible polarizations and wave vectors, and yields the expectation value

$$\langle \alpha_r(\mathbf{k}), \alpha_{r'}'(\mathbf{k}'), \ldots | a_r(\mathbf{k}) | \alpha_r(\mathbf{k}), \alpha_{r'}'(\mathbf{k}'), \ldots \rangle = \alpha_r(\mathbf{k}), \qquad (4.123)$$

analogous to Equation 4.120.

The expectation of the second-quantized electric field given by Equation 4.122, with the help of Equation 4.123, is given by

$$\langle \alpha_r(\mathbf{k}), \alpha_{r'}'(\mathbf{k}'), \ldots | \mathbf{E}(\mathbf{x}, t) | \alpha_r(\mathbf{k}), \alpha_{r'}'(\mathbf{k}'), \ldots \rangle$$
$$= i \sum_{\mathbf{k}, r} \sqrt{\frac{2\pi \hbar \omega_k}{L^3}} \boldsymbol{\epsilon}_r(\mathbf{k}) \left\{ \alpha_r(\mathbf{k}) e^{i(\mathbf{k} \cdot \mathbf{x} - \omega_k t)} - \alpha_r^*(\mathbf{k}) e^{-i(\mathbf{k} \cdot \mathbf{x} - \omega_k t)} \right\}. \qquad (4.124)$$

Recall that the coefficient $\sqrt{\frac{2\pi\hbar\omega_k}{L^3}}$ is the electric field strength of a single photon of frequency ω_k. As such, $a_r(\mathbf{k})$ is a dimensionless complex amplitude that when multiplied by the electric field due to a single photon yields the classical electric field amplitude. The conclusion is that the expectation value of the coherent state is the classical electric field – and $a_r(\mathbf{k})$, the eigenvalue of the raising operator – is the dimensionless complex amplitude that acts as the multiplier to get the field.

> **Problem 4.3-1(a):** Determine the expectation value of the Hamiltonian of a harmonic oscillator, $\langle \alpha | H | \alpha \rangle$, if it is in a coherent state.
>
> **(b):** Calculate $\langle \alpha' | \alpha \rangle$ and determine from your result if coherent states are orthogonal.
>
> **(c):** Calculate $\langle \alpha | x | \alpha \rangle$, $\langle \alpha | p | \alpha \rangle$ and $\langle \alpha | e^{-iHt/\hbar} | \alpha \rangle$.
>
> **(d):** Determine the time evolution of a coherent state using $e^{-iHt/\hbar} |\alpha\rangle$ and describe if the answer makes sense.

4.4 A Molecule in a Cavity with Photons

Consider a one-dimensional cavity along x so that the electric field vanishes at $x = 0$ and $x = L$. These boundary condition, when applied to Equation 4.59, demand that $a_r(k_x) = a_r(-k_x)$ – thus making a standing wave. Summing just over positive k_x, because the standing wave includes both positive and negative wave vectors, the quantized field takes the form

$$\mathbf{E}(x, t) = i \sum_{k \geq 0, r} \sqrt{\frac{4\pi\hbar\omega_k}{L(L_y L_z)}} \boldsymbol{\epsilon}_r(\mathbf{k}) \left\{ a_r(k) e^{-i\omega_k t} - a_r^\dagger(k) e^{i\omega_k t} \right\} \sin kx, \quad (4.125)$$

where we have defined $k = k_x$. By virtue of the boundary conditions,

$$k_n = n\pi/L, \quad (4.126)$$

where n is a positive nonzero integer. If $L_y, L_z \ll L$, the energies of the modes that are perpendicular to x are not excited, making this system essentially one-dimensional. As such, we ignore the transverse directions. Equation 4.125 is the one-dimensional quantized cavity field analogous to Equation 4.122.

Let's now place an atom or molecule in the middle of the cavity at $x = L/2$ and for the sake of simplicity, assume that it is a two-state system with the two eigen energies 0 and e. Furthermore, we assume that the atom is initially in its ground state and that the cavity is filled with N photons of energy $\hbar\omega = e - 0 = e$, and the photon field corresponds to the lowest-energy mode, or n=1. Assume further that all the photons are polarized along y. Then, the wavefunction of the system initially is of the form $|0; N_\omega\rangle$. Note that we can reference a mode by its angular frequency since it uniquely describes a particular mode because of the special way we have constructed the system. The only other permitted state with the same total energy of $N\hbar\omega$ is $|1; (N-1)_\omega\rangle$. Thus, the most general orthogonal states of the system can be expressed as a superposition of the form

$$|\psi_+(\phi)\rangle = \cos\phi |0; N_\omega\rangle + e^{i\delta} \sin\phi |1; (N-1)_\omega\rangle \qquad (4.127)$$

and

$$|\psi_-(\phi)\rangle = \sin\phi |0; N_\omega\rangle - e^{i\delta} \cos\phi |1; (N-1)_\omega\rangle, \qquad (4.128)$$

where the coefficients are expressed in trigonometric form to impose normalization.

The perturbation potential in the dipole approximation of an electron in the molecule is given by $-eyE(L/2, t))$, which using Equation 4.125 for the lowest mode is

$$V_I(t) = -iey\sqrt{\frac{4\pi\hbar\omega}{L(L_y L_z)}} \left\{ a_y(k)e^{-i\omega t} - a_y^\dagger(k)e^{i\omega t} \right\}. \qquad (4.129)$$

Recall that the photons are polarized along the y axis so $\boldsymbol{\epsilon}_r(\boldsymbol{k}) = \hat{y}$ and $\omega = \omega_1 = ck_1 = \pi c/L$ so $\sin(kL/2) = 1$ in Equation 4.125. Note that we have purposefully placed the molecule at the center of the cavity where the field amplitude is the highest.[1] Equation 4.129 is a time-dependent potential; but, in the Schrödinger picture – the one we are using, the wavefunctions are time dependent and the operators are not. In the discussion that follows, we will move the time dependence to the wavefunctions.

As the student can easily verify, there are only two nonzero matrix elements of $V_I(t)$, which are $\langle 0; N_\omega | V_I | 1; (N-1)_\omega \rangle$ and $\langle 1; (N-1)_\omega | V_I | 0; N_\omega \rangle = \langle 0; N_\omega | V_I | 1; (N-1)_\omega \rangle^*$. Defining the off diagonal matrix element

$$V \equiv \langle 0; N_\omega | V_I | 1; (N-1)_\omega \rangle = ie \langle 0|y|1\rangle \sqrt{\frac{4\pi N\hbar\omega}{L(L_y L_z)}} \qquad (4.130)$$

---

[1] Had we placed the molecule near the cavity walls, where the field vanishes, there would be no dipole coupling but the gradient of the field there is nonzero, so the molecule would couple through the quadrupole moment. Since this is a higher-order term, the coupling would be weaker than for dipole case.

and recalling that the unperturbed Hamiltonian is diagonal with two degenerate energies

$$E_0 = N\hbar\omega, \tag{4.131}$$

the full Hamiltonian in matrix form is given by

$$H = \begin{pmatrix} E_0 & V \\ V^* & E_0 \end{pmatrix}. \tag{4.132}$$

The eigenvalues of Equation 4.132 are calculated in the usual way from the determinant $\det(H - \mathbb{1}\lambda) = 0$, which yields

$$E_\pm = E_0 \pm |V|. \tag{4.133}$$

The stationary state energy eigenvectors are calculated from the condition $H|\psi_\pm\rangle = E_\pm|\psi_\pm\rangle$, which after normalization yields

$$|\psi_\pm\rangle = \frac{1}{\sqrt{2}}|0;N_\omega\rangle \mp \frac{i}{\sqrt{2}}|1;(N-1)_\omega\rangle. \tag{4.134}$$

Note that Equation 4.134 is of the form given by Equation 4.127 with $\phi = \mp\pi/4$ and $\delta = \pi/2$. Finally, applying the time evolution operator to Equation 4.134, which includes both the radiation field and the molecule, yields

$$|\psi_\pm(t)\rangle = \frac{1}{\sqrt{2}}|0;N_\omega\rangle e^{\mp iN\omega t} \mp \frac{i}{\sqrt{2}}|1;(N-1)_\omega\rangle e^{\mp iN\omega t} \tag{4.135}$$

If the molecule is in the ground state with N photons in the cavity at $t = 0$, the wavefunction will be in a superposition of states $|\psi_+(t)\rangle$ and $|\psi_-(t)\rangle$ given by

$$|\psi(t)\rangle = \frac{1}{\sqrt{2}}\left(|\psi_+(t)\rangle + |\psi_-(t)\rangle\right) \tag{4.136}$$

where clearly, $|\psi(0)\rangle = |0;N_\omega\rangle$. Then, the probability of finding the molecule in the state $|0;N_\omega\rangle$, $\mathscr{P}_{|0;N_\omega\rangle}$, is given by

$$\mathscr{P}_{|0;N_\omega\rangle} = |\langle 0;N_\omega|\psi(t)\rangle|^2 = \cos^2(N\omega t). \tag{4.137}$$

Thus, the probability oscillates at angular frequency $N\omega$, so the frequency increases with the number of photons. Such behavior is expected given the fact that the excitation rate depends on the number of photons as does the stimulated emission rate. Since these processes compete with each other, a larger number of photons causes the population to transfer between the two states more rapidly. These are called Rabi oscillations and the two-level model is that of Jaynes and Cummings.

> **Problem 4.4-1(a):** In the two-level model of a molecule, we argued that the only two allowed states are $|0;N_\omega\rangle$ and $|1;(N-1)_\omega\rangle$ since they have the same energy. However, other states with the same energy include $|0;(N-2)_\omega,1_{2\omega}\rangle$, $|1;(N-3)_\omega,1_{2\omega}\rangle$, etc. If the system starts in the state $|0;N_\omega\rangle$, is it possible for it to make a transition to these other states trough $V_I(t)$ given by Equation 4.129? Be sure to give a good argument.
>
> **(b):** If the molecule has three states with equal energy spacing and at $t=0$ starts in the state $|0;N_\omega\rangle$, in what states can the system be found at later times through transitions mediated by $V_I(t)$? Give examples of pathways that could lead to such states.
>
> **Problem 4.4-2(a):** If the "molecule" is a harmonic oscillator, where its energy spacing is $\hbar\omega$, and the system starts in state $|0;N_\omega\rangle$, what are the possible states that are occupied at later times?
>
> **(b):** Describe what the Hamiltonian looks like in matrix form, being sure to point out which elements vanish and how the nonzero ones are related to each other. Be sure to label the row and columns. Hint: the upper left corner is labelled $|0;N_\omega\rangle$, $|0;N_\omega\rangle$.

4.5 Nonlinear Susceptibilities and Dipole Moment Expectation Value

Our goal is to calculate the nonlinear susceptibility of a quantum system. As is always the case, the first step requires a calculation of the induced dipole moment as a function of the electric field, which is then differentiated according to Equation 1.33 to get the nonlinear susceptibilities.

All of the information about a quantum system is contained in the wavefunction and any particular observable is simply the expectation value of an operator that represents the observable. Thus, we need to calculate the wavefunction of a molecule in an electric field, from which we calculate the expectation value of the dipole moment, which we differentiate with respect to the electric field to get the susceptibility.

We therefore define $|\psi_n(\mathbf{E})\rangle$ as the nth energy eigenstate of an atom or molecule in the presence of electric field $\mathbf{E}(t)$, where

$$\mathbf{E}(t) = \sum_{i=1}^{\text{\# of fields}} \mathbf{E}^{\omega_i}(t). \qquad (4.138)$$

At zero temperature the atom/molecule is in its ground state, so the polarization is given by

$$\mathbf{P}(\mathbf{E}) = \langle \psi_0(\mathbf{E}) | \mathbf{P} | \psi_0(\mathbf{E}) \rangle, \qquad (4.139)$$

where $|\psi_0(\mathbf{E})\rangle$ is the perturbed ground state. The susceptibility is then given by

$$\chi^{(n)}_{ijk\ldots}(-\omega_\sigma;\omega_1,\omega_2,\ldots) = \frac{1}{D'} \frac{\partial^n}{\partial E^{\omega_1}_j \partial E^{\omega_2}_k \ldots} \langle \psi_0(\mathbf{E}) | P_i | \psi_0(\mathbf{E}) \rangle \Big|_{\mathbf{E}=0} \qquad (4.140)$$

where D' is the frequency dependent degeneracy denominator introduced earlier in Chapter 2.

4.5.1 Time-Dependent Perturbation Theory

Our strategy is to determine the wavefunction of the molecule in the presence of the electric field using the zero-field wavefunctions as a basis. The field dependence will be calculated using dipole coupling between the field and the molecule as a time-dependent perturbation.

Let H_0 be the molecular Hamiltonian without perturbation, that is, when the electric field is not present, the time evolution of a state $|\psi\rangle$ is given by

$$i\hbar \frac{\partial}{\partial t} |\psi\rangle = H_0 |\psi\rangle. \qquad (4.141)$$

We consider as a perturbation

$$V(t) = -\boldsymbol{\mu} \cdot \mathbf{E}(t), \qquad (4.142)$$

which is due to the interaction energy of the electric dipole moment of the system and the applied electric field, where

$$\mathbf{E}(t) = \frac{1}{2} \sum_p \mathbf{E}^{\omega_p} e^{-i\omega_p t} \qquad (\omega_{-p} = -\omega_p), \qquad (4.143)$$

where $\boldsymbol{\mu}$ is the dipole moment of the atom/molecule.

We define the unperturbed eigenstates as

$$|\psi^{(0)}_n(t)\rangle = |u^{(0)}_n\rangle e^{-i\Omega_n t}, \qquad (4.144)$$

where $\Omega_n = E_n/\hbar$ and $|u^{(0)}_n\rangle$ are the stationary state vectors of the zero-field Hamiltonian. We can then write the total Hamiltonian, H, as the sum of the molecular Hamiltonian and the perturbation potential multiplied by a parameter λ that tracks the order of the perturbation, which is given by

$$H = H_0 + \lambda V. \tag{4.145}$$

The perturbed states can be expressed as a sum of ascending orders of correction that are labeled by λ^m, or

$$|\psi_n(t)\rangle = \sum_{m=0}^{\infty} \lambda^m |\psi_n^{(m)}(t)\rangle, \tag{4.146}$$

where m is an integer representing the order of correction to to the wavefunction.

Substituting Equation 4.146 into the Schrodinger equation $H|\psi_n(t)\rangle = i\hbar \frac{\partial}{\partial t}|\psi_n(t)\rangle$, and keeping all terms of order m, we get

$$i\hbar \frac{\partial}{\partial t}|\psi_n^{(m)}(t)\rangle = H_0|\psi_n^{(m)}(t)\rangle + V(t)|\psi_n^{(m-1)}(t)\rangle. \tag{4.147}$$

Using the fact that the eigenvectors, $|\psi_n^{(m)}(t)\rangle$, can be expanded in terms of unperturbed states $|\psi_l^{(0)}(t)\rangle$ with coefficients $a_{ln}^{(m)}(t)$,

$$|\psi_n^{(m)}(t)\rangle = \sum_l a_{ln}^{(m)}(t)|\psi_l^{(0)}(t)\rangle \tag{4.148}$$

the ground state wavefunction in the presence of an electric field can be expressed as

$$|\psi_0^{(m)}(t)\rangle = \sum_l a_l^{(m)}(t)|\psi_l^{(0)}(t)\rangle, \tag{4.149}$$

where $a_l^{(m)}(t) \equiv a_{l0}^{(m)}(t)$. Substituting Equation 4.149 into Equation 4.147 we get

$$i\hbar \left(\sum_l \dot{a}_l^{(m)}|\psi_l^{(0)}(t)\rangle - \sum_l i\Omega_l a_l^{(m)}|\psi_l^{(0)}(t)\rangle \right)$$
$$= \sum_l E_l a_l^{(m)}|\psi_l^{(0)}(t)\rangle + V(t)\sum_l a_l^{(m-1)}|\psi_l^{(0)}(t)\rangle. \tag{4.150}$$

Operating on Equation 4.150 from left with $\langle u_p^{(0)}|$, we get

$$i\hbar \dot{a}_p^{(m)} e^{-i\Omega_p t} + \hbar \Omega_p a_p^{(m)} e^{-i\Omega_p t} = \hbar \Omega_p a_p^{(m)} e^{-i\Omega_p t}$$
$$+ \sum_l \langle u_p^{(0)}|V(t)|u_l^{(0)}\rangle a_l^{(m-1)} e^{-i\Omega_l t}. \tag{4.151}$$

By defining Ω_{lp} and $V_{pl}(t)$ as

$$\Omega_{lp} = \Omega_l - \Omega_p$$
$$V_{pl}(t) = \langle u_p^{(0)} | V(t) | u_l^{(0)} \rangle \qquad (4.152)$$

we can write $\dot{a}_p^{(m)}$ as

$$\dot{a}_p^{(m)} = \frac{1}{i\hbar} \sum_l V_{pl}(t) a_l^{(m-1)}(t) e^{-i\Omega_{lp}t}. \qquad (4.153)$$

Integration of Equation 4.153 gives

$$a_p^{(m)}(t) = \frac{1}{i\hbar} \sum_l \int_{-\infty}^{t} V_{pl}(t) a_l^{(m-1)}(t) e^{-i\Omega_{lp}t} dt. \qquad (4.154)$$

Since in the unperturbed state the system is in its ground state, or $a_l^{(0)} = \delta_{l,0}$, Equation 4.154 with the help of Equation 4.142 and Equation 4.143 yields

$$a_p^{(1)}(t) = \frac{1}{i\hbar} \sum_l \int_{-\infty}^{t} \boldsymbol{\mu}_{pl} \cdot \frac{1}{2} \sum_q \mathbf{E}^{\omega_q} e^{-i\omega_q t} \delta_{l,0} e^{-i\Omega_{lp}t} dt. \qquad (4.155)$$

The integral at the negative infinity limit in Equation 4.155 vanishes if we make the Bohr frequency Ω_{p0} have a small negative imaginary part, which then is evaluated in the limit when the Bohr frequency is real, yielding

$$a_p^{(1)}(t) = \frac{1}{2\hbar} \sum_q \frac{\boldsymbol{\mu}_{p0} \cdot \mathbf{E}^{\omega_q}}{\Omega_{p0} - \omega_q} \exp\left[i\left(\Omega_{p0} - \omega_q\right)t\right]. \qquad (4.156)$$

The significance of the complex nature of Ω_p will be discussed shortly.

The coefficient $a_p^{(2)}$ is derived by substituting Equation 4.156 into Equation 4.154, and gives

$$a_r^{(2)}(t) = \frac{1}{4\hbar^2} \sum_{q,s} \sum_v \frac{[\boldsymbol{\mu}_{rv} \cdot \mathbf{E}^{\omega_s}][\boldsymbol{\mu}_{v0} \cdot \mathbf{E}^{\omega_q}]}{(\Omega_{r0} - \omega_q - \omega_s)(\Omega_{v0} - \omega_q)} \exp\left[i\left(\Omega_{r0} - \omega_q - \omega_s\right)t\right], \qquad (4.157)$$

and $a_p^{(3)}$ is calculated using Equation 4.157 in Equation 4.154, or

$$a_p^{(3)}(t) = \frac{1}{8\hbar^3} \sum_{q,r,s} \sum_{m,n} \frac{[\boldsymbol{\mu}_{pm} \cdot \mathbf{E}^{\omega_s}][\boldsymbol{\mu}_{mn} \cdot \mathbf{E}^{\omega_q}][\boldsymbol{\mu}_{n0} \cdot \mathbf{E}^{\omega_r}]}{(\Omega_{p0} - \omega_q - \omega_r - \omega_s)(\Omega_{m0} - \omega_q - \omega_r)(\Omega_{n0} - \omega_q)}$$
$$\times \exp\left[i\left(\Omega_{p0} - \omega_q - \omega_r - \omega_s\right)t\right]. \qquad (4.158)$$

We assume that the system remains in the ground state in the presence of the field; but, because the field is time-dependent, the ground state wavefunction will evolve with time as

$$|\psi_0(t)\rangle = |0\rangle + \lambda \sum_p a_p^{(1)}(t)|p\rangle e^{-i\Omega_{p0}t} + \lambda^2 \sum_p a_p^{(2)}(t)|p\rangle e^{-i\Omega_{p0}t} + \ldots, \quad (4.159)$$

where we have used the fact that

$$\left|\psi_p^{(0)}(t)\right\rangle = \left|u_p^{(0)}\right\rangle e^{-i\Omega_{p0}t}. \quad (4.160)$$

Note that since only energy differences are important, Ω_p and Ω_{p0} are equivalent.

Using Equation 4.159, the expectation value of the dipole moment will be of the form

$$\begin{aligned}\langle\boldsymbol{\mu}\rangle(t) &= \left(\langle 0| + \lambda \sum_p a_p^{(1)*}(t)\langle p|e^{+i\Omega_{p0}^*t} + \ldots\right)\boldsymbol{\mu} \\ &\times \left(|0\rangle + \lambda \sum_p a_p^{(1)}(t)|p\rangle e^{-i\Omega_{p0}t} + \lambda^2 \sum_p a_p^{(2)}(t)|p\rangle e^{-i\Omega_{p0}t} + \ldots\right).\end{aligned}$$
(4.161)

Here, we allow Ω_{p0} to be complex. The meaning of this will be clarified later.

Given Equation 4.161, the nth-order nonlinear susceptibilities will be related to dipole moment to order λ^n.

4.5.2 First-Order Susceptibility

Using Equation 4.161, the dipole moment to first order in λ is given by

$$\langle\boldsymbol{\mu}\rangle^{(1)}(t) = \sum_p a_p^{(1)*}(t)\langle p|e^{+i\Omega_{p0}^*t}\boldsymbol{\mu}|0\rangle + \langle 0|\boldsymbol{\mu}\sum_p a_p^{(1)}(t)|p\rangle e^{-i\Omega_{p0}t}. \quad (4.162)$$

We evaluate Equation 4.162 using Equation 4.156. However, the calculation is made simpler if we first recast $a_p^{(1)*}$, as follows.

Taking the complex conjugate of Equation 4.156, we get

$$a_p^{(1)*}(t) = \frac{1}{2\hbar}\sum_q \frac{\boldsymbol{\mu}_{0p}^* \cdot \mathbf{E}(-\omega_q)}{\Omega_{p0}^* - \omega_q}\exp\left[-i\left(\Omega_{p0}^* - \omega_q\right)t\right], \quad (4.163)$$

where we have used $\boldsymbol{\mu}_{p0}^* = \boldsymbol{\mu}_{0p}$. But, recall that q sums over positive and negative frequencies with $\omega_{-q} = -\omega_q$. Thus, if we invert the sum by taking

$q \to -q$, the sum remains unchanged since we are still summing over all q. This re-indexing of Equation 4.163 leads to

$$a_p^{(1)*}(t) = \frac{1}{2\hbar} \sum_q \frac{\boldsymbol{\mu}_{0p} \cdot \mathbf{E}^{\omega_q}}{\Omega_{p0}^* + \omega_q} \exp\left[-i\left(\Omega_{p0}^* + \omega_q\right)t\right]. \qquad (4.164)$$

Substituting Equations 4.164 and 4.156 into Equation 4.162 gives

$$\langle \boldsymbol{\mu} \rangle^{(1)}(t) = \frac{1}{2\hbar} \sum_p \sum_q \left(\frac{\boldsymbol{\mu}_{0p} \cdot \mathbf{E}^{\omega_q}}{\Omega_{p0}^* + \omega_q} \boldsymbol{\mu}_{p0} + \boldsymbol{\mu}_{0p} \frac{\boldsymbol{\mu}_{p0} \cdot \mathbf{E}^{\omega_q}}{\Omega_{p0} - \omega_q} \right) e^{-i\omega_q t}, \qquad (4.165)$$

where we have used the shorthand notation $\boldsymbol{\mu}_{nm} = \langle n | \boldsymbol{\mu} | m \rangle$. But the polarization is given by

$$\mathbf{P}(t) = \sum_\omega \left(\frac{\mathbf{P}^\omega}{2} e^{-i\omega t} + \frac{\mathbf{P}^{-\omega}}{2} e^{i\omega t} \right) = N \langle \boldsymbol{\mu} \rangle^{(1)}(t), \qquad (4.166)$$

so projecting out the ω Fourier component yields

$$\mathbf{P}^{\omega_\sigma} = \frac{N}{\hbar} \sum_p \left(\frac{\boldsymbol{\mu}_{0p} \cdot \mathbf{E}^{\omega_\sigma}}{\Omega_{p0}^* + \omega_q} \boldsymbol{\mu}_{p0} + \boldsymbol{\mu}_{0p} \frac{\boldsymbol{\mu}_{p0} \cdot \mathbf{E}^{\omega_\sigma}}{\Omega_{p0} - \omega_\sigma} \right). \qquad (4.167)$$

The first-order susceptibility is calculated using Equation 4.140 and yields

$$\chi_{ij}^{(1)}(-\omega_\sigma; \omega_\sigma) = -N \sum_p \left(\frac{\mu_{0p}^j \mu_{p0}^i}{\mathcal{E}_{p0}^* + \hbar\omega_\sigma} + \frac{\mu_{0p}^i \mu_{p0}^j}{\mathcal{E}_{p0} - \hbar\omega_\sigma} \right), \qquad (4.168)$$

where μ_{0p}^k is the kth cartesian component of $\boldsymbol{\mu}_{0p}$, $\mathcal{E}_{p0} = \hbar\Omega_{p0}$, and where we have used the fact that the definition of $\chi_{ij}^{(1)}$ assigns i to the direction of the polarization $\mathbf{P}^{\omega_\sigma}$, and j to the direction of the applied field $\mathbf{E}^{\omega_\sigma}$.

In deriving Equation 4.168, we neglected to explain how the Bohr frequencies can be complex if the Hamiltonian is real; and, we never checked if the wavefunction given by Equation 4.149 is normalized. The quick answer is that the Bohr frequencies must be real for a Hermitian Hamiltonian; but, we add a small imaginary part to allow for energy loss. As a result, the Hamiltonian is still *approximately* Hermitian. With regards to normalization, as luck would have it, the wavefunctions turn out not to be normalized; but, this has no effect on the linear susceptibility. However, the nonlinear susceptibilities are a different story.

In the two sections that follow, we briefly discuss complex Bohr frequencies and normalization.

Complex Bohr Frequencies

The complex Bohr frequency Ω_{n0} can be expressed in terms of the transition energy E_{n0} as

$$\hbar\Omega_{n0} = E_{n0} - i\frac{\hbar\Gamma}{2}, \tag{4.169}$$

where $\hbar\Gamma$ is the imaginary part and describes the rate of energy lost from the system. Since nonlinear optics deals with the exchange of energy between photons and materials, Γ represents other source of losses such as non-radiative losses when a molecule de-excites without emitting a photon. Collisional losses, where the energy is transferred into kinetic energy, is one well-known example.

Recall that the time evolution of an energy eigenvector is of the form

$$|\psi_n(t)\rangle = |u_n\rangle e^{-i\Omega_{n0}t}, \tag{4.170}$$

where $|u_n\rangle$ is the stationary state vector of energy eigenstate n. Substituting Equation 4.170 into Equation 4.169 yields

$$|\psi_n(t)\rangle = |u_n\rangle e^{-i\frac{E_{n0}}{\hbar}t} \cdot e^{-\frac{\Gamma}{2}t}. \tag{4.171}$$

Thus, the probability of finding the system in state n is given by

$$\langle \psi_n(t)|\psi_n(t)\rangle = e^{-\Gamma t}. \tag{4.172}$$

so the probability decays to $1/e$ over a time $1/\Gamma$.

As long as the decay time constant is long compared with the Bohr frequencies, the Hamiltonian is *approximately* Hermitian, and the quantum calculations can be applied in the usual way. Thus, the addition of a small imaginary part to the energy is a phenomenological way to add energy dissipation to a pure quantum theory that does not allow for dissipation.

Normalizing the Wavefunctions

Equation 4.148 is not normalized, so all expectation values need to be divided by a normalization factor. For the first-order corrections to the ground-state wavefunctions, this normalization factor is given by

$$\begin{aligned}\left\langle \psi_0^{(1)}(t) \middle| \psi_0^{(1)}(t) \right\rangle &= \sum_{p,l} \left(\langle p|a_p^{(1)}(t) + \langle 0| \right) e^{iHt/\hbar} \cdot e^{-iHt/\hbar} \left(|0\rangle + a_l^{(1)}(t)|l\rangle \right) \\ &\approx \langle 0|0\rangle + \left(\sum_p a_p^{(1)}(t)\langle p|0\rangle + \text{c.c.} \right) = 1 + a_0^{(1)}(t) + a_0^{(1)*}(t),\end{aligned}$$

$$\tag{4.173}$$

where we have kept terms to order $a_p^{(1)}(t) \propto \boldsymbol{\mu}_{0p} \cdot \mathbf{E}^{\omega_q}$ (see Equation 4.156) since we are interested in the corrections to $\chi^{(1)}$ due to normalization.

Normalizing the expectation value of the dipole operator given by Equation 4.161 using Equation 4.173 to order $a_p^{(1)}(t)$ leads to

$$\begin{aligned}
\langle\boldsymbol{\mu}\rangle(t) &= \frac{\langle 0|\boldsymbol{\mu}|0\rangle + \left(\langle 0|\boldsymbol{\mu}\sum_p a_p^{(1)}(t)|p\rangle e^{-i\Omega_{p0}t} + \text{c.c.}\right) + \ldots}{1 + a_0^{(1)}(t) + a_0^{(1)*}(t)} \\
&= \langle 0|\boldsymbol{\mu}|0\rangle + \left(-a_0^{(1)}(t)\langle 0|\boldsymbol{\mu}|0\rangle + \sum_p a_p^{(1)}(t)\langle 0|\boldsymbol{\mu}|p\rangle e^{-i\Omega_{p0}t} + \text{c.c.}\right) + \ldots,
\end{aligned}$$
(4.174)

where we expanded the denominator in a series using $1/(1+x) \approx 1-x$ for small x. The net result of normalization is that it subtracts the $p = 0$ term from the sum given by last term in parentheses in Equation 4.174.

To determine the effect of normalization on the linear susceptibility, we consider the $p = 0$ term in Equation 4.168, which obviously vanishes. Thus, for the linear susceptibility, neglecting normalization has no affect on the result. This will not be the case for higher-order susceptibilities. The calculations that follow will take into account normalization, but, the messy details will not be given. The method employed will be similar to the one above for the linear susceptibility.

4.5.3 Nonlinear Susceptibilities and Permutation Symmetry

The second- and third-order (and higher order) susceptibilities can be calculated using a similar approach as the first-order one by first finding $\langle\boldsymbol{\mu}\rangle^{(n)}(t)$ from $\langle\boldsymbol{\mu}\rangle(t)$ using n^{th}-order λ terms and differentiating the result according to Equation 4.140 after selecting the ω_σ Fourier component. The second-order susceptibility is then given by

$$\begin{aligned}
\chi_{ijk}^{(2)}(-\omega_\sigma;\omega_1,\omega_2) = &-\frac{Ne^3}{\hbar^2}\mathcal{P}_\mathcal{I}\sum_{m,n}\left[\frac{r_{0n}^i r_{nm}^j r_{m0}^k}{(\Omega_{n0}-\omega_1-\omega_2)(\Omega_{m0}-\omega_2)}\right. \\
&+ \frac{r_{0n}^j r_{nm}^i r_{m0}^k}{(\Omega_{n0}+\omega_1)(\Omega_{m0}-\omega_2)} \\
&\left.+ \frac{r_{0n}^j r_{nm}^k r_{m0}^i}{(\Omega_{n0}+\omega_1)(\Omega_{m0}+\omega_2+\omega_1)}\right],
\end{aligned}$$
(4.175)

where we have used $\boldsymbol{\mu} = -e\mathbf{r}$, and $\mathcal{P}_I$ is the 'intrinsic permutation operator'. It dictates that we take an average over all permutations of ω_1 and ω_2 with simultaneous permutations of the Cartesian components. For example, the first term in brackets in Equation 4.175 under permutation yields

$$\frac{1}{2}\left[\frac{r^i_{0n}r^j_{nm}r^k_{m0}}{(\Omega_{n0}-\omega_1-\omega_2)(\Omega_{m0}-\omega_2)} + \frac{r^i_{0n}r^k_{nm}r^j_{m0}}{(\Omega_{n0}-\omega_2-\omega_1)(\Omega_{m0}-\omega_1)}\right]. \quad (4.176)$$

Equation 4.175 can be rewritten using the full permutation operator $\mathcal{P}_F$. When ω is far from resonance ω_σ can be permuted with ω_n with simultaneous permutation of the Cartesian components, so the first order susceptibility can be rewritten as

$$\chi^{(1)}_{ij}(-\omega_\sigma;\omega_\sigma) = \frac{Ne^2}{\hbar}\mathcal{P}_F \sum_m{}' \left(\frac{r^i_{0m}r^j_{m0}}{(\Omega_{m0}-\omega_\sigma)}\right), \quad (4.177)$$

where the prime on the sum indicates that it excludes the ground state as required for normalization.

Next we revisit the second-order susceptibility and consider normalization. Taking into account normalization, which was not done in arriving at Equation 4.175, after some manipulations we get

$$\chi^{(2)}_{ijk}(-\omega_\sigma;\omega_1,\omega_2) = -\frac{Ne^3}{2\hbar^2}\mathcal{P}_F \sum_{m,n}{}' \left(\frac{r^i_{0n}\bar{r}^j_{nm}r^k_{m0}}{(\Omega_{m0}-\omega_\sigma)(\Omega_{n0}-\omega_2)}\right), \quad (4.178)$$

where again, the prime indicates that the sum excludes the ground state, and $\bar{r}^j = r^j - r^j_{00}$. Equation 4.175 for $\chi^{(2)}_{ijk}$ indeed reduces to Equation 4.178 upon the full permutation operation.

And similarly the third order susceptibility is given by

$$\chi^{(3)}_{ijkh}(-\omega_\sigma;\omega_1,\omega_2,\omega_3) = \frac{Ne^4}{6\hbar^3}\mathcal{P}_F\left[\sum_{m,n,l}{}'\right.$$
$$\frac{r^i_{0l}\bar{r}^j_{ln}\bar{r}^k_{nm}r^h_{m0}}{(\Omega_{l0}-\omega_\sigma)(\Omega_{n0}-\omega_1-\omega_2)(\Omega_{m0}-\omega_1)}$$
$$\left.-\sum_{m,n}{}'\frac{r^i_{0m}r^j_{m0}r^k_{0n}r^h_{n0}}{(\Omega_{m0}-\omega_\sigma)(\Omega_{n0}-\omega_1)(\Omega_{n0}+\omega_2)}\right]$$
$$(4.179)$$

Problem 4.5-1: Using time dependent perturbation theory, we found that

$$\dot{a}_p^{(m)}(t) = \frac{1}{i\hbar} \sum_l \langle u_p^{(0)} | V(t) | u_l^{(0)} \rangle a_l^{(m-1)}(t) e^{-i\omega_{lp} t}, \quad (4.180)$$

where $V(t) = e\mathbf{r} \cdot \mathbf{E}$. Express the electric field in terms of photon creation and annihilation operators and solve for $a_r^{(2)}(t)$ using the approximation $\mathbf{k} \cdot \mathbf{r} = 0$.

Problem 4.5-2: Start from Equation 4.178, and calculate the second-order susceptibility of a system made of aligned 1D molecules (only matrix elements of x contribute) in the two-state approximation (i.e. only state 0 and 1 contribute to the sum) and show that it can be expressed in the compact form

$$\chi^{(2)} = -\frac{3e^3 N}{\hbar^2} \cdot \frac{\omega_{10}^2 |x_{10}|^2 (x_{11} - x_{00})}{\left(\omega_{10}^2 - \omega^2\right)\left(\omega_{10}^2 - 4\omega^2\right)}. \quad (4.181)$$

How does this result compare with the classical nonlinear harmonic oscillator result given by Equation 2.59?

Problem 4.5-3a: Write a Python module that has as its input a vector that represents the eigenenergies and an array of matrix elements of the position operator; and as an output has the off-resonance polarizability, hyperpolarizability and second hyperpolarizability. Use Equations 4.177 through 4.179 without the N factor. Since these are off-resonance, all the angular frequencies are set to zero and the energies are real.

(b): Calculate the position matrix and energy vector for a harmonic oscillator using 20 states, and use these to calculate the polarizability and hyperpolarizabilities in the code you wrote in Part (a). Do the results make sense? Explain.

(c): Now generalize your code to take as input a complex energy vector, complex matrix elements and photon frequencies; and, as output, have it return the complex polarizability and hyperpolarizabilities. Test your code to get the dispersion of the real and imaginary parts of the polarizability of a harmonic oscillator.

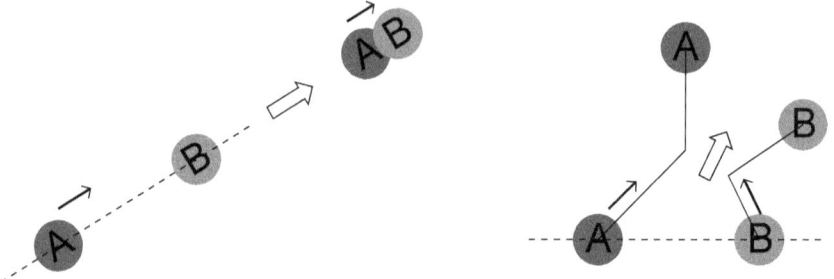

Figure 4.2: Particle trajectories in an inelastic (left) and elastic (right) collision.

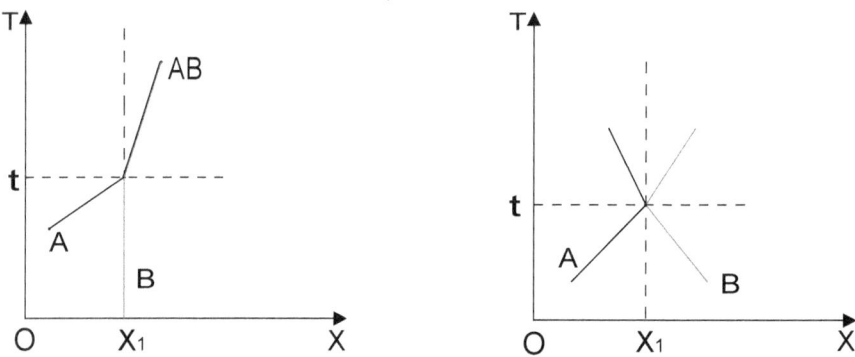

Figure 4.3: Space-time diagram for process shown in Figure 4.2.

4.6 Using Feynman-like Diagrams

4.6.1 Introduction

A **Feynman diagram** is a pictorial aid to evaluating the nonlinear susceptibilities, and provides a simple picture of the physical process of photons being absorbed or emitted by a molecule. It accurately embodies the underlying physics and allows one to write down immediately the mathematical expression associated with any particular process to any order of perturbation.

Figure 4.2 shows an inelastic and elastic collision of two particles in space and Figure 4.3 shows the same two processes on a space-time diagram, which is amenable to be diagrammatically evaluated. Both are classical descriptions of collisions. Figure 4.4 shows a Feynman diagram for a collision between two charged particles such electrons through an electromagnetic interaction mediated by the exchange of a photon. It is understood that the vertical and horizontal axes represent time and space, so the axes will be omitted.

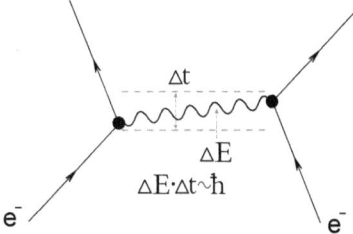

Figure 4.4: A Feynman diagram that represents the interaction between two electrons (solid lines) that is mediated by the exchange of a photon (wiggly line).

The incident electron on the left in Figure 4.4 emits a photon for no apparent reason and then deflects. This is a virtual process because energy is not conserved between the two dashed lines. A second incident electron absorbs the photon at a latter time. The net effect is that both electrons deflect as a result of the collision. Though energy is not conserved in the time interval between the dashed lines, as far as the outside world is concerted, energy is conserved.

The uncertainty principle is mother-nature's way of preventing this embarrassing violation of energy conservation from being observed. Recall that $\Delta E \Delta t \geq \hbar/2$. Thus, if the energy is measured over a time interval Δt, it's uncertainty is $\Delta E \geq \hbar/2\Delta t$. Thus, energy conservation is broken but it is not observed because its magnitude is smaller than the uncertainty of the measurement. As such, the area between the dashed lines is an unobservable. Momentum conservation is also violated; but is unobservable in the volume containing the photon because the momentum uncertainty is too large.

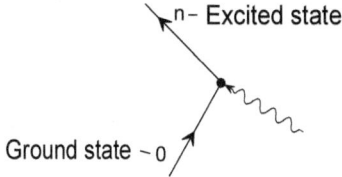

Figure 4.5: A molecule (solid line) in it's ground state absorbs a photon and is excited to state n.

Atoms and molecules can be represented in a similar way. Figure 4.5

shows a molecule absorbing a photon, which kicks it into a state of higher energy. A similar diagram can be drawn to represent the emission of a photon and the two diagrams can be combined to show the absorption of a photon followed by an emission. Note that we will not be concerned with making the slopes of the molecule and photon lines accurate because they do not have an effect of how the diagrams are evaluated.

4.6.2 Elements of Feynman Diagrams

The elements and physical meaning of Feynman diagrams as applied to nonlinear optics follow.

- a line represents a molecule
- a wavy line represents a photon
- a vertex represents absorption or emission of a photon
- a diagonal line represents a specific state
- a vertical line represents many possible states which do not necessarily conserve energy through a vertex

A vertex often represents a virtual process, so as we have already discussed, energy does not have to be conserved between vertices. However, energy is conserved between the initial state – at a time prior to the first vertex, and the final state – after the last vertex.

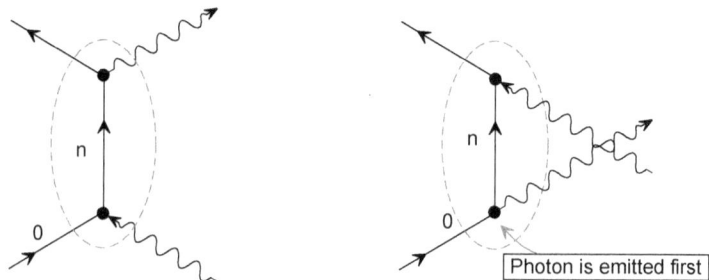

Figure 4.6: The two Feynman diagrams for the linear susceptibility.

As a first example, we consider the linear susceptibility. The first step is to draw all possible distinct diagrams, which contribute equally to $\chi^{(1)}_{ij}(-\omega_\sigma;\omega_\sigma)$. The fact that the photon is emitted before a photon is absorbed,

Nonlinear Optics: A student's perspective

as shown in the second figure, may seem strange, but, the uncertainty principle prevents it from being observed. The observer "sees" the process outside the dashed red regions, so they appear identical – the incident photon interacts with the molecule and is emitted.

The righthand diagram shows the value of adding points at the vertices to represent an interaction. Though the two photons cross, they interact only at the vertices that include the molecule.

Recall that

$$\chi_{ij}^{(1)}(-\omega_\sigma;\omega_\sigma) = N \sum_p \left(\frac{\mu_{0p}^j \mu_{p0}^i}{\mathcal{E}_{p0}^* + \hbar\omega_\sigma} + \frac{\mu_{0p}^i \mu_{p0}^j}{\mathcal{E}_{p0} - \hbar\omega_\sigma} \right), \qquad (4.182)$$

where μ_{0p}^k is the kth cartesian component of μ_{0p}, $\mathcal{E}_{p0} = \hbar\Omega_{p0}$, and where we have used the fact that the definition of $\chi_{ij}^{(1)}$ assigns i to the direction of the polarization $\mathbf{P}^{\omega_\sigma}$, and j to the direction of the applied field $\mathbf{E}^{\omega_\sigma}$. We will show in the next section how the two terms in Equation 4.182 are related to the two diagrams shown in Figure 4.6. In each case, we will need to step back to the perturbation calculation to understand which elements of the diagram correspond to which mathematical operation.

4.6.3 Rules of Feynman Diagram

This section describes how Feynman diagrams are used to evaluate nonlinear susceptibilities. The two components of a Feynman diagram are vertices and propagators, which represent parts of the integrals that are evaluated in time-dependent perturbation theory, which to first-order are of the form

$$\int_{t_1}^{t_2} dt\, e^{-i\Omega_{n0}t} \langle f|\boldsymbol{\mu}\cdot\mathbf{E}|i\rangle. \qquad (4.183)$$

The nth-order perturbation involves a product of n integrals.

Second quantization provides a guide for drawing the diagrams. A vertex is the point at which an absorption or emission takes place, so the number of photons changes and the molecule makes a transition between states and is represented by the matrix element $\langle f|\boldsymbol{\mu}\cdot\mathbf{E}|i\rangle$. This is just the probability amplitude of a transition between the initial and final states.

A propagator describes how the system evolves in time and originates in the time integral. The electric fields are time harmonic at frequency ω and the time evolution of the molecular wave functions are at frequency Ω_{n0} so integration yields energy denominators of the form $1/(\Omega_{n0}\pm\omega)$. The Fyenman diagram will be evaluated by working forward in time, evaluating the propagators and the intervening vertices from bottom up. The details and rules follow.

Vertex Rule

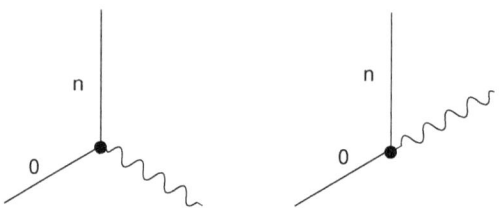

Figure 4.7: A vertex where a photon is absorbed (left) or emitted (right) by a molecule in its ground state to create a virtual superposition of states (represented by a vertical line). These diagrams must be part of a larger one since they do not individually conserve energy.

A vertex represents the probability amplitude of absorbing or emitting a photon that is accompanied by a dipole transition between states of the molecule. It is of the form $\langle f|\boldsymbol{\mu}\cdot\mathbf{E}|i\rangle$, where $\langle f|$ is the final state just above the vertex; and, $|i\rangle$ is the initial state just below the vertex. The state vectors include both molecule and photon states. Recall that $\boldsymbol{\mu}$ is an operator that acts on molecular states and $\mathbf{E}$ is an operator that acts on the state of the electromagnetic field. In second-quantized language $\mathbf{E} \propto a - a^\dagger$ so for the photon part an absorption contributes $\langle 0|\mathbf{E}|1\rangle \propto \langle 0|a|1\rangle = 1$ to the transition amplitude and for an emission the contribution is $\langle 0|\mathbf{E}|1\rangle \propto \langle 1|a^\dagger|0\rangle = 1$. It may seem peculiar that a molecule in its ground state can emit a photon, but this prejudice is based on our steadfast commitment to energy conservation. Since this is a virtual process, the energy ledger does not need to balanced until all interactions are accounted for.

If the molecule is initially in its ground state $|0\rangle$ with one incident photon, and makes a transition by virtual excitation of the molecule to state $|n\rangle$ with zero photons as shown in the left-hand part of Figure 4.7, the dipole transition amplitude of this absorption process is given by $\langle n;0|\boldsymbol{\mu}\cdot\mathbf{E}|0;1\rangle \propto \langle n|ex_i|0\rangle = -ex^i_{n0}$, where i is the field polarization direction and the dipole moment of the molecule is given by $\mu_i = -ex_i$. Similarly, the vertex in the righthand Feynman diagram shown in Figure 4.7 gives the dipole transition amplitude $\langle n;1|\boldsymbol{\mu}\cdot\mathbf{E}|0;0\rangle \propto \langle n|ex_i|0\rangle$ for the emission process.

Each vertex contributes to the Feynman diagram with the factor

ex_{n0}^i, for both an emitted or absorbed photon, where i is the polarization of the photon. Note that the field amplitude does not appear because the susceptibilities are calculated by differentiation with respect to the field.

Propagator Rule

Since the electric field operator is proportional to $ae^{-i\omega t} - a^+ e^{i\omega t}$, the integrand in Equation 4.183 includes a term of the form $e^{-i(\Omega_{n0} \pm \omega)t}$. Thus integration yields a factor proportional to $\frac{1}{\Omega_{n0} \pm \omega}$,

$$\int dt e^{-i\Omega_{n0}t} \langle f | \boldsymbol{\mu} \cdot \mathbf{E} | i \rangle \propto \int e^{-i(\Omega_{n0} \pm \omega)t} dt \propto \frac{1}{\Omega_{n0} \pm \omega}. \quad (4.184)$$

In the case of the outgoing photon, the complex Bohr frequency appears as its complex conjugate. Details of the rules for complex conjugations will be described in more detail when the higher-order susceptibilities are discussed. Clearly, the product of the two vertices and propagator in Figure 4.6 yields the two terms in Equation 4.182.

The propagators are evaluated across each vertex, adding the Bohr frequency corresponding to the two states on either side of the vertex to all outgoing photons and subtracting the incident frequencies. This rule will be made more clear when we consider the second-order susceptibility. But, even for the linear susceptibility, we see that the last vertex gives a bit of trouble, which we can appreciated by considering any one of the diagrams for $\chi^{(1)}$. The propagator taking the molecule to the ground state after the last photon is absorbed or emitted will be of the form $1/(\Omega_{00} + \omega - \omega) = 1/0$, which diverges.

The divergence of the last propagator can be traced to the last step in calculating the susceptibilities, in which a delta function in the energies ensures energy conservation. For the first-order case, refer to Equations 2.135, 2.136, and 2.137 to refresh your memory on why this is so. The simple resolution to the divergence is to disregard the last propagator, because that delta function gets integrated over when taking the Fourier transform in the case where there are a finite number of monochromatic waves.

Next, we calculate the propagator for the segment of the Feynman diagram to second-order as shown in Figure 4.8 between the two crosses, which gives

$$\frac{1}{\Omega_{n0} - \omega_1 + \Omega_{mn} - \omega_2} = \frac{1}{\Omega_{m0} - \omega_1 - \omega_2}. \quad (4.185)$$

In summary, the **Propagator Rule** gives the time integrals in the perturbation expansion. To calculate the the full propagator of a Feynman diagram,

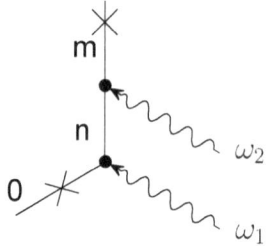

Figure 4.8: Applying the propagator rule to second-order in the field between the two crosses.

one calculates the propagator from the first molecule segment (where the system is in the ground state) to each molecule segment up the chain and taking the product of all the propagators. The propagator to the final molecular state when it returns to the ground state is ignored. The process of calculating propagators will be made more clear in the example using sum frequency generation, as described in Section 4.6.4.

4.6.4 Feynman Diagrams and Sum Frequency Generation

In the previous sections, we derived the mathematical expressions of the nonlinear susceptibilities using perturbation theory, which even to second-order was a time-consuming task. In this section, we will see how we can write down all those mathematical expressions immediately by evaluating the Feynman Diagrams.

Figure 4.9 shows all possible diagrams for sum frequency generation. It is a simple task to show that there are a total of 6 diagrams (3! = 6).

Let's examine the first diagram which is drawn in more detail in Figure 4.10 (left). Applying the vertex and propagator rules from the bottom up yields

1. $|0\rangle$ to $|n\rangle$:

    - from the vertex rule, we have $-ex_{n0}^{j}$.
    - from the propagator rule, we have $\frac{1}{\hbar(\Omega_{n0}-\omega_1)}$.

2. $|n\rangle$ to $|m\rangle$:

    - from the vertex rule, we have $-ex_{mn}^{k}$.
    - from the propagator rule, we have $\frac{1}{\hbar(\Omega_{m0}-\omega_1-\omega_2)}$, where $\Omega_{m0} = \Omega_{mn} + \Omega_{n0}$. Note that in the denominator the energy difference

Nonlinear Optics: A student's perspective

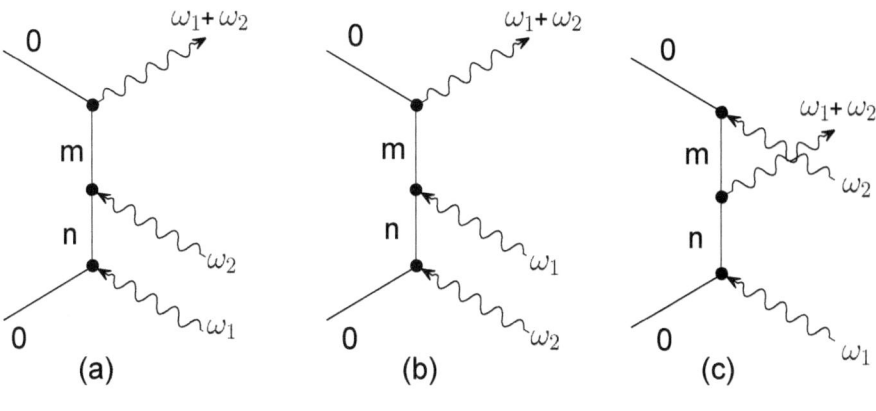

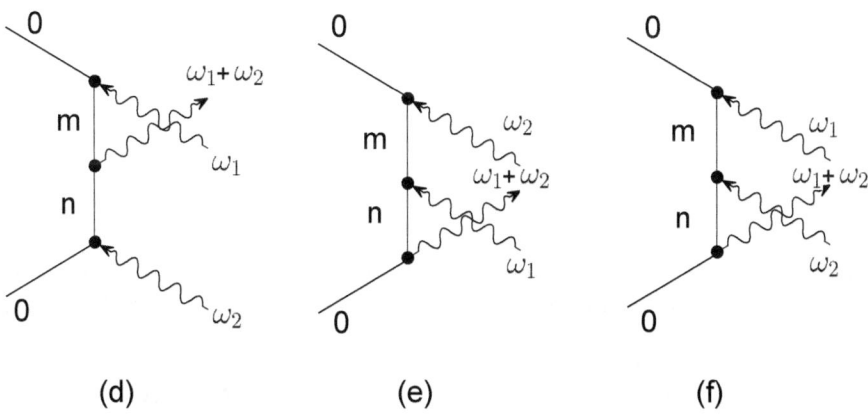

Figure 4.9: All possible Feynman diagrams for sum frequency generation.

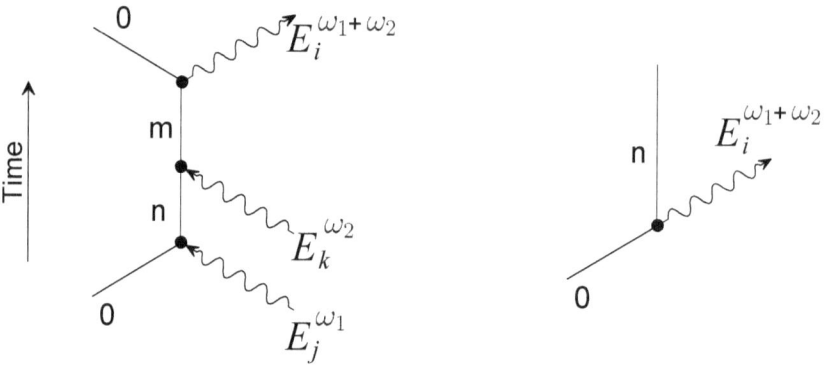

Figure 4.10: (left) A more detailed drawing of one Feynman diagram for sum frequency generation and (right) an outgoing photon in the propagator.

is always between the ground state and the upper excited state in the diagram.

3. $|m\rangle$ to $|0\rangle$:
    - from the vertex rule, we have $-ex^i_{0m}$.
    - from the propagator rule, since the transition is from the ground state back to the ground state, there is no energy denominator, because it was already used to enforce energy conservation.

We multiply all those terms together and sum over the interior indices, which give us the contribution to β for that diagram

$$\beta = -\frac{e^3}{\hbar^2}\sum_{mn}{}' \frac{x^i_{0m}\bar{x}^k_{mn}x^j_{n0}}{(\Omega_{m0}-\omega_1-\omega_2)(\Omega_{n0}-\omega_1)}. \tag{4.186}$$

Equation 4.186 is only for the first diagram. To get the total hyperpolarizability β, we need to add all the other five diagrams, which gives the result

$$\beta = -\frac{e^3}{\hbar^2}\sum_{mn}{}' \frac{x^i_{0m}\bar{x}^k_{mn}x^j_{n0}}{(\Omega_{m0}-\omega_1-\omega_2)(\Omega_{n0}-\omega_1)} + \text{the other five diagrams}. \tag{4.187}$$

If the outgoing photon is in the propagator, Ω_{m0} is changed to Ω^*_{m0}. For example, in Figure 4.10 (right) the propagator rule gives $\frac{1}{\Omega^*_{n0}+\omega_1+\omega_2}$. All the other propagators above the first one with a complex conjugate are include complex conjugation of the form Ω^*. Of course, the center matrix element gets a "bar" to enforce normalization and the ground state is excluded from the sum.

4.7 Virtual States and Virtual Excitations

We saw how virtual states and excitations don't conserve energy but the time scales of the processes are so short that the uncertainty in the measurement exceeds the amount of energy conservation violation, so the offense against nature is not observed.

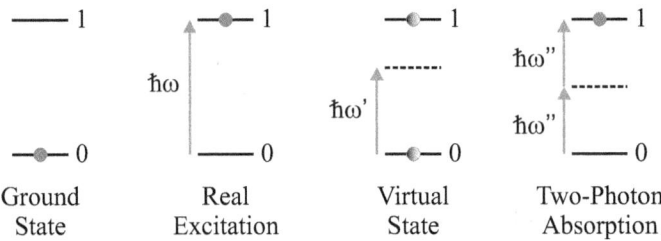

Figure 4.11: Real excitations occur when the photon energy matches the energy difference between two states of the material system. If the photon energy is less than this energy difference, the system cannot be excited and the state of the material system can be thought of as being in a superposition of states. Two-photon absorption can be mediated by an excitation through a virtual state.

Figure 4.11 shows a two-state quantum system that is excited by a photon under several conditions. When the photon energy matches the energy difference between its states, the material system that is initially in its ground state is excited into the upper state and the photon is destroyed. This is called a real excitation because the material's final state is an anergy eigenstate. The the initial state of the universe is $|0; 1_{\hbar\omega}\rangle$ and the final state is $|1; 0_{\hbar\omega}\rangle$, where $E_{10} = \hbar\omega$. As we saw in Section 4.2.3, the transition amplitude for an excitation is given by $\langle 0; 1_{\hbar\omega} | \mathbf{r} \cdot \mathbf{E} | 1; 0_{\hbar\omega}\rangle$

Next, consider the case where the photon's energy $\hbar\omega'$ is less than the energy difference between the two states. No real excitation occurs; rather, the system is excited into a superposition of states. This superposition includes many energy eigenstates . Since the photon energy does not match any of the material energy differences, energy is not conserved in the process; but, the time interval is small enough that the violation of energy conservation is not observed before the system returns to it's final state. This superposition is represented by the vertical molecule line in the Feynman diagram and the excitation is said to be a *virtual state*. Each energy eigenstates $|n\rangle$ is the superposition is said to undergo a *virtual excitation*.

Two photons are absorbed through the two-photon absorption process as long as the energy of the two photons combined matches the energy difference between the two states of the material. This two-step process leads to a real excitation, but the first photon can be viewed as being absorbed by a virtual, from which the second photon is absorbed. This intermediate state does not conserve energy, and can be viewed as being a superposition of virtual expiations.

Problem 4.6-1(a): Use the Feynamn-like diagram in Figure 4.12(a) to calculate its contribution to $\chi^{(-)}_{----}(-;--\ldots)$, where you are to fill in the blanks. Note that the field at frequency $\omega_2 - \omega_1$ is the field that is being observed.

(b): Cascading is the process by which two or more molecules interact with light and each other through a nonlinear response of order less than or equal to n and results in a process of order greater than n. As an example, a third-order response can result from two molecules, each of which interacts with the light through a second-order process as shown in Figure 4.12(b). Evaluate this diagram. Note that you have the freedom to decide how to account for distance between the molecules and their relative orientations.

Problem 4.6-2(a): The third harmonic spectrum is measured for an unknown material and resonance enhancement is observed when the incident fundamental frequencies correspond to the energies $E = 0.8\text{eV}$, 1.3eV and 2.4eV. What are the energy levels and symmetries of the excited states if the ground state is symmetric. Choose a system with the minimum number of states that gives this behavior.

(b): If the magnitude of the susceptibility is the same at each of the three resonances in the above spectrum, how would the spectrum appear (peak placement and magnitude) if the potential were perturbed to leave it slightly asymmetric? Make a sketch of $|\chi^{(3)}|$.

Problem 4.6-3(a): The spectrum shown in Figure 4.13 is observed when an intense beam of light at frequency ω_1 is launched into a sample. You have already determined that a $\chi^{(2)}$ process is responsible. Propose the simplest **two** hypotheses for the source of the three peaks in the spectrum.

(b): What experimental parameters (such as laser intensity, etc.) would you vary that would differentiate between the two hypotheses?

Nonlinear Optics: A student's perspective

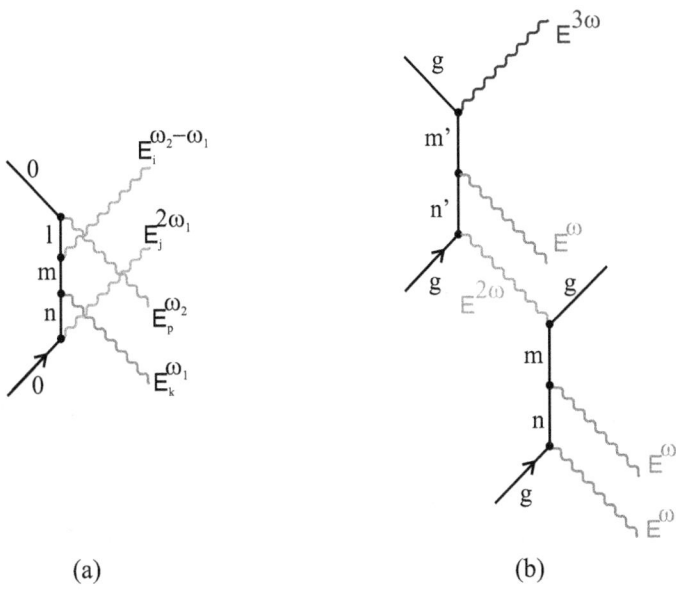

(a) (b)

Figure 4.12: (a) Second hyperpolarizability and (b) the cascading second hyperpolarizability that results from two interacting molecules through the first hyperpolarizability.

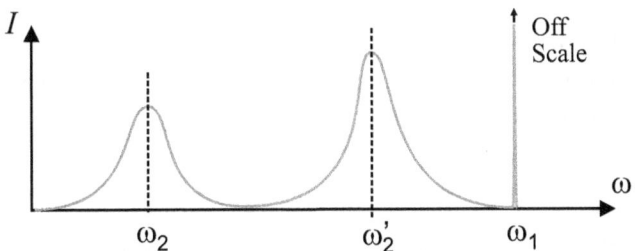

Figure 4.13: The output spectrum observed when an intense beam at frequency ω_1 is launched into a sample with $\chi^{(2)} \neq 0$.

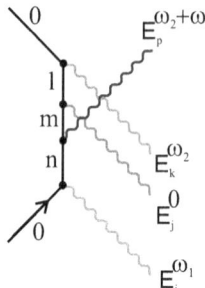

Figure 4.14: A Feynman diagram for a process that includes a static electric field.

> **Problem 4.6-4(a):** Evaluate the Feynman diagram shown in Figure 4.14.
>
> **(b):** Draw a diagram of an experiment that you would build to measure this process including two laser sources, detectors, filters that pass one color and reflect another, electrodes, etc.
>
> **(c):** Plot the measured sum frequency power as a function applied voltage and explain how you would get $\chi^{(3)}$ from the data (constants of proportionality are unimportant). Also describe what information the y-intercept and the slope would give and under what conditions it would vanish.

4.8 Broadening Mechanisms

As we saw in Equation 4.169, decay from an excited state to a lower state can be approximated by making the Hamiltonian slightly non-Hermitian with a small imaginary part of the energy

$$E_{n0} \rightarrow E_{n0} - \frac{i\Gamma_{n0}}{2}, \qquad (4.188)$$

where Γ_{n0} is the inverse decay time constant. Γ_{n0} is the width of the peak in the plot of the imaginary part of α, as shown in Figure 4.15. The decay width can be directly measured with linear absorption spectroscopy. This is an example of homogeneous broadening, where the lifetime is the same for each molecule in the ensemble.

Inhomogeneous broadening comes from molecules that are found in varying environments. As an example, consider a monomolecular gas in which the

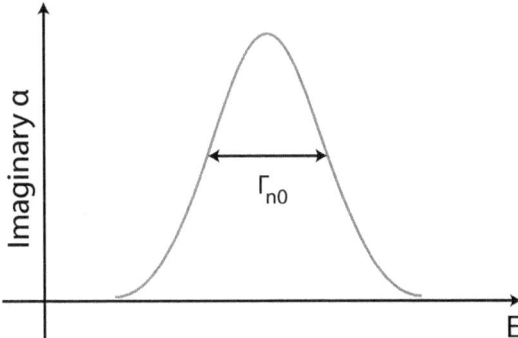

Figure 4.15: A plot of the imaginary part of the polarizability α as a function of energy peaks for state n of width Γ_{n0}.

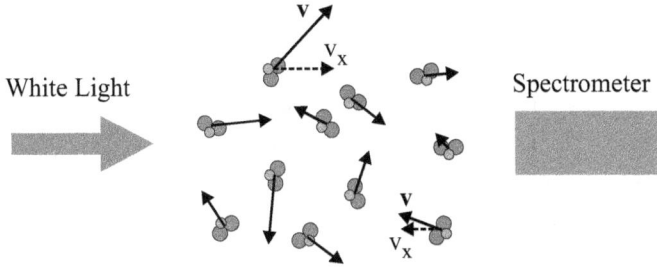

Figure 4.16: The light transmitted by a monomolecular gas is measured with a spectrometer.

velocities of the molecules obey the Boltzmann distribution. The probability of finding a molecule with velocity $\mathbf{v}$ is $P(\mathbf{v})$, which is given by

$$P(\mathbf{v}) \propto \exp\left[\frac{m\mathbf{v}^2}{2kT}\right], \tag{4.189}$$

where m is the mass of the molecule, T is the temperature and k is Boltzmann's constant.

Now consider an experiment which measures the amount of light that passes through a monomolecular gas as shown in Figure 4.16. If the molecule is approaching the laser, it will "see" blue-shifted light, so will absorb light of lower energy from the beam, resulting in a peak that is shifted to lower energy. Thus, the peak will be shifted in proportion to the x-component of the velocity v_x.

Figure 4.17a shows the imaginary part of the polarizability, which is proportional to the absorbance, versus energy for three particles – one that

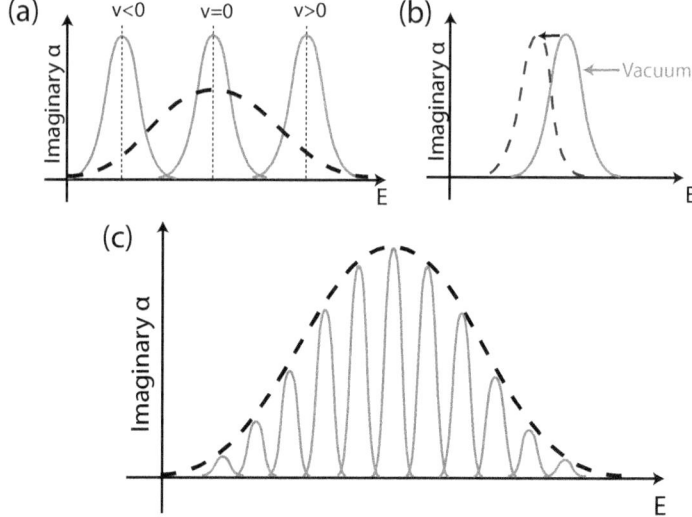

Figure 4.17: a) The absorption spectrum of three different molecules with different velocities (red), and the Boltzmann distribution of peak positions (dashed curve). b) The spectrum shifts when nearby molecules interact with each other through their coulomb fields. c) The sum of the individual contributions in an ensemble (red peaks) weighted by the Boltzmann factor leads to the aggregate inhomogeneously-broadened peak (dashed curve).

moves away from the spectrometer, $v < 0$, one with zero velocity, $v = 0$, and one that is moving toward the spectrometer, $v > 0$. The dashed curve represents the velocity distribution, which is used as a weighting factor in summing the individual peaks to determine the aggregate peak from all the molecules in the gas. We call the result the inhomogeneously-broadened spectrum, as shown in Figure 4.17c. Note that the peaks are packed much more closely in reality, giving the smooth dashed aggregate curve.

Another example of inhomogeneous broadening comes from collision between pairs of molecules. During the close encounter, the electric field due to one molecule on the other one varies in time, growing as they approach each other and decreasing when they recede, leading to a time-dependent Stark shift of the eigen energies, hence the peak position. Figure 4.17b shows the shift at one instant of time. The superposition of all the shifts, weighted according to the distribution of shifts from a scattering calculation, yields the inhomogeneous width.

In general, a quantum system is never isolated because it will interact with the environment. For example, when a molecule is in contact with a temperature bath, the energy has a Boltzman distribution. In the quantum picture, the molecules in vacuum are all in their ground states, but because of the temperature bath, other states may be populated as well. When the quantum system is not isolated and interacts with the environment, it is characterized by both a quantum probability and an ensemble average weighted according to a statistical mechanical probability. In such cases, the density matrix is the appropriate tool for taking into account our ignorance of the details of the system.

4.9 Introduction to Density Matrices

The density operator, ρ, in Dirac notation is given by

$$\rho = \sum_s p(s) |s(t)\rangle \langle s(t)|, \qquad (4.190)$$

where summation is over states of the system, $|s(t)\rangle$, which are assumed to be normalized and $p(s)$ is the probability of finding the system to be in a state $|s\rangle$. $p(s)$ is usually associated with the proportion of a population of particles in an ensemble that are in state s. The states $|s\rangle$ are not necessarily orthogonal and may or may not span the full Hilbert space. The consequence of this property will be described later. We stress that $p(s)$ is not necessarily a quantum probability. For example, when the system is in thermal equilibrium with a bath of temperature T, $p(s)$ would be proportional to $e^{-E_s/kT}$ where E_s is the energy of state s.

Equation 4.190 can be represented in terms of the energy eigenstate basis $\{|n\rangle\}$ since $|s\rangle$ can be written in the energy basis using superposition, or

$$|s\rangle = \sum_n c_n^{(s)} |n\rangle, \qquad (4.191)$$

where we have dropped the time dependence to remove clutter. Substituting Equation 4.191 into Equation 4.190 yields

$$\rho = \sum_s p(s) \sum_n \sum_{n'} c_{n'}^{(s)*} c_n^{(s)} |n'\rangle \langle n| \equiv \sum_n \sum_{n'} \rho_{n'n} |n'\rangle \langle n|, \qquad (4.192)$$

where $\rho_{n'n} = \sum_s p(s) c_{n'}^{(s)*} c_n^{(s)}$ is the matrix form of the density operator ρ in Equation 4.190. The density operator is a quantity that is independent of basis, whereas the density matrix is written here in the energy basis, as is common practice. Other bases can also be used when convenient. Equation 4.190 is more elegant and more transparent when deriving general properties of the density matrix.

Density matrices describe a quantum systems that interact with the environment, where the interaction is taken into account via the probability term $p(s)$. The equilibrium condition determines which states are populated. If $|n\rangle$ is the eigenvector of the Hamiltonian,

$$H|n\rangle = E_n |n\rangle, \qquad (4.193)$$

then any given state $|s\rangle$ can be expanded as a superposition of energy eigenvectors, $|n\rangle$,

$$|s\rangle = \sum_n a_n |n\rangle, \qquad (4.194)$$

where a_n is the *quantum* probability of finding the particle in state n and should not be confused with $p(s)$. a_n is the quantum amplitude and $p(s)$ is due to the interaction of the system with the outside world. We can calculate the quantum time evolution of state $|s\rangle$ via,

$$|s(t)\rangle = e^{-iHt/\hbar} |s\rangle = \sum_n a_n e^{-iE_n t/\hbar} |n\rangle. \qquad (4.195)$$

Density matrices can be used to calculate the expectation value of any observable, A, by taking

$$\langle A \rangle = Tr\left[\rho A\right]. \qquad (4.196)$$

Equation 4.196 takes into account both thermal and quantum effects. The meaning of Equation 4.196 becomes apparent by expressing it as

$$\begin{aligned}
Tr[\rho A] &= Tr\left[\sum_u p(u)|u\rangle\langle u|A\right] \\
&= \sum_u \sum_v \langle v|p(u)|u\rangle\langle u|A|v\rangle \\
&= \sum_{u,v} p(u)\langle u|A|v\rangle\langle v|u\rangle \\
&= \sum_u p(u)\langle A\rangle_{uu} \\
&= \langle A\rangle,
\end{aligned} \quad (4.197)$$

where we have used closure, or

$$\sum_u |u\rangle\langle u| = 1. \quad (4.198)$$

Thus, the expected value of an observable is the sum over the quantum expectation values of that observable weighted by the proportion of the population of particles that populate those states.

The index v in Equation 4.197 spans the Hilbert space, , which for a continuous variable is infinite. The dimension of the Hilbert space for a discrete systems such as spin 1/2 particles is 2: spin up, $|\uparrow\rangle$, and spin down, $|\downarrow\rangle$. The important point of the density matrix is that the sum in Equation 4.190 extends over the states of the particles in the system as it is initially prepared. All the eigenstates may not be populated or the each particle in the system can be in a different superposition eigenstates that are not orthogonal to each other. Consequently, the summation may span more or less terms that the dimension of the Hilbert space . Examples will follow.

As a first example we consider a system of a spin half particle in a uniform magnetic field, $\mathbf{B} = B_0\hat{z}$, in equilibrium with a bath at temperature T. The particle can be spin up or spin down. The energy of a particle with spin S_z is,

$$H = -\mu\cdot\mathbf{B} = -aS_z B_0, \quad (4.199)$$

where a is a constant and S_z is the spin component in z direction. The Hamiltonian in matrix form is

$$H = \begin{pmatrix} -a\hbar B_0/2 & 0 \\ 0 & a\hbar B_0/2 \end{pmatrix}, \quad (4.200)$$

and the probabilities of finding finding a particle with spin $\pm\hbar$ is,

$$p_\pm = \frac{e^{\pm a\hbar B_0/2kT}}{e^{a\hbar B_0/2kT} + e^{-a\hbar B_0/2kT}}. \quad (4.201)$$

Inserting Equation 4.201 in Equation 4.190 yields the density matrix,

$$\begin{aligned}\rho &= p_+|+\rangle\langle+| + p_-|-\rangle\langle-| \\ &= \begin{pmatrix} p_+ & 0 \\ 0 & p_- \end{pmatrix}. \end{aligned} \quad (4.202)$$

Now, using Equations 4.200 and 4.202 in Equation 4.196, the expectation value of the Hamiltonian can be calculated,

$$\langle H \rangle = Tr[H\rho] = -\frac{a\hbar B_0}{2}(p_+ - p_-) = -\frac{a\hbar B_0}{2}\tanh\frac{a\hbar B_0}{2kT}. \quad (4.203)$$

When the system is at zero temperature, the population is in its ground state, so Equation 4.203 yields $\langle H \rangle = -a\hbar B_0/2$. Similarly, when the temperature becomes infinite, $\langle H \rangle = 0$. This makes sense because at infinite temperature both states are equally populated so that the average of the energies is zero.

As a second example, we describe a situation in which the number of states exceeds the dimension of the Hilbert space. Consider,

$$p(|+\rangle) = \frac{1}{4}, \quad p\left(\frac{|+\rangle + |-\rangle}{\sqrt{2}}\right) = \frac{1}{4}, \quad p(|-\rangle) = \frac{1}{2}. \quad (4.204)$$

So, there are 3 states in the population, which is larger than the dimensionality of the Hilbert space, which is 2. The density matrix is then given by

$$\begin{aligned}\rho &= \frac{1}{4}|+\rangle\langle+| + \frac{1}{4}\left(\frac{|+\rangle + |-\rangle}{\sqrt{2}}\right)\left(\frac{\langle+| + \langle-|}{\sqrt{2}}\right) + \frac{1}{2}|-\rangle\langle-| \\ &= \begin{pmatrix} 3/8 & 1/8 \\ 1/8 & 5/8 \end{pmatrix}. \end{aligned} \quad (4.205)$$

The off-diagonal terms in the density matrix come from the quantum superposition of states. They are called coherence terms because they are non-zero only when the wave function is composed of a superposition of states. The trace of the density matrix given by Equation 4.205 is equal to unity. This is a general characteristic of density matrices.

For pure states, represented by a single vector in Hilbert space, the density matrix can be written as

$$\rho = |s(t)\rangle\langle s(t)|. \quad (4.206)$$

In this case,

$$\rho = \rho^\dagger, \quad (4.207)$$

so a pure state density matrix must be Hermitian. Additionally, since the wavefunction is normalized, $\rho^2 = |s(t)\rangle \langle s(t)| \cdot |s(t)\rangle \langle s(t)| = |s(t)\rangle \langle s(t)| = \rho$. These two properties make the density matrix for a pure state a projector.[2]

As an example, for a spin 1/2 particle, the most general real state vector is

$$|s\rangle = \cos(\theta)|+\rangle + \sin(\theta)|-\rangle, \qquad (4.208)$$

whence

$$\rho = \begin{pmatrix} \cos^2\theta & \sin\theta\cos\theta \\ \sin\theta\cos\theta & \sin^2\theta \end{pmatrix}. \qquad (4.209)$$

It is straightforward to show that this density matrix is a projector.

The time evolution of the density matrix operator can be studied in the Heisenberg representation, which for any operator A, yields

$$\frac{dA}{dt} = \frac{\partial A}{\partial t} - \frac{i}{\hbar}[A, H]. \qquad (4.210)$$

For the density matrix operator,

$$\frac{d\rho}{dt} = -\frac{i}{\hbar}[H, \rho], \qquad (4.211)$$

where we assume that the density matrix does not depend explicitly on time. The evolution of each component of ρ is given by

$$\frac{d\rho_{nm}}{dt} = -\frac{i}{\hbar}\langle n|H\rho - \rho H|m\rangle = -\frac{i}{\hbar}(E_n - E_m)\rho_{nm}, \qquad (4.212)$$

so, solving Equation 4.212, we see that the time evolution of the density matrix is given then by

$$\rho_{nm} = \rho_{nm}^{(0)} e^{-iE_{nm}t/\hbar}, \qquad (4.213)$$

where $\rho_{nm}^{(0)}$ is the density matrix at $t = 0$.

In matrix form, the expectation value of an operator A, $\langle A \rangle$, is calculated from the density matrix according to

$$\langle A \rangle = \sum_{n,m} \rho_{nm} A_{mn}. \qquad (4.214)$$

Using Equation 4.213, the time-dependent expectation value of the off-diagonal components, $n \neq m$, are found to oscillate at the Bohr frequency. For diagonal terms, $n = m$, the expectation value is time independent.

---

[2] see any textbook on quantum mechanics.

4.9.1 Phenomenological Model of Damping

When a system interacts with the environment, it will settle down into a steady state where the populations are described by the equilibrium density matrix. When the system interacts with light, it will be knocked out of equilibrium, then will relax back to equilibrium at a rate given by a phenomenological parameter γ. When the system is not in equilibrium, the time evolution of the density matrix can be modeled by

$$\frac{d\rho_{nm}}{dt} = -\frac{i}{\hbar}[H,\rho]_{nm} - \gamma_{nm}\left(\rho_{nm} - \rho_{nm}^{eq}\right), \qquad (4.215)$$

where ρ_{nm}^{eq} is the equilibrium density matrix and the second term (corresponding to $\partial\rho/\partial t$), represents the relaxation rate back to equilibrium. Since γ represents the decay rate, it is a real quantity, which implies that $\gamma_{nm} = \gamma_{mn}$. Thus, an underpopulated (depleted) state's population will return back to equilibrium at the same rate as it decays when overpopulated.

In general, a system can be in a superposition of states; but, in thermal equilibrium, the populations are given by $\rho(E_n) \propto e^{-\beta E_n}$. Thus, thermal agitation can excite population, but will not on average lead to a superposition. Consequently, coherence is lost and the off-diagonal elements vanish, or $\rho_{nm}^{eq} \propto \delta_{n,m} e^{-E_n/kT}$. The off-diagonals elements of γ_{nm} therefore give the rates at which coherence between states n and m if the system is prepared in a superposition of states. The diagonal terms describe the changing populations.

An alternative way to model time evolution of the diagonal elements ρ_{nn} is to determine the rates at which the populations are exchanged between states. The phenomenological model then takes the form

$$\dot{\rho}_{nn} = -\frac{i}{\hbar}[H,\rho]_{nn} + \sum_{m>n}\Gamma_{nm}\rho_{mm} - \sum_{m<n}\Gamma_{mn}\rho_{nn}. \qquad (4.216)$$

Γ_{mn} for terms with $m > n$ represents the decay rate from higher-energy states to state n and $n > m$ gives the decay rate out of state n to lower-energy states. Thus, the sum over m includes all states that are a source of decays into state n (second term on the right-hand side of Equation 4.216), and all the states to which state n decays (the last term in Equation 4.216).

Next we show how γ in Equation 4.215 is related to Γ in Equation 4.216. Consider a pure state, $|\psi(t)\rangle$, for which $\rho = |\psi(t)\rangle\langle\psi(t)|$. Its time dependence is given by

$$|\psi(t)\rangle = e^{-iHt/\hbar}\sum_n a_n|n\rangle = \sum_n e^{-iE_n t/\hbar}a_n|n\rangle. \qquad (4.217)$$

Nonlinear Optics: A student's perspective

To include the decay of population, we can add a small imaginary part to the energy eigenstates given by Equation 4.188, so the density matrix becomes

$$\rho = \sum_{nm} a_m a_n^* e^{-i\omega_{mn} t} e^{-(\Gamma_m + \Gamma_n)t/\hbar} |m\rangle \langle n|, \quad (4.218)$$

whence

$$\rho_{nn} = |a_n|^2 e^{-\Gamma_n t}, \quad (4.219)$$

which indicates that population is leaving state n to lower states at a decay rate Γ_n. Taking the time derivative of Equation 4.219 yields

$$\dot{\rho}_{nn} = -\Gamma_n \rho_{nn}. \quad (4.220)$$

Since Γ_n here is the decay from state n, comparing Equation 4.220 with Equation 4.216 leads to the conclusion that

$$\Gamma_n = \sum_{m<n} \Gamma_{mn}. \quad (4.221)$$

On the other hand, associating the coefficient of $|m\rangle \langle n|$ in Equation 4.218 with ρ_{mn}, both $\sum_{m<n} \Gamma_{nm} \rho_{nn}$ and $\gamma_{nm}\left(\rho_{mm} - \rho_{nn}^{equil.}\right)$ contribute to ρ_{nm}. Thus

$$\gamma_{nm} = \frac{1}{2}(\Gamma_n + \Gamma_m) + \gamma_{nm}^{col}, \quad (4.222)$$

where the last term is due to collisional dephasing, where collisions between molecules washes out coherence but does not result in a change of population.

To illustrate the use of the density matrix, consider a two-level atom with dipole operator μ in the matrix form given by

$$\mu = \begin{pmatrix} 0 & \mu_{ab} \\ \mu_{ba} & 0 \end{pmatrix}, \quad (4.223)$$

where a and b are the ground and excited states. The diagonals vanish because atoms are centrosymmetric and have no dipole moment. Since the dipole moment operator is Hermitian, $\mu_{ab} = \mu_{ba}^*$. The expectation value of this dipole operator is then given by

$$\langle \mu \rangle = Tr[\rho\mu] = Tr\left[\begin{pmatrix} \rho_{aa} & \rho_{ab} \\ \rho_{ba} & \rho_{bb} \end{pmatrix}\begin{pmatrix} 0 & \mu_{ab} \\ \mu_{ab}^* & 0 \end{pmatrix}\right] = \rho_{ab}\mu_{ab}^* + \rho_{ab}^*\mu_{ab}. \quad (4.224)$$

Equation 4.224 indicates that a centrosymmetric atom has a dipole moment that is induced when it is in contact with a thermal bath.

4.10 Calculating the Polarizability with the Density Matrix

The polarizability and hyperpolarizabilities can be calculated with density matrices using perturbation theory. We illustrate the method for the polarizability. The more complex nonlinear case follows the same procedure, so will not be presented here.

As we did for the Schrädinger Equation, we start with the Hamiltonian

$$H = H_0 + \lambda V(t), \qquad (4.225)$$

and treat the light coupling term $V(t)$ as a perturbation. Equation 4.215 then gives the time evolution of the density matrix, yielding

$$\dot{\rho}_{nm} = -\frac{i}{\hbar}(E_n - E_m)\rho_{nm} - \frac{i}{\hbar}[V(t), \rho]_{nm} - \gamma_{nm}\left(\rho_{nm} - \rho_{nm}^{eq}\right). \qquad (4.226)$$

The density matrix can be expanded in a series of the order of the perturbation

$$\rho = \rho^{(0)} + \lambda \rho^{(1)} + \cdots + \lambda^n \rho^{(n)} + \cdots. \qquad (4.227)$$

Keeping the zeroth order terms of Equation 4.226 after substitution of Equation 4.227 leads to

$$\dot{\rho}_{nm}^{(0)} = -i\omega_{nm}\rho_{nm}^{(0)} - \gamma_{nm}\left(\rho_{nm}^{(0)} - \rho_{nm}^{eq}\right), \qquad (4.228)$$

and to first order in λ gives

$$\begin{aligned}\dot{\rho}_{nm}^{(1)} &= -i\omega_{nm}\rho_{nm}^{(1)} - \frac{i}{\hbar}\left[V(t), \rho^{(0)}\right]_{nm} - \gamma_{nm}\rho_{nm}^{(1)} \\ &= -(i\omega_{nm} + \gamma_{nm})\rho_{nm}^{(1)} - \frac{i}{\hbar}\left[\sum_p \left(V_{np}\rho_{pm}^{(0)} - \rho_{np}^{(0)}V_{pm}\right)\right],\end{aligned} \qquad (4.229)$$

where we have used closure, which is given by Equation 4.198.

As we previously saw, thermal fluctuations wash out the zeroth-order off-diagonal elements of the density matrix, so only the diagonal terms in the summation in Equation 4.229 remain. Thus

$$\dot{\rho}_{nm}^{(1)} = -(i\omega_{nm} + \gamma_{nm})\rho_{nm}^{(1)} - \frac{i}{\hbar}V_{nm}\left(\rho_{mm}^{(0)} - \rho_{nn}^{(0)}\right), \qquad (4.230)$$

where the last term vanishes for $n = m$. With no perturbation, $V(t) = 0$ and by definition $\rho_{nm} = \rho_{nm}^{eq}$, which also vanishes when $n \neq m$. This implies that

$$\dot{\rho}_{nm}^{(0)} = 0 \quad \text{for} \quad n \neq m, \qquad (4.231)$$

and

$$\dot{\rho}_{nn}^{(0)} = 0 \quad \text{for} \quad n = m, \tag{4.232}$$

which indicates that both diagonal and off-diagonal components are consistent with the zeroth order result.

Rearranging Equation 4.229, we get the first order equation

$$\dot{\rho}_{nm}^{(1)} + (i\omega_{nm} + \gamma_{nm})\rho_{nm}^{(1)} = -\frac{i}{\hbar}V_{nm}\left(\rho_{mm}^{(0)} - \rho_{nn}^{(0)}\right), \tag{4.233}$$

which can be solved using Green's function as follows. Defining the operator $\hat{O}_t$ as

$$\hat{O}_t = \frac{\partial}{\partial t} + i\omega_{nm} + \gamma_{nm}, \tag{4.234}$$

Equation 4.233 can be written in the form

$$\hat{O}_t \rho_{nm}^{(1)} = f(t). \tag{4.235}$$

The general solution of Equation 4.235 is the homogeneous solution of $\rho_{nm,hom}^{(1)}$, for which $f(t) = 0$, plus the inhomogeneous solution, where $f(t) \neq 0$, so

$$\rho_{nm}^{(1)} = \rho_{nm,hom}^{(1)} + \rho_{nm,inhom}^{(1)}. \tag{4.236}$$

We can easily find a solution for the homogeneous equation, yielding

$$\rho_{nm,hom}^{(1)} = Ae^{(-i\omega_{nm} + \gamma_{nm})t}, \tag{4.237}$$

where A is the integration constant. To solve the inhomogeneous equation, we define a Green's function, $G(t - t')$, so that,

$$\hat{O}_t G(t - t') = \delta(t - t'). \tag{4.238}$$

Once we have the Green's function, then we can find the inhomogeneous solution for the density matrix by simple integration

$$\rho_{nm,inhom}^{(1)}(t) = \int_{-\infty}^{\infty} G(t - t')f(t')dt'. \tag{4.239}$$

As a check, we can operate with $\hat{O}_t$ on Equation 4.239 with the use of Equation 4.238, yielding

$$\hat{O}_t \rho_{nm,inhom}^{(1)}(t) = \int_{-\infty}^{\infty} f(t')\hat{O}_t G(t - t')dt' = f(t), \checkmark \tag{4.240}$$

where we have used the fact that

$$\int f(t')\delta(t'-t)dt' = f(t). \qquad (4.241)$$

To solve a particular differential equation requires a Green's function that can be constructed from a superposition of the homogeneous solutions. For a first order differential equation, the Green's function is simply a step function. We construct the Green's function by setting $A = 0$ for $t' < t$ and $A = 1$ in Equation 4.237 when $t' > t$. These two regions are connected with a unit step where its derivative is a delta function. Thus

$$G(t'-t) = \begin{cases} 0 & t'-t < 0 \\ \exp\left[-(i\omega_{nm}+\gamma_{nm})t\right] & t'-t > 0. \end{cases} \qquad (4.242)$$

Equivalently,

$$G(t'-t) = \Theta(t'-t)e^{-(i\omega_{nm}+\gamma_{nm})t}, \qquad (4.243)$$

where $\Theta(t'-t)$ is a step function, with $\Theta(t'-t) = 0$ for $t'-t < 0$ and $\Theta(t'-t) = 1$ when $t'-t \geq 1$. Using Equation 4.243, the solution for $\rho_{nm}^{(1)}(t)$ in Equation 4.233 is obtained,

$$\rho_{nm}^{(1)}(t) = \frac{-i}{\hbar}\left[\int_t^\infty dt' e^{-(i\omega_{nm}+\gamma_{nm})t}V_{nm}\left(\rho_{mm}^{(0)}-\rho_{nn}^{(0)}\right)\right] + Ae^{-(i\omega_{nm}+\gamma_{nm})t}. \qquad (4.244)$$

The expectation value of the dipole moment with

$$V = -\mu \cdot \mathbf{E} = -\mu \cdot \frac{\mathbf{E_0}}{2}e^{-i\omega_p t} + \text{c.c.}, \qquad (4.245)$$

and the first order density matrix

$$\rho_{nm} = \rho_{nm}^{(0)} + \lambda \rho_{nm}^{(1)}. \qquad (4.246)$$

eventually leads to the susceptibility

$$\chi_{ij}^{(1)}(-\omega_p;\omega_p) = \frac{N}{\hbar}\sum_{n,m}\frac{\mu_{mn}^i \mu_{nm}^j}{\omega_{nm}-\omega_p-i\gamma_{nm}}\left(\rho_{mm}^{(0)}-\rho_{nn}^{(0)}\right). \qquad (4.247)$$

Manipulating the index in Equation 4.247 gives an expression for $\chi^{(1)}$, which is similar to what we previously had, or

$$\chi_{ij}^{(1)}(-\omega_p;\omega_p) = \frac{N}{\hbar}\sum_{n,m}\rho_{mm}^{(0)}\left[\frac{\mu_{mn}^i \mu_{nm}^j}{\omega_{nm}-\omega_p-i\gamma_{nm}} + \frac{\mu_{nm}^i \mu_{mn}^j}{\omega_{nm}+\omega_p+i\gamma_{nm}}\right]. \qquad (4.248)$$

If $\rho_{mm} = \delta_{m0}$, the expression for $\chi^{(1)}$ in Equation 4.248 is the same as the results for vacuum, given by Equation 4.168, where the molecule is originally in its ground state. However, in reality, the system can start from any other state. Therefore, $\rho_{mm}^{(0)}$ is a weighting factor that describes the distribution of molecules in different initial states.

4.11 Parity

In this section we use parity to characterize the symmetry of the Hamiltonian and investigate its effect on the nonlinear response. For a Hamiltonian given by

$$H = \frac{\mathbf{p}^2}{2m} + V(\mathbf{r}), \tag{4.249}$$

the energy eignestates $|n\rangle$ are solutions of

$$H|n\rangle = E_n |n\rangle. \tag{4.250}$$

The parity operator π acts to invert the coordinate system, so $\mathbf{r} \to -\mathbf{r}$ and $\mathbf{p} \to -\mathbf{p}$. Formally,

$$\pi \mathbf{r} \pi^{-1} = -\mathbf{r} \quad \text{and} \quad \pi \mathbf{p} \pi^{-1} = -\mathbf{p}. \tag{4.251}$$

When the potential is symmetric,

$$\pi V(\mathbf{r}) \pi^{-1} = V(-\mathbf{r}) = V(\mathbf{r}), \tag{4.252}$$

then

$$\pi H(\mathbf{r}) \pi^{-1} = \frac{-\mathbf{p}^2}{2m} + V(-\mathbf{r}) = \frac{\mathbf{p}^2}{2m} + V(\mathbf{r}) = H, \tag{4.253}$$

so the parity operator π commutes with the Hamiltonian, or

$$[H, \pi] = 0. \tag{4.254}$$

The parity operator is thus a simultaneous observable with the Hamiltonian, so the energy eigenvectors are also parity eigenvectors. Since clearly $\pi\pi = 1$, the parity eigenvalues are ± 1, yielding

$$\pi |n\rangle = \pm |n\rangle. \tag{4.255}$$

As a result, the wave function can be characterized as either odd $|o_i\rangle$ or even $|e_i\rangle$ wave functions,

$$\begin{aligned} \pi |o_i\rangle &= -|o_i\rangle, \\ \pi |e_i\rangle &= +|e_i\rangle. \end{aligned} \tag{4.256}$$

In position space,

$$\langle \mathbf{x} | \pi | o_i \rangle = \langle -\mathbf{x} | o_i \rangle = -\langle \mathbf{x} | o_i \rangle, \tag{4.257}$$

which implies that the odd energy eigenfunctions are spatially asymmetric, or $\psi_{oi}(-\mathbf{x}) = -\psi_{oi}(\mathbf{x})$. The even functions, on the other hand, remain unchanged under the parity operation.

We can use the wave functions of a parity-invariant Hamiltonian to determine the first order susceptibility. Recall that the first-order susceptibility is given by

$$\chi^{(1)}(\omega_p) = Ne^2 \sum_n{}' |x_{0n}|^2 \left[\frac{1}{(E_{n0} - \hbar\omega_p)} + \frac{1}{(E^*_{n0} + \hbar\omega_p)} \right], \qquad (4.258)$$

so the strength of the contribution to state n is given by the magnitude of $|x_{0n}|^2$. This matrix element can be expressed as

$$\begin{aligned} x_{0n} = \langle 0|x|n\rangle &= \langle 0|\mathbb{1}x\mathbb{1}|n\rangle = \langle 0|\pi^{-1}\pi x \pi^{-1}\pi|n\rangle = -\langle 0|\pi^{-1}x\pi|n\rangle \\ &= -p_0 p_n \langle 0|x|n\rangle, \end{aligned} \qquad (4.259)$$

where p_0 and p_n are the parities of states 0 and n. Equation 4.259 implies that $p_0 p_n = -1$ if the matrix element is nonzero, or that the two states must be of opposite parity.

According to Equation 4.258 the linear susceptibility shows a resonance when the photon energy matches the transition energy, which results in a peak in the linear absorption spectrum at frequency $E_{i0}/\hbar$. Usually, the ground state of a molecule whose electrons are bound with a symmetric potential is of even parity. In this case only transitions to odd-parity excited states are allowed. These states are called one-photon states. On the other hand the even-parity states have zero transition moment from the even-parity ground state, thus there is no resonance. If a molecule had only one excited state, and if it was the same parity as the ground state, it would be invisible under dipole transitions.

States of the same parity as the ground state are called two-photon states because a transition from the ground state to a two photon state can only be reached by a two-step process involving two or more photons, with two-photon absorption being the lowest-order process that is symmetry-allowed. As we will see later, transitions to two-photon states are observed in $\chi^{(3)}$ processes.

The contribution to the second order susceptibility from states 0 and 1, $\chi^{(2)}_{2L}$, is given by

$$\chi^{(2)}_{2L} \propto Ne^2 |x_{01}|^2 \Delta x_{10}, \qquad (4.260)$$

in which case both $|x_{01}|^2$ and Δx_{10} must be nonzero if state 1 is to contribute. For a centrosymmetric potential, when the first excited state is odd, $|x_{01}|^2 =$

$|\langle e_0|x|o_1\rangle|^2 \neq 0$. However, $\Delta x_{10} = x_{oo} - x_{ee} = \langle e|x|e\rangle - \langle o|x|o\rangle = 0$ because the ground and exited state dipole moments each vanish given that the parity of the states are necessarily the same, and this results in $\chi^{(2)}_{2L} = 0$.

In the more general case that includes 3-level terms,

$$\chi^{(2)} \propto Ne^2 \sum_n |x_{0n}|^2 \Delta x_{n0} + Ne^2 \sum_{n,m} x_{0n} x_{nm} x_{m0}, \qquad (4.261)$$

where the first sum vanishes term-by-term according to the same argument as above. And $x_{0n} x_{nm} x_{m0}$ in the second term is always zero due to the fact that two of the three excited states are either both even or odd, which forces one of the matrix elements to vanish. Therefore, $\chi^{(2)}$ is always zero for a centrosymmetric potential, which we have already seen is required from general arguments.

In a two-level model, the third order susceptibility is of the form

$$\chi^{(3)}_{2L} \propto -|x_{01}|^2(|x_{01}|^2 - |\Delta x_{10}|^2). \qquad (4.262)$$

Since the dipole moment is zero for a centrosymmetric potential, then $\chi^{(3)}_{2L} \propto |x_{01}|^4$. As we saw above, the transition moment to an excited state of opposite parity is nonzero. Thus, one strategy to make a good $\chi^{(3)}$ material is to use a centrosymmetric material that has a large value of $\chi^{(1)}$.

For potentials with no symmetry, we can always express any wave function as a superposition of even- and odd-symmetry functions. In general we can write the nth state vector as

$$|n\rangle = \sin \alpha_n |n_e\rangle + \cos \alpha_n |n_o\rangle, \qquad (4.263)$$

where $\sin \alpha_n$ and $\cos \alpha_n$ are normalization coefficients. Thus, the susceptibilities for asymmetric potentials can be expressed in terms of the the phase factors α_n, which describe how close a state is to being of even parity ($\alpha_n = \pi/2$) or odd parity ($\alpha_n = 0$).

It is instructive to use a two-level model to study parity. In this case there are two basis vectors, which each can be expressed as a superposition of a symmetric and antisymmetric part. If the state vectors are to be orthogonal, they need to be of the form

$$|0\rangle = \sin \alpha |e\rangle + \cos \alpha |o\rangle \qquad (4.264)$$

and

$$|1\rangle = -\cos \alpha |e\rangle + \sin \alpha |o\rangle. \qquad (4.265)$$

The process by which the separation into even and odd parity states is implemented can be understood by considering the wavefunctions, which can be represented by a Fourier series. The even- and odd-parity parts of the wavefunctions are determined simply by summing the even-parity terms (i.e. the cosines) to get the even-parity contribution and summing the odd-parity terms (i.e. the sines) to get the odd-parity contributions.

We can get insights into the role of parity yet avoid unimportant complications by assuming that $\langle e|x|o\rangle$ is real, or $\langle e|x|o\rangle = \langle o|x|e\rangle$. Then, the first-order susceptibility – which is proportional to $|x_{01}|^2$, is found using Equations 4.264 and 4.265, yielding

$$\chi^{(1)} \propto |x_{01}|^2 = 4|\langle e|x|o\rangle|^2 \cos^2 2\alpha. \tag{4.266}$$

$\chi^{(1)}$ is thus maximum when $\alpha = n\pi/2$, where n is an integer. This makes good sense since $\alpha = \pi/2$ corresponds to a ground state that is symmetric (or even parity) and the excited state is of odd parity. When $\alpha = \pi$, the ground state is of odd parity and the excited state is of even parity. In all cases, the first-order susceptibility is maximum when the ground and excited state are of opposite parity. Thus, the Hamiltonian must commute with the parity operator if it is to yield the maximum $\chi^{(1)}$.

On the other hand, the first-order susceptibility vanishes when $\cos^2 2\alpha = 0$, or $\alpha = (2n+1)\pi/4$ for integer n. In this case, both the ground and excited states are made of equal parts of even- and odd-parity state vectors. Thus, potentials optimized for $\chi^{(1)}$ are centrosymmetric while highly asymmetric potentials that give an even mixture of even and odd character yields small $\chi^{(1)}$.

$\chi^{(2)}$ for a two-level system is a function of both $|x_{01}|^2$ and $\Delta x_{10} = x_{11} - x_{00}$. Using Equations 4.264 and 4.265, the ground and excited state dipole moments are proportional to the diagonal position matrix elements

$$x_{00} = \sin 2\alpha \langle e|x|o\rangle \tag{4.267}$$

and

$$x_{11} = -\sin 2\alpha \langle e|x|o\rangle, \tag{4.268}$$

where again we have assumed that $\langle e|x|o\rangle$ is real. Equations 4.267 and 4.268 give

$$\Delta x_{10} = x_{11} - x_{00} = 2\sin 2\alpha \langle e|x|o\rangle. \tag{4.269}$$

The two-state model of the second-order susceptibility, given by Equation 4.260 with the help of Equations 4.266 and 4.269, is of the form

$$\chi^{(2)} \propto \langle e|x|o\rangle \sin 2\alpha \cdot \cos^2 2\alpha. \tag{4.270}$$

We can find the maximum of Equation 4.270 by differentiating it with respect to α and finding the zeros, which gives

$$\alpha = \frac{1}{2}\sin^{-1}\left(\pm\sqrt{\frac{1}{3}}\right). \tag{4.271}$$

The second-order susceptibility can be positive or negative, so the positive and negative angles in Equation 4.271 correspond to maximum positive and negative second-order susceptibilities. $\chi^{(2)}$ vanishes when $\alpha = n\pi/4$, where n even is the case of an asymmetric molecule and n odd is for a molecule with an equal mixture of both parities. Thus, the optimum is at a point in between these two extremes.

The three-level model of the third-order susceptibility is given by Equation 4.262. It is straightforward to show that $\chi^{(3)}$ is at its negative maximum when the potential is centrosymmetric and $\alpha = n\pi/2$. Odd n corresponds to an even-parity ground state and odd-parity excited state while even n gives an odd-parity ground state and even-parity excited state. The third-order susceptibility vanishes when the ground state and excited state wavefunctions are a superposition of equal parts of even and odd-parity states. A positive peak is also found and is the topic of the homework problem.

> **Problem 4.11-1(a):** Evaluate the two-state model for the third-order susceptibility using the state vectors given by Equations 4.264 and 4.265 and determine which values of α give extrema. Be sure to sketch the third-order susceptibility as a function of α.
>
> **(b):** Determine the ratio of the largest positive and negative susceptibilities and describe what proportion of the state vectors are of even or odd parity in each case.

4.12 Sum Rules

Now we demonstrate a simple approach to understanding the magnitude of the nonlinear response using the broadest of fundamental principles. To optimize a nonlinear response, one has to optimize an infinite number of material parameters that appear in infinite sums - a difficult if not impossible task. If a quantum system can be approximated by a two-level model, then a deeply-colored molecule, which necessarily has large $|x_{01}|$, will have a large linear susceptibility.

We can calculate the upper limit of the first order susceptibility using the sum rules. The sum rules, which are derived directly from the Schrödinger equation without approximation, state that

$$\sum_n |x_{n0}|^2 E_{n0} = N_{el} \frac{\hbar^2}{2m}, \qquad (4.272)$$

where N_{el} is the number of electrons in the system and m is the mass of the electron. First we simplify Equation 4.258 by assuming that the light is in the off-resonant regime, so that we can approximate $\omega_p \approx 0$. Assuming $E_{n0} \sim E_{n0}^*$, yields

$$\chi^{(1)} = Ne^2 \sum_n \frac{2|x_{0n}|^2}{E_{n0}}, \qquad (4.273)$$

where N is the number density of molecules.

All terms in the sum in Equation 4.273 are zero or positive and the energy denominator insures that each term, on average, gets smaller with increasing n. Multiplying each term in Equation 4.273 by E_{n0}/E_{n0} we get

$$\chi^{(1)} = Ne^2 \sum_n \frac{2|x_{0n}|^2 E_{n0}}{E_{n0}^2}. \qquad (4.274)$$

Using the fact that $1/E_{n0} \leq 1/E_{10}$, we can replace E_{n0} by E_{10} in the denominator, leading to

$$\chi^{(1)} \leq Ne^2 \sum_n \frac{2|x_{0n}|^2 E_{n0}}{E_{10}^2}. \qquad (4.275)$$

Factoring $1/E_{01}^2$ out from the sum yields

$$\chi^{(1)} \leq 2\frac{Ne^2}{E_{10}^2} \sum_n |x_{0n}|^2 E_{n0}, \qquad (4.276)$$

and applying the sum rules gives

$$\chi^{(1)} \leq N \cdot \frac{2e^2\hbar^2}{m} \cdot \frac{N_{el}}{E_{10}^2} \equiv \chi_{max}^{(1)}. \qquad (4.277)$$

Equation 4.277 gives the fundamental limit for $\chi^{(1)}$. More precisely, the fundamental limit is of the xx tensor component. We assume that the xx component is the largest one. If not, we can rotate the coordinate system so that it is. In this case, E_{10} corresponds to the lowest-energy excitation along x for light polarized along x. A similar derivation can be carried out for $\chi^{(2)}$ and $\chi^{(3)}$, but requires additional assumptions because the terms in the sums are not positive definite as they are for the linear susceptibility. Thus, for a given number of electrons and energy scale, there is a fundamental limit to all susceptibilities. More precisely, the fundamental limits are for the microscopic polarizability α and hyperpolarizabilities β, γ, etc. These limits can be used to assess molecules and aid in the design of better materials.

The definition of the intrinsic polarizability α_{int} given by

$$\alpha_{int} = \frac{\alpha}{\alpha_{max}} \quad (4.278)$$

facilitates the comparison between molecules. The magnitude of α_{int} does not exceed unity.

Problem 4.12-1: Use the commutator $[x,[x,H]] = -\hbar^2/m$ to derive the sum rules given by Equation 4.272. *Hint: Take matrix elements of the commutator and use closure . Do this first for one electron as practice, then generalize the result to N_{el} electrons by making the position operator $x = \sum_{i=1}^{N_{el}} x_i$ and generalizing the Hamiltonian in the same way.*

Problem 4.12-2: Show that the intrinsic polarizability of a harmonic oscillator is given by $\alpha_{int} = 1$.

Problem 4.12-3: Write a Python module that randomly picks the energies and position matrix elements x_{0n} under the constraint that the 1-electron sum rules are obeyed. Then calculate the intrinsic polarizability from these parameters. Repeat the calculation 10,000 times and plot a histogram of the intrinsic polarizabilities to verify that their magnitudes never exceed unity. This is an example of what is called a Monte Carlo calculation because of the random choices of energies and position matrix elements. Investigate how the shape of the histogram depends on the method of randomization used in your Monte Carlo algorithm.

4.13 Local Fields

Linear Local Field

We seek to determine the electric field at a particular molecule in a material that is made of an ensemble of similar molecules, such as an H_2O molecule

230

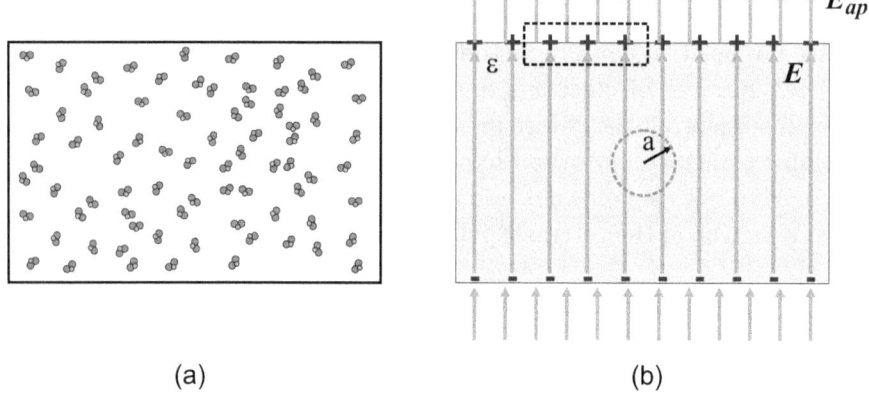

Figure 4.18: (a) A material made of molecules (b) is modeled as a uniform dielectric of permittivity ϵ. A uniform electric field is applied, which induces charges on the surfaces. The dashed circle represents a spherical volume that contains one molecule and the dashed rectangle is the side view of a Gaussian box.

in a bottle of water, to which an electric field is applied. For now, we treat the electric field as static – an approximation that can later be relaxed. At any point within the material, the total electric field $\mathbf{E}$ is the sum of the applied electric field $\mathbf{E}_{app}$ and the fields due to the induced moments (dipole, quadrupole, etc.) $\mathbf{E}_{ind}$. However, the electric field at any molecular site includes the electric field due to that molecule, which needs to be removed.

We attack the problem by making the continuum approximation, so that the molecules as shown in Figure 4.18a are approximated by a uniform dielectric, as shown in Figure 4.18b. The dielectric constant of the medium, ϵ, is thus a scalar. We will later see that this approximation can be relaxed. When a uniform electric field is applied, charges accumulate at the surface, thus partially screening the electric field inside the material.

The relationship between the applied external field $\mathbf{E}_{app}$ and the field inside the dielectric $\mathbf{E}$ is simple to determine from continuity of the electric displacement, or $\mathbf{D}_{app} = \mathbf{D}$. Given that $\mathbf{D} = \epsilon \mathbf{E}$, we get

$$\mathbf{E} = \mathbf{E}_{app}/\epsilon. \qquad (4.279)$$

Our goal is to find a relationship between the field inside the dielectric $\mathbf{E}$ and the local electric field $\mathbf{F}$ – the field that acts on the molecule.

The dashed circle in Figure 4.18b represents a spherical volume within the material that on average contains one molecule. It too is represented by

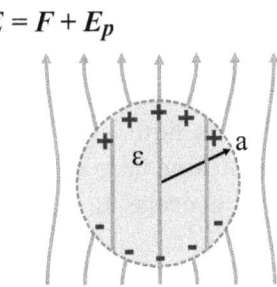

Figure 4.19: An electric field **F** applied to a dielectric sphere induces charges on its surface, leading to a dipole field **E**$_P$. The total field **E** is a sum of the applied field and the dipole field.

a smooth dielectric. The polarization within the material is given by Equation 2.77, which can be written in terms of the dielectric constant using Equation 2.79, yielding

$$\mathbf{P} = \frac{\epsilon - 1}{4\pi}\mathbf{E}. \tag{4.280}$$

Using Equation 4.280, the dipole moment of the spherical volume of radius a is

$$\mathbf{p} = \frac{4}{3}\pi a^3 \cdot \mathbf{P} = \frac{4}{3}\pi a^3 \cdot \frac{\epsilon-1}{4\pi}\mathbf{E} = \frac{\epsilon-1}{3}a^3\mathbf{E}. \tag{4.281}$$

We now apply a trick, and that is to determine the applied electric field required to induce the dipole moment on the spherical chunk of material given by Equation 4.281. That field will be the local electric field that we seek. Figure 4.19 shows the induced charge on a dielectric sphere in response to an applied uniform electric field **F**, which gives a dipole field **E**$_P$. The total field **E** is then the superposition of the applied field and dipole field

$$\mathbf{E} = \mathbf{F} + \mathbf{E}_P. \tag{4.282}$$

Applying continuity of the normal component of the electric displacement and the tangential component of the electric field at the surface of the sphere, it is straightforward to show that the induced moment is given by[10]

$$\mathbf{p} = \frac{\epsilon-1}{\epsilon+2}a^3\mathbf{F}. \tag{4.283}$$

Setting equal to each other the righthand sides of Equations 4.281 and 4.283 yields

$$\mathbf{F} = \frac{\epsilon+2}{3}\mathbf{E}. \tag{4.284}$$

This is called the Lorentz-Lorenz local field.

Before moving on, we address the objection of treating a molecule as a dielectric sphere. In the case of a liquid, the molecule is buffeted around, so after many collisions, all orientations are equally represented. Thus, even a long and thin molecule will on average be spherical. Experiments show that this approximation works well in gasses and liquids.[13]

We can picture the molecule as residing in a cavity, to which a local electric field is applied. If the field is time varying and takes the form of a plane wave of wave length λ, Equation 4.284 can be used when the cavity radius a is much smaller than the wavelength of the field ($a \ll \lambda$), as shown in Figure 4.20. Under this condition, called the dipole approximation, the field is spatially uniform in the cavity and it's amplitude is the one we calculated assuming static fields. Its time dependence is calculated simply by multiplying the static field by a time harmonic function.

The dielectric constant will then depend on the frequency of the fields and the local field is expressed as

$$\mathbf{F}(\omega) = f(\omega)\mathbf{E}(\omega), \qquad (4.285)$$

where

$$f(\omega) = \frac{\epsilon(\omega)+2}{3}. \qquad (4.286)$$

Here, $f(\omega)$ is the local field factor and in general it is a tensor when the dielectric is anisotropic. Even so, if one of the principle axes of the dielectric tensor is aligned with the applied electric field, the Lorentz-Lorenz local field model still holds, and takes the form

$$F_i = \frac{\epsilon_{ii}+2}{3}E_i, \qquad (4.287)$$

where summation notation does not apply. If none of the principle axes are aligned with the field, then the field can be projected onto the three principle axes, and the local electric field calculated for each component. The total local field is then the vector sum of the three components $\mathbf{F} = F_x\hat{x} + F_y\hat{y} + F_z\hat{z}$, where each component is given by Equation 4.287.

Local field models can take into account other effects, such as the the interaction between molecules and composite systems, such as molecules of Type A being dissolved in a solution of Molecules of Type B. The Onsager local field model takes these cases into account. Such more complex local field models will not be discussed here.

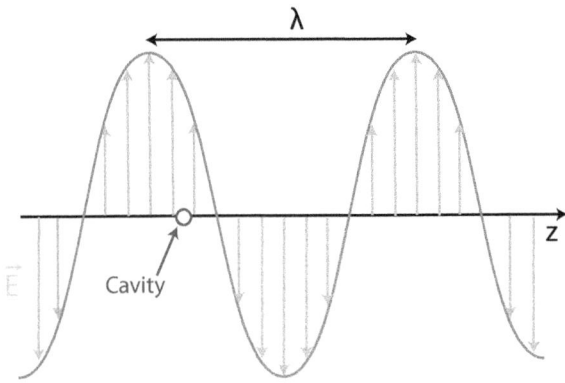

Figure 4.20: If the cavity radius is much smaller than the plane wave's wavelength, the electric field will be spatially uniform inside the cavity that contains a molecule. The Lorentz-Lorenz local field model for static fields then applies.

We can evaluate the local field factor using the relationship between the dielectric function and the refractive index $\epsilon = n^2$. For visible light in glass, for example, $n \sim 1.5$, thus $f(\omega) \sim 1.42$ and $\mathbf{F}(\omega) > \mathbf{E}(\omega)$.

In the opposite extreme is a material made of aligned one-dimensional molecules that are are rigidly held in place, as in a molecular crystal[14] or dye-doped and electric-field poled polymer[15]. These materials are called electrets because they emanate a permanent electric field in analogy to the magnetic field of a magnet. For such an anisotropic material, and ignoring the static electric field,[3] the most general form of the ith component of the local field is

$$F_i(\omega) = \sum_j f_{ij}(\omega) E_j(\omega), \qquad (4.288)$$

where $f(\omega)$ is a tensor.

To apply this model to nonlinear optics, we recall that the induced dipole moment is related to the applied local field and the vacuum hyperpolarizability

$$p^{(n)}_{i'}(\omega_\sigma) = \zeta^{(n)}_{i'jk\ldots}(-\omega_\sigma; \omega_1, \omega_2, \cdots) F_j(\omega_1) F_k(\omega_2) \cdots, \qquad (4.289)$$

where $\zeta^{(n)}_{i'jk\ldots}$ is the nth order molecular susceptibility in vacuum. Substituting

---
[3] we are only interested in changes of the local field in response to an applied electric field

Equation 4.288 into Equation 4.289 yields

$$p^{(n)}_{i'}(\omega_\sigma) = \zeta^{(n)}_{i'jk\ldots}(-\omega_\sigma;\omega_1,\omega_2,\cdots)f_{jj'}(\omega_1)f_{kk'}(\omega_2)\cdots$$
$$\times E_{j'}(\omega_1)E_{k'}(\omega_2)\cdots, \qquad (4.290)$$

where summation convention holds. Multiplying both sides by $f_{ii'}(\omega_\sigma)$, and summing over i' gives

$$f_{ii'}(\omega_\sigma)p^{(n)}_{i'}(\omega_\sigma) = \zeta^{(n)}_{i'jk\ldots}(-\omega_\sigma;\omega_1,\omega_2,\cdots)f_{ii'}(\omega_\sigma)$$
$$\times f_{jj'}(\omega_1)f_{kk'}(\omega_2)\cdots E_{j'}(\omega_1)E_{k'}(\omega_2)\cdots. \qquad (4.291)$$

Defining $p^{(n)*}_{i'}(\omega_\sigma)$ as the dressed dipole moment, and $\zeta^{(n)*}_{i'jk\ldots}(-\omega_\sigma;\omega_1,\omega_2,\cdots)$ as the dressed n^{th} order nonlinear molecular susceptibility, Equation 4.291 becomes

$$p^{(n)*}_{i'}(\omega_\sigma) = f_{ii'}(\omega_\sigma)p^{(n)}_{i'}(\omega_\sigma),$$
$$\zeta^{(n)*}_{i'jk\ldots}(-\omega_\sigma;\omega_1,\omega_2,\cdots) = \zeta^{(n)}_{i'jk\ldots}(-\omega_\sigma;\omega_1,\omega_2,\cdots)$$
$$\times f_{ii'}(\omega_\sigma)f_{jj'}(\omega_1)f_{kk'}(\omega_2)\cdots. \qquad (4.292)$$

The dressed polarization can thus be expressed in terms of the dressed molecular susceptibility and the applied fields,

$$p^{(n)*}_{i'}(\omega_\sigma) = \zeta^{(n)*}_{i'jk\ldots}(-\omega_\sigma;\omega_1,\omega_2,\cdots)E_{j'}(\omega_1)E_{k'}(\omega_2)\cdots. \qquad (4.293)$$

Equation 4.293 as derived above includes an implicit assumption that the dielectric constant is a function of only the linear susceptibility $\chi^{(1)}$, which is a good approximation but not precisely true. In particular, if the local electric field is due only to the linear response, then Equation 4.293 is exact for any nonlinearity. However, as given by Equation 4.304 and later derived, the local field factor of the linear susceptibility is not squared as implied by Equation 4.292.

The dressed susceptibilities are the quantities that are measured by experiments and local field models are used to relate them to vacuum values. These vacuum values, in turn, can be calculated from quantum theory as described in this chapter. The polarization, the bulk observable, is the sum over the dressed dipole moments of the molecules in a fixed volume

$$P^{(n)}_i(\omega_\sigma) = \frac{1}{V}\sum p^{(n)*}_i(\omega_\sigma). \qquad (4.294)$$

Since

$$P_i^{(n)}(\omega_\sigma) = \chi_{ijk\cdots}^{(n)}(-\omega_\sigma;\omega_1,\omega_2,\cdots)E_j(\omega_1)E_k(\omega_2)\cdots, \qquad (4.295)$$

substituting Equation 4.293 into Equation 4.294 and then substituting the result into Equation 4.295 yields

$$\chi_{ijk\cdots}^{(n)}(-\omega_\sigma;\omega_1,\omega_2,\cdots) = \frac{N}{V}\zeta_{ijk\cdots}^{(n)*}(-\omega_\sigma;\omega_1,\omega_2,\cdots), \qquad (4.296)$$

where N is the number of dipoles per volume. We have assumed that all the dipoles are aligned in taking the sum.

If the molecules are not aligned, then $\zeta_{ijk\cdots}^{(n)*}$ in Equation 4.296 is replaced by the orientational average of the molecular susceptibility $\langle\zeta_{ijk\cdots}^{(n)*}\rangle$, which is calculated using Equation 4.294. Thus, Equation 4.296 becomes

$$\chi_{ijk\cdots}^{(n)}(-\omega_\sigma;\omega_1,\omega_2,\cdots) = \frac{N}{V}\langle\zeta^{(n)*}(-\omega_\sigma;\omega_1,\omega_2,\cdots)\rangle_{ijk\cdots}, \qquad (4.297)$$

where the orientational averaging can be implemented using an orientational distribution function.[16] Note that $\mathscr{L}$ is sometimes used to express the local field tensor, which is of the form

$$\mathscr{L}_{ii'jj'kk'\cdots}^{(n)}(-\omega_\sigma;\omega_1,\omega_2,\cdots) = f_{ii'}(\omega_\sigma)f_{jj'}(\omega_1)f_{kk'}(\omega_2)\cdots. \qquad (4.298)$$

The local field used for the third-order susceptibility $\chi^{(3)}$ is proportional to f^4. The Lorentz-Lorenz field factor for visible light in glass is $f \simeq 1.42$, leading to a correction factor of $f^4 \approx 4$. Therefore, the local field factor results in an enhancement of the bulk nonlinear susceptibility relative to the sum over the microscopic units. Local fields can therefore be used to enhance the nonlinear optical response. For example, metal nanoparticles can act as strong electric field intensifiers that can theoretically lead to an enhancement of 8 orders of magnitude when using silver spheres at the surface plasmon resonance. Surface plasmons are discussed in Section 4.14.

The local electric field will depend on the shape of the cavity used in the calculation, as you will show in Problem 4.13-2. Since the shape of the cavity used in the derivations is a computational tool, not a real physical cavity, it is peculiar that the result should depend on the shape. To understand what is going on, recall that a sphere is symmetric to rotation by any angle and to the interchange of any two axes. It is thus the simplest geometry for three dimensions. Alternatively we could have chosen a cylinder; but, for a finite length, the electric fields are not possible to calculate analytically. If we had

Table 4.1: The local field factors in one, two, and three dimensions; and, the relationship between the field inside and outside an infinite sheet, cylinder and sphere.

Dimension	Geometry	Local Field	Inside Field
1D		$\mathbf{F} = \epsilon \mathbf{E}$	$\mathbf{E} = \frac{1}{\epsilon}\mathbf{F}$
2D	$E = F + E_p$	$\mathbf{F} = \frac{\epsilon+1}{2}\mathbf{E}$	$\mathbf{E} = \frac{2\mathbf{F}}{\epsilon+1}$
3D		$\mathbf{F} = \frac{\epsilon+2}{3}\mathbf{E}$	$\mathbf{E} = \frac{3\mathbf{F}}{\epsilon+2}$

used numerical methods to overcome the limitations imposed by an analytical analysis, would we get the same local field factor? Recall that we argued that the sphere was a reasonable approximation even for a long narrow molecule since it is on average spherical when interacting stochastically with its environment, so maybe the microscopic geometry of the system does require a spherical cavity. If you are curious about the matter, try a numerical simulation of a finite cylinder.

An infinite cylindrical cavity gives a difference result than the sphere. Consider the geometry where the field is applied perpendicular to its axis. As in the case of the sphere, the analytical solution gives a uniform interior electric field. However, by virtue of the cylinder being infinite, the electric field is translationally-invariant along its axis, z. If the field is independent of the z coordinate, the system is effectively two-dimensional, so this local field model applies only in two dimensions. And therein is the resolution;

Nonlinear Optics: A student's perspective

the infinite cylinder removes one degree of freedom, and thus gives the two-dimensional local field factors. Similarly, if we used a cavity in the form of a thin infinite sheet with the field applied perpendicular to the sheet, this would give the one-dimensional local field factor. Table 4.1 summarizes the local field factors for one, two and three dimensions as well as the relationship between the interior and exterior fields.

> **Problem 4.13-1:** Determine the induced dipole moment of a dielectric sphere of dielectric constant ϵ in a uniform electric field **F** to verify Equation 4.283.
>
> **Problem 4.13-2:** The local field derivation in three dimensions used a spherical piece of dielectric. In two-dimensions, one can use a circle instead, which is the cross-section of an infinitely-long cylinder (the fields are invariant along the cylinder's axis thus removing one degree of freedom). For an electric field that is applied perpendicular to the axis of the cylinder, determine the local field factor by paralleling the technique developed in this section. Do the same in one-dimension, where the dielectric is an infinite planar thin sheet. Do you see a pattern in the local field factors as a function of dimensionality?

Non-linear Local Field

Finally, we tackle the nonlinear local field, using a self-consistent method. At heart is the fact that the dielectric function is a nonlinear function of the electric field, which can be written as

$$\epsilon = 1 + 4\pi\chi^{(1)} + 4\pi\chi^{(2)}E + 4\pi\chi^{(3)}E^2 + \cdots \equiv \epsilon_0 + \epsilon_1 E + \epsilon_2 E^2 + \ldots. \quad (4.299)$$

Once again, we assume a static scalar field, which can later be generalized.

We will solve this problem self consistently using a method similar to the one introduced in Section 2.4.1. The most confusing part of applying the self-consistent method is in identifying which relationship is fixed in stone and which one we are after. In the case of local electric fields, Equations 4.285 and 4.286 define the local field and Equation 4.299 is the dielectric function, which depends on the electric field E. Our goal is to find the relationship between the susceptibilities $\chi^{(1)}$, $\chi^{(2)}$, etc. and the polarizabilities/hyperpolarizabilities α, β, etc. in terms of the linear local field factors $(\epsilon_0 + 2)/3$.

The local electric field is given by substituting Equation 4.285 into 4.285

and 4.286, yielding

$$F = \frac{1}{3}\left((\epsilon_0 + 2)E + \epsilon_1 E^2 + \epsilon_2 E^3 + \ldots\right), \tag{4.300}$$

$$F^2 = \frac{1}{9}\left((\epsilon_0 + 2)^2 E^2 + 2\epsilon_1(\epsilon_0 + 2)E^3 + \ldots\right), \tag{4.301}$$

and

$$F^3 = \frac{1}{27}\left((\epsilon_0 + 2)^3 E^3 + \ldots\right), \tag{4.302}$$

where we have kept terms to third order in the electric field. But $p = \alpha F + \beta F^2 + \gamma F^3$, so using Equations 4.300-4.302, we get

$$\begin{aligned} P &= N'p \\ &= N'\alpha\left(\frac{\epsilon_0+2}{3}\right)E + N'\left(\frac{\alpha}{3}\epsilon_1 + \beta\left(\frac{\epsilon_0+2}{3}\right)^2\right)E^2 + N'\left(\frac{\alpha}{3}\epsilon_2 + \frac{2\beta}{9}\epsilon_1(\epsilon_0+2)\right)E^3, \end{aligned} \tag{4.303}$$

where $N' = N/V$. We can compare the coefficients of powers in the electric field of $P = \chi^{(1)}E + \chi^{(2)}E^2 + \chi^{(3)}E^3$ with Equation 4.303 to get the susceptibilities.

The linear susceptibility trivially yields

$$\chi^{(1)} = N'\alpha \frac{\epsilon_0 + 2}{3}, \tag{4.304}$$

which relates α to $\chi^{(1)}$ in terms of the linear local field factor. However, since $\epsilon_0 = 1 + 4\pi\chi^{(1)}$, we can eliminate ϵ_0 from Equation 4.304, yielding

$$\chi^{(1)} = \frac{3N'\alpha}{3 - 4\pi N'\alpha}. \tag{4.305}$$

We arrived at Equation 4.305 through a self-consistent method since ϵ_0 depends on $\chi^{(1)}$ so it is a function that depends on itself.

Next we move on to the second-order term, which yields

$$\chi^{(2)} = \frac{N'\alpha}{3}\epsilon_1 + N'\beta\left(\frac{\epsilon_0 + 2}{3}\right)^2. \tag{4.306}$$

Substituting $\epsilon_1 = 4\pi\chi^{(2)}$ into Equation 4.306 and solving for $\chi^{(2)}$ yields

$$\chi^{(2)} = N'\beta\left(\frac{\epsilon_0 + 2}{3}\right)^3. \tag{4.307}$$

Equation 4.307 is of the same form as Equation 4.296, so the two methods in this case give the same result and generalizing to an anisotropic material and including dispersion is straightforward as described in the previous section.

To third order, we get

$$\chi^{(3)} = \frac{N'\alpha}{3}\epsilon_2 + \frac{2N'\beta}{3}\left(\frac{\epsilon_0+2}{3}\right)\epsilon_1 + N'\gamma\left(\frac{\epsilon_0+2}{3}\right)^3. \quad (4.308)$$

Substituting $\epsilon_1 = 4\pi\chi^{(2)}$ and $\epsilon_2 = 4\pi\chi^{(3)}$ into Equation 4.308, then substituting Equation 4.307 into the result to eliminate $\chi^{(2)}$, and solving for $\chi^{(3)}$ yields

$$\chi^{(3)} = N'\gamma\left(\frac{\epsilon_0+2}{3}\right)^4 + \frac{8\pi}{3}(N'\beta)^2\left(\frac{\epsilon_0+2}{3}\right)^5. \quad (4.309)$$

The first term in Equation 4.309 is of the same form as Equation 4.296 but the second term is the contribution from the nonlinear part of the local field. The local field model derived here gives the continuum version of cascading. The Feynman diagram in Figure 4.12b shows the microscopic version of cascading between two molecules, which also gives a contribution in proportion to β^2 as you have verified in Problem 4.6-1(b) on page 208.

A liquid is isotropic because the molecules are randomly oriented, so the hyperpolarizability of each sphere that represents an average molecule vanishes. In this case, there is no net local field contribution from the second term in Equation 4.309. However, as molecules in a liquid are buffeted about, fluctuations can lead to instances in time when the molecules are partially aligned, leading to tiny bursts of second harmonic light that leave the material and can be detected. This process is called hyper Rayleigh scattering (HRS), the nonlinear analog of light scattering from density fluctuations in a material. Though this effect is too small to be of concern to local field contributions, HRS is used as a technique for determining the hyperpolarizability of molecules that cannot be aligned with an electric field,[17] such as noncentrosymmetric octupolar molecules with no dipole moment.[18]

In a material where the molecules are orientationally ordered, the second term in Equation 4.309 might dominate, so a third-order nonlinear-optical response can result even if the material is made of molecules that have no second hyperpolarizability γ. It is worth doing a hand-waving calculation of typical order-of-magnitude values of β and γ to determine when each term is important. Figure 4.21 shows an example of an ordered material, the volume swept out by a molecule V_2 and the volume per molecule in the material V_1, which is related the number density $N' = 1/V_1$.

Neglecting the numerical factors and linear local fields for the purpose of getting ballpark numbers, the quantity of interest that assesses the relative

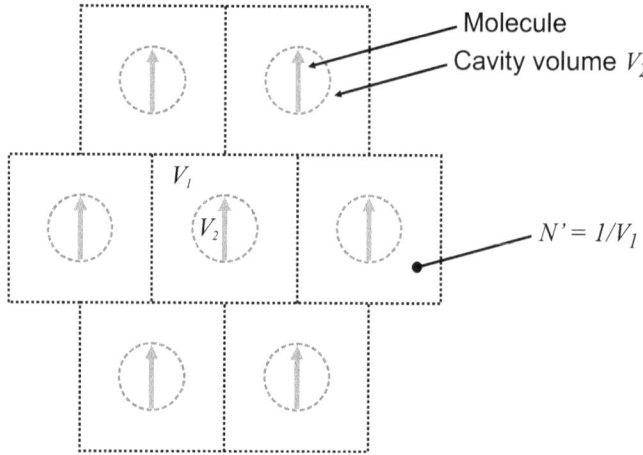

Figure 4.21: In a dye-doped polymer or molecular crystal, the volume swept out by a molecule V_2 is less than the volume occupied per molecule $V_1 = 1/N'$.

contribution of the two terms is $(N'\beta)^2/N'\gamma$. The quantum model of the second order susceptibility is given by the product of three transition moments and is divided by two energies as given by Equation 4.175, so we can abstractly write this as $\beta \sim e^3 x^3/E^2$. In this notation, the third-order susceptibility from Equation 4.179 can be expressed as $\gamma \sim e^4 x^4/E^3$. The ratio between the two terms then takes the form

$$\frac{\beta^2 N'}{\gamma} = \frac{e^2 x^2}{V_1 E}, \qquad (4.310)$$

where we have used the fact that $N' = 1/V_1$.

The position matrix element x is related to the size of the molecule and with the wave of the hand can be assumed to be limited by the size of the molecule, or $x \sim \sqrt[3]{V_2}$. Finally, modelling the molecule as an electron in a box of size x, it's energy is on the order of $\hbar^2/mx^2$, where again, we are ignoring factors of 2. m here is the mass of the electron. Putting it all together, Equation 4.310 becomes

$$\frac{\beta^2 N'}{\gamma} = \frac{m e^2 V_2^{4/3}}{V_1 \hbar^2}, \qquad (4.311)$$

where we have used the fact that $x \sim V_2^{1/3}$. Finally, since $V_1 \gtrsim V_2$, lets define $V_1 = \xi V_2$ where $1/\xi$ is the fraction of the total volume of the material taken by the molecules. Equation 4.312 then becomes

$$\frac{\beta^2 N'}{\gamma} = \frac{me^2 V_2^{1/3}}{\xi \hbar^2} \sim \frac{L[\text{Å}]}{\xi}, \qquad (4.312)$$

where we have evaluated the constants and expressed the length $V_2^{1/3} \approx L[\text{Å}]$ as the molecule's size in angstroms.

Thus, if the molecules are tightly packed as it would be in a crystal, $\xi \approx 1$ and the contribution from the β^2 term is of the same order as the γ term. As the molecule is made larger, the β term dominates. In a dye-doped polymer where $\xi \approx 10^{-3}$, the β^2 term becomes negligible. Note that the calculation here is quite rough, so the numbers are to be taken lightly. However, this shows that when the packing density is high, the β^2 term might be large and needs to be included in the analysis. When the density is low enough, the β^2 term will be negligible.

To this point, we have only considered static fields. As before, if the fields are time harmonic but of wavelength that is long compared with the size of the molecule, the static results hold but the field frequency needs to be included. The mixed terms, such as β^2 to messy to describe here.

> **Problem 4.13-3:** Start with Equation 4.304 or Equation 4.305 to derive what is called the Clausius-Mossotti relation, or
>
> $$\frac{\epsilon_0 - 1}{\epsilon_0 + 2} = \frac{4\pi}{3} N' \alpha. \qquad (4.313)$$
>
> Equation 4.313 is a way to get the polarizability of a molecule from a measure of the dielectric constant ϵ_0 or the refractive index $n_0 = \sqrt{\epsilon_0}$.
>
> **Problem 4.13-4:** Fill in the steps between Equation 4.306 to Equation 4.307. Also, fill in the missing steps to arrive at Equation 4.309. Note that the Clausius-Mossotti relation given by Equation 4.313 might be helpful.

4.14 Surface Plasmons in Nanospheres

A surface plasmon in a metallic sphere is a collective motion of the electrons that run around the surface in response to an electric field. Since the response behaves classically, we will draw upon the classical spring model to describe the material. The linear susceptibility of the linear spring is given by Equation 2.30. In a metal, the charges are free to move without impediment, which is the case when there is no spring, or $\omega_0 = 0$. Equation 2.30 then becomes

$$\chi^{(1)} = \frac{e^2 N}{m} \cdot \frac{2i\Gamma\omega - \omega^2}{\omega^4 + 4\Gamma^2\omega^2} \equiv \frac{\omega_p^2}{4\pi} \cdot \frac{2i\Gamma\omega - \omega^2}{\omega^4 + 4\Gamma^2\omega^2} = \frac{1}{4\pi} \frac{2i\frac{\Gamma}{\omega} - 1}{\left(\frac{\omega}{\omega_p}\right)^2 + 4\left(\frac{\Gamma}{\omega_p}\right)^2}, \quad (4.314)$$

where Equation 4.314 defines the plasma frequency ω_p, or

$$\omega_p = \sqrt{\frac{4\pi e^2 N}{m}}. \quad (4.315)$$

The plasma frequency is a material property that depends on the density of electrons and varies from metal to metal due to differences in their electron densities.

The dielectric constant is then of the form

$$\epsilon = 1 + \frac{2i\frac{\Gamma}{\omega} - 1}{\left(\frac{\omega}{\omega_p}\right)^2 + 4\left(\frac{\Gamma}{\omega_p}\right)^2} \approx 1 - \left(\frac{\omega_p}{\omega}\right)^2 + 2i \cdot \left(\frac{\omega_p}{\omega}\right)^2 \cdot \frac{\Gamma}{\omega}, \quad (4.316)$$

where we have assumed that $\Gamma \ll \omega \sim \omega_p$. Up to this point, we have been dealing only with the material. Next we use the fact that the material is in the shape of a sphere. The spherical geometry is embodied in Equation 4.283, which relates the induced dipole moment of a sphere to the applied electric field. If the object is of a different shape, this relationship would be different.

The geometry of the sphere and the metal's properties as given by the plasma model are combined by substituting Equation 4.316 into Equation 4.283, yielding

$$\mathbf{p} = \frac{-\left(\frac{\omega_p}{\omega}\right)^2 + 2i \cdot \left(\frac{\omega_p}{\omega}\right)^2 \cdot \frac{\Gamma}{\omega}}{3 - \left(\frac{\omega_p}{\omega}\right)^2 + 2i \cdot \left(\frac{\omega_p}{\omega}\right)^2 \cdot \frac{\Gamma}{\omega}} a^3 \mathbf{F}. \quad (4.317)$$

Note that the real part of the denominator vanishes when $\omega^2 = \omega_p^2/3$, leaving an imaginary term that is small since $\omega \gg \Gamma$. As such, $\omega^2 = \omega_p^2/3$ is the resonance condition, and is called the surface plasmon resonance frequency of a sphere, giving an induced dipole moment of

$$\mathbf{p} = \left(1 - \frac{1}{2\sqrt{3}i} \frac{\omega_p}{\Gamma}\right) a^3 \mathbf{F}. \quad (4.318)$$

To appreciate the significance of Equation 4.318, recall that the coefficient of **F** is the polarizability α, and that at it's largest, is given by the volume of the sphere. The factor in parenthesis is thus the enhancement factor. When $\omega_p/\Gamma \approx 100$, the magnitude of the electric dipole is enhanced over the static value given by $\omega_p = 0$ by a hundredfold. The radiation scattered at this resonance is proportional to the square of the dipole movement, so is enhanced ten thousand times.

Similarly, the electric field just outside the sphere at its surface due to the induced dipole moment is given by

$$\mathbf{E}_P(r=a) = \frac{\mathbf{p}}{a^3} = \left(1 - \frac{1}{2\sqrt{3}i}\frac{\omega_p}{\Gamma}\right)\mathbf{F}. \quad (4.319)$$

Thus, the electric field is enhanced at the surface a hundredfold over the applied field. Both the field near the surface and the induced dipole movement is mostly imaginary at the surface plasma resonance because the induced dipole moment is out of phase with the field.

We can calculate the relationship between the applied external electric field and the electric field inside a dielectric for various geometries, as you will do in Problem 4.14-2. You will find that the surface plasmon resonance depends on the object's shape, so the intentional control of the geometry can be used to adjust the frequency of the surface plasmon resonance for a given material.

Table 4.1 summarizes the relationship between the applied field and the field inside the dielectric. For the three-dimensional case, at the surface plasmon resonance, the electric field inside the sphere **E** is related to the applied electric field **F** according to

$$\mathbf{E} = \frac{\omega_p}{2\sqrt{3}i\Gamma}\mathbf{F}. \quad (4.320)$$

Since $\omega_p \gg \Gamma$, the field inside the metal particle can be quite large. Since the nth-order nonlinear response of a sphere is proportional to E^n, the nonlinearity is enhanced even more than the local electric field.

Figure 4.22 shows a plot of the electric field inside a particle as a function of the frequency of the applied field for one, two and three dimensions using the expressions from Table 4.1 and the dielectric constant from Equation 4.316. The resonance structure shifts to lower frequency as the particle shape is changed from a sphere (3D) to a cylinder (2D). For the sheet (1D), the field inside is independent of frequency and is real. This illustrates how the plasmon resonance for a particular metal can be changed simply by varying the particle's shape. A metal particle can be used to intensify the electric

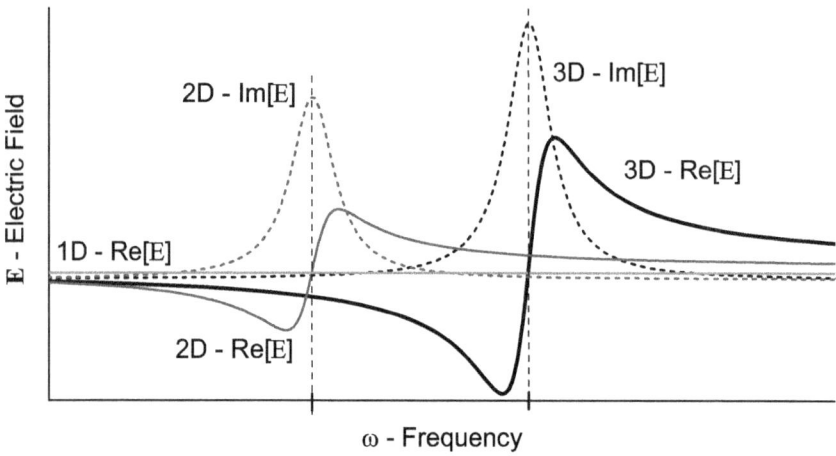

Figure 4.22: The position of the surface plasmon peak changes with the shape of the metal particle.

field near a molecule to enhance its nonlinearity, varying the shape of the metal nano-particle so that the electric field is enhanced the most when the nonlinearity of the nearby molecule is the largest.

The fact that the higher-order nonlinearities are enhanced more than lower-order ones by a surface plasmon resonance can be used to the engineer's advantage when designing an optical device. Optical loss is the bane of devices because it removes light before it is acted upon. Thus, loss needs to be made as low as possible while the nonlinearity needs to be as large as possible. Since the surface plasmon resonance enhances the nonlinear response more that the loss, a material that does not meet the criteria for making a device might become usable when a surface plasmon resonance is used.

Sadly, there is no free lunch, and subtle issues make this line of reasoning not as straightforward as it may appear. Recall that Equation 4.277 gives the fundamental limit of the linear susceptibility. A similar limit holds for the nonlinear susceptibilities. A molecule near a metal sphere may appear to break the fundamental limit due to local field enhancements. However, if the nanoparticle and molecule are taken together as the quantum system – and all the electrons are counted, the nonlinear response will be well below the limit. Thus, it is best to make larger molecules with more electrons, which will have a nonlinear response that is many orders of magnitude greater, than to waste electrons by using them to enhance the local electric field.[19]

Problem 4.14-1: Substitute reasonable values for the parameters that define the plasma frequency in Equation 4.314. Is the plasma frequency you get reasonable?

Problem 4.14-2a: Determine the electric field inside an infinitely-long dielectric cylinder of dielectric constant ϵ in terms of **F**, the uniform electric field in which the cylinder is emersed. The uniform field is perpendicular to the cylinder's axis. Do the same case for a field that is applied perpendicular to a thin dielectric slab of infinite surface area.

(b): What is the frequency of the surface plasmon resonance for each shape. Do you see a pattern as a function of dimensionality?

Chapter 5

Using the OKE to Determine Mechanisms

This chapter focuses on how the various mechanisms of the Optical Kerr Effect in a liquid affects the third-order susceptibility tensor; and, how measurements of the $\chi^{(3)}$ tensor can be used to study the underlying mechanisms. We begin with a general description of the mechanisms and end with a detailed discussion of how molecular reorientation and electronic mechanisms can be separated using polarized nonlinear refractive index measurements.

5.1 Intensity Dependent Refractive Index

We start with a calculation of the refractive index experienced by a weak beam in the presence of a strong pump. Such interactions can be used to control the propagation of one beam with another, a phenomena that is at the heart of optical switching and computing. Assume $|E(\omega)|^2 \gg |E(\omega')|^2$, where $E(\omega)$ and $E(\omega')$ are the pump and probe field, respectively. Since we are interested in measuring the probe beam, we focus on the polarization at

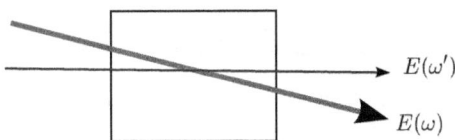

Figure 5.1: A pump beam of light influences the propagation of a probe beam.

frequency ω'

$$P^{(3)}_{\omega'} = \left(\frac{3}{2}\chi^{(3)}\left(-\omega';\omega',-\omega',\omega'\right)E_{\omega'}E_{\omega'}{}^* + \frac{3}{4}\chi^{(3)}\left(-\omega';\omega,-\omega,\omega'\right)E_{\omega}E_{\omega}{}^*\right)E_{\omega'}. \tag{5.1}$$

We can use a color filter or beam block that only allows the beam at ω' to pass through to a detector. For weak probe fields, Equation 5.1 can be expressed to first order in $E_{\omega'}$, yielding

$$P^{(3)}_{\omega'} = \frac{3}{2}\chi^{(3)}\left(-\omega';\omega,-\omega,\omega'\right)|E_{\omega}|^2 E_{\omega'}, \tag{5.2}$$

which is the lowest order term that describes two-beam mixing. Note that the polarization in Equation 5.2 is proportional to the intensity of the strong pump beam.

To observe ω', we project out the ω' Fourier component from Maxwell's wave equation. We obtain essentially the same result as when we did using a similar approach in Section 2.2. This leads to,

$$k^2_{\omega'} = \varepsilon(\omega')\frac{\omega'^2}{c^2} + 4\pi \frac{\omega'^2}{c^2}\frac{3}{2}\chi^{(3)}|E_{\omega}|^2, \tag{5.3}$$

or

$$k^2_{\omega'} \equiv \varepsilon_{\text{eff}}\frac{\omega'^2}{c^2}, \tag{5.4}$$

where $\varepsilon_{\text{eff}} = \varepsilon(\omega') + 6\pi\chi^{(3)}|E_{\omega}|^2$. Then the effective refractive index is

$$n_{\text{eff}} = \sqrt{\varepsilon_{\text{eff}}} = n(\omega')\sqrt{1 + \frac{6\pi\chi^{(3)}}{n^2(\omega')}|E_{\omega}|^2}, \tag{5.5}$$

where the refractive index is evaluated at ω', the frequency of the probe beam.

For a small nonlinear phase shift $\chi^{(3)}|E_{\omega}| \ll 1$, the square root in Equation 5.5 can be expanded in a series, yielding

$$n_{\text{eff}} \approx n(\omega') + \frac{3\pi\chi^{(3)}}{n(\omega')}|E_{\omega}|^2. \tag{5.6}$$

This is similar to the one-beam result given by Equation 2.109. Recalling that

$$I_{\omega} = \frac{c}{8\pi}n(\omega)|E_{\omega}|^2, \tag{5.7}$$

and substituting Equation 5.7 into Equation 5.6, the effective intensity-dependent refractive index becomes

$$n_{\text{eff}} \approx n(\omega') + \frac{24\pi^2\chi^{(3)}\left(-\omega';\omega,-\omega,\omega'\right)}{cn(\omega)n(\omega')}I_{\omega}. \tag{5.8}$$

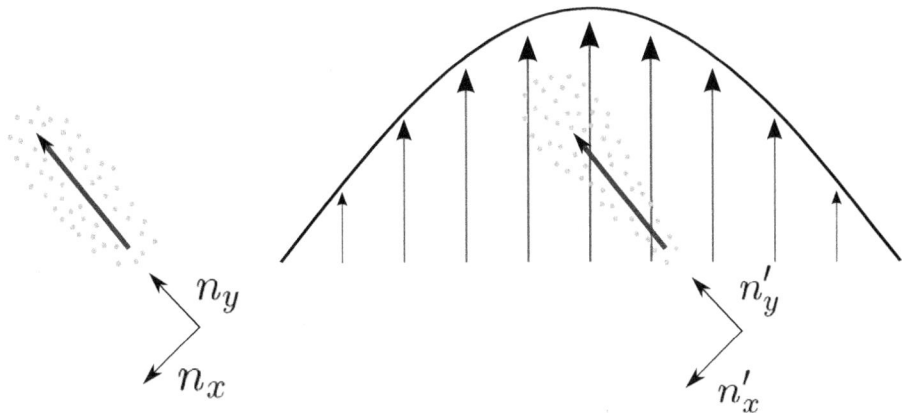

Figure 5.2: Electronic mechanism of $\chi^{(3)}$. A molecule with a nuclear framework represented by the thick arrow, and electrons represented by the surrounding dots. The electron cloud is deformed by the electric field, changing the optical refractive index of the molecule. Note that the degree of deformation is greatly exaggerated.

The most general form of the intensity-dependent refractive index can be written as

$$n = n_0 + n_2 I + n_4 I^2 + ..., \tag{5.9}$$

where $n_4 \propto \chi^{(5)}$, and $n_l \propto \chi^{(l+1)}$. The coefficient to the intensity in Equation 5.8 is thus n_2.

5.1.1 Mechanisms of $\chi^{(3)}$

Electronic

Figure 5.2 shows a molecule and its electron cloud at an arbitrary initial orientation. As an electric field is applied, changing the shape of the electron cloud, which in turn alters the refractive index. The change in refractive index is measured by using the strong electric field of a very powerful laser pulse, and then probing the refractive index change with a weak laser. This is essentially a measurement of the Kerr susceptibility, as quantified by $n_2(-\omega'; \omega, -\omega', \omega)$.

The electronic response is calculated using time-dependent perturbation theory as described in Section 4.5

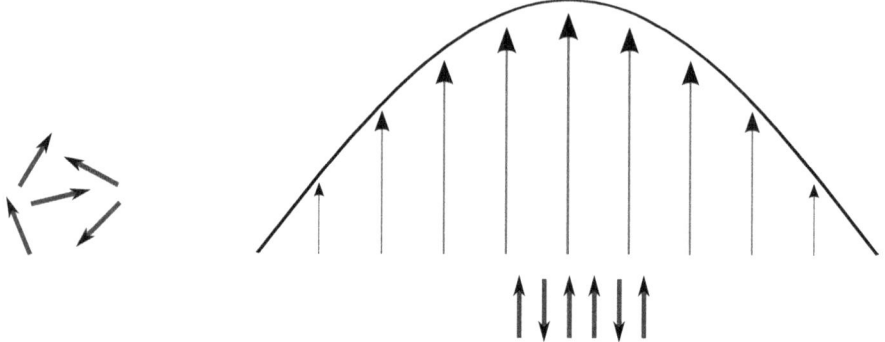

Figure 5.3: Reorientational mechanism of $\chi^{(3)}$. The nuclei of each molecule are shown as arrows.

Molecular Reorientation

A set of randomly oriented molecules is shown in Figure 5.3. The applied electric field is in the optical range, so oscillates too quickly for the molecules to keep up. The molecules will align axially along the polarization axis of the electric field as shown in Figure 5.3. This leads to an increase in the refractive index along the beams polarization and a decrease perpendicular to it, as is illustrated in Problem 5.1.1.

> **Problem 5.1.1-1:** A stationary molecule is oriented at an angle θ to the polarization of an optical field as shown in Figure 5.4a. The optical field induces a dipole moment. If the only nonzero component of the polarizability of the molecule is $\alpha_{z'z'} \equiv \alpha$, calculate the polarizability tensor of the molecule in the lab frame and the induced dipole moment along the z-axis. (A rotation matrix may be required.)
>
> **(b):** Calculate the torque on the molecule as a function of θ and α.
>
> **(c):** If a static electric field E_0 is applied when the molecule is initially at rest at an angle $\theta_0 \ll 1$ to the z axis, determine the polarizability tensor in the lab frame as a function of time. Take the moment of inertia of the molecule to be I. What is the frequency of oscillation for a small molecule? Use reasonable values and describe if the the result makes sense.

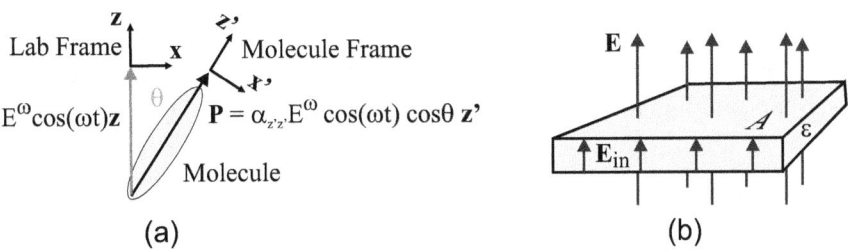

Figure 5.4: (a) An optical field is polarized along the $\hat{z}$ axis and induces a dipole moment in a molecule along its long axis $\hat{z}'$. (b) The electric field inside and outside of a dielectric slab is polarized perpendicular to its surface as shown.

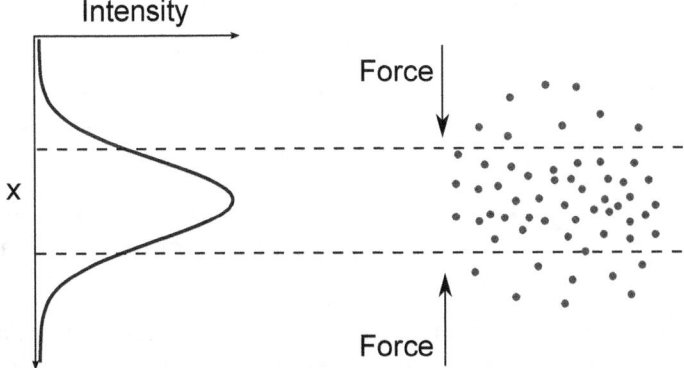

Figure 5.5: Electrostriction mechanism of $\chi^{(3)}$ in solution. The effective width of the light beam is shown by the dashed lines and the dots represent molecules in solution, which are attracted to regions of higher intensity.

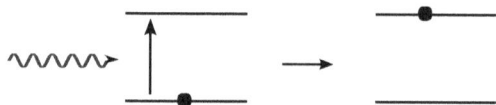

Figure 5.6: Saturated absorbtion mechanism of $\chi^{(3)}$, where the molecule in its excited state has a different refractive index than in its ground state.

Electrostriction

Electrostriction means squeezing (stricton) due to an electric field (electro) and can operate in liquid solution, the gas phase and in solids.

A Gaussian beam, as shown in Figure 5.5, with full width half maximum shown by the dashed lines, travels through a solution of molecules and creates an inward force on the molecules due to the electric field gradient, which then move into the region of the beam. More molecules in a region results in a density change and thus a change in refractive index. In a solid material the force is on the material itself. For instance, rhythmic mechanical contraction and relaxation can be created by successively pulsing light into the material such that acoustical waves are produced.

> **Problem 5.1.1-2:** An optical field illuminates a thin dielectric sheet with the electric field perpendicular to its surface as shown in Figure 5.4b. Calculate the electric field inside the sheet as a function of the external field and the dielectric constant ϵ. Use this result to calculate the surface charge density on the bottom and top surfaces as well as the force on each surface, making sure to note its direction.
>
> **(b):** If Young's modulus of the material is complex and of the form $E = E_R + iE_I$, calculate the strain as a function of time if the applied field is sinusoidal in time.
>
> **(c):** If the change in refractive index Δn of the material is related to the strain u through the stress-optical coefficient $b > 0$ according to $\Delta n = -bu$, determine the intensity-dependent refractive index n_2 as defined in Equation 5.9.

Saturated Absorbtion

In the electronic mechanism, the electron cloud deforms but no real excitations to excited states take place because the energy of the photon is much

Nonlinear Optics: A student's perspective

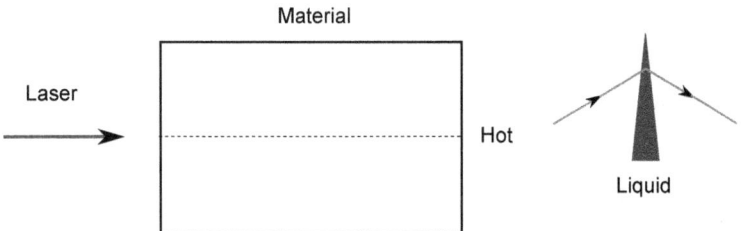

Figure 5.7: Some light from a laser beam is absorbed by a material, depositing heat in its path, and thus creates a thermal gradient. The temperature gradient acts like a prism, bending the light.

smaller than the transition energy between states in the molecule. However in saturated absorbtion, as shown in Figure 5.6, a molecule originally in its ground state is excited to a state of higher energy when the photon energy matches the transition energy, thus changing the refractive index.

Thermal Effects

When a laser beam travels through a material as shown in Figure 5.7, even a small amount of absorption deposits heat that leads to an increase in temperature near the beam. The resulting temperature gradient causes a change in the density, which induces a refractive index change. Generally, the thermal mechanism leads to a negative refractive index change because heating usually decreases the density.

An interesting example of a device that converts a continuous light beam to a pulsed one is a liquid placed in a thin prism cell. The prism at first bends the light, but through photothermal heating the refractive index decreases and the degree of bending decreases. As a result, the beam moves away from the heated region, allowing the liquid to cool. Subsequently, the cooled liquid is back to its original state causing the beam to deflect. The competing processes of deflection and cooling leads to pulsations whose frequency depends on the intensity. The continuous laser beam is thus converted into a pulsed source.

Table 5.1 shows the strength of $\chi^{(3)}$ and the response time for each of the mechanisms. The magnitude of $\chi^{(3)}$ and the response times vary by 10 to 14 orders of magnitude; but, the ratio of $\chi^{(3)}$ to the response time varies by ony a few orders. A device using the aforementioned mechanisms ideally has a large nonlinearity and also operates quickly. However, according to Table 5.1, if the device is fast $\chi^{(3)}$ is small so the device is less efficient and therefore

Mechanism	$\chi^{(3)}\left(\frac{cm^3}{erg}\right)$	τ Response time (sec)	$\chi^{(3)}/\tau$
Electronic	10^{-14}	10^{-15}	10
Reorientation	10^{-12}	10^{-12}	1
Electrostriction	10^{-12}	10^{-9}	0.001
Saturated Absorbtion	10^{-8}	10^{-8}	1
Thermal	10^{-4}	10^{-3}	0.1

Table 5.1: The magnitude of $\chi^{(3)}$ and the response times of several mechanisms.

requires a higher light intensity to operate. Thus the different mechanisms are essentially equivalently effective if this ratio is the figure of merit for a device.

5.2 Tensor Nature of $\chi^{(3)}_{ijkl}$ in Liquids

The third-order susceptibility tensor $\chi^{(3)}_{ijkl}$ has 81 independent elements. Material symmetry can significantly decrease the number of independent components and can make some of the components vanish. Liquids are centrosymmetric so have vanishing $\chi^{(2)}$, making the third-order susceptibility the lowest-order nonvanishing nonlinearity. Such centrosymmetric systems are thus ideal for studying third-order processes.

We proceed in four steps to determine the minimal number of independent $\chi^{(3)}$ tensor components. First, we will apply inversion to show that only 21 of the 81 components are nonzero. Then will apply rotations by 90° so show that the tensor elements are are interrelated, leaving only 4 independent ones. Applying rotations by 45° reduces the independent set to two. This procedure applies to any fourth-rank tensor. Since this chapter is concerned with the intensity-dependent refractive index, we conclude this section by applying intrinsic permutation symmetry to the two degenerate frequencies, which reduces the number of independent parameters to two.

5.2.1 Inversion Symmetry

The effects of centrosymmetry can be studied with the inversion operator, $\hat{I}$, which changes the signs of all vector components; and $\hat{I}_i$, which changes the sign of only the ith component. In a liquid,

$$\hat{I}_i \chi^{(3)} = \chi^{(3)} \tag{5.10}$$

because a liquid is invariant under inversion along an arbitrary axis. It thus follows that a liquid is also invariant to inversion $\hat{I}$. Being the same in all directions makes a liquid isotropic. A liquid is also equivalent from place to place, so is homogeneous. For the purposes of studying the tensor properties the isotropic nature is the most relevant.

Consider the constitutive relationship

$$P_i^{(3)} = \chi_{iijk}^{(3)} E_i E_j E_k \tag{5.11}$$

with $i \neq j \neq k$. Applying inversion through j, Given Equation 5.10, Equation 5.11 yields

$$P_i^{(3)} = \chi_{iijk}^{(3)} E_i (-E_j) E_k = -\chi_{iijk}^{(3)} E_i E_j E_k. \tag{5.12}$$

Equations 5.11 and 5.12 imply that $\chi_{iijk}^{(3)} = -\chi_{iijk}^{(3)}$, which holds only when $\chi_{iijk}^{(3)} = 0$. However, if we apply inversion to the constitutive equation corresponding to $\chi_{iijj}^{(3)}$, then we get

$$P_i^{(3)} = \chi_{iijj}^{(3)} (E_i)(-E_j)(-E_j) = +\chi_{iijj}^{(3)} E_i E_j E_j. \tag{5.13}$$

Inversion through i, on the other hand, gives

$$-P_i^{(3)} = \chi_{iijj}^{(3)} (-E_i)(E_j)(E_j) = -\chi_{iijj}^{(3)} E_i E_j E_j. \tag{5.14}$$

Inversion through k trivially leaves the equation invariant. Thus, a thrid-order susceptibility of the form $\chi_{iijj}^{(3)}$ is an allowed tensor component. It might vanish for other reasons, for example when the potential is that of a *linear spring*, but not due to symmetry.

The same argument holds for third-order susceptibilities of the form $\chi_{jiij}^{(3)}$ and $\chi_{jiji}^{(3)}$. For a given value of i and j, there are $\frac{4!}{2!2!} = 6$ ways of distributing the two indices, i.e. $iijj$, $ijij$, etc. And, there are three possible pairs of indices, $(i,j) = (1,2), (1,3),$ or $(2,3)$, so this makes a total of 18 tensor elements. The case $i = j$ adds three more nonzero elements, so there are a total of 21 independent nonzero tensor components that are identified from inversion symmetry.

> **Problem 5.2.1-1:** Apply all possible inversion operations on the third-order susceptibility tensors of the form $\chi_{jiij}^{(3)}$, $\chi_{jiji}^{(3)}$ and $\chi_{jjjj}^{(3)}$ to show that they are allowed. Also show that the tensor components of the form $\chi_{ijjk}^{(3)}$ must vanish. This covers all the possible cases.

5.2.2 90° Rotation Symmetry

Next we can use the property that a rotation of the liquid or the coordinate system leaves the nonlinear susceptibility unchanged. Rotating the system 90° about any particular axis interchanges the two indices, i.e. $i \rightleftarrows j$, which adds the following conditions:

$$\chi_{1111} = \chi_{2222} = \chi_{3333}, \tag{5.15}$$

where the two equalities come from a rotation about the 3-axis followed by a rotation about the 1-axis. Note that we drop the superscript "3" since it is understood that we are dealing with a third-order susceptibility. Similarly, we have for the off-diagonals

$$\chi_{1212} = \chi_{1313} = \chi_{2323}, \tag{5.16}$$

by rotating about the 1-axis then the 3-axis. Following this procedure for the rest of the components, we get that all $\chi^{(3)}_{iijj}$ are the same for any pair of i and j as are $\chi^{(3)}_{ijji}$ and $\chi^{(3)}_{ijij}$.

Summarizing the results, we get four independent components, given by

$$\chi_{1111} = \chi_{2222} = \chi_{3333}, \tag{5.17}$$

$$\chi_{1122} = \chi_{1133} = \chi_{2211} = \chi_{2233} = \chi_{3311} = \chi_{3322}, \tag{5.18}$$

$$\chi_{1212} = \chi_{1313} = \chi_{2323} = \chi_{2121} = \chi_{3131} = \chi_{3232}, \tag{5.19}$$

$$\chi_{1221} = \chi_{1331} = \chi_{2112} = \chi_{2332} = \chi_{3113} = \chi_{3223}. \tag{5.20}$$

Using the fact that there are only four independent components as given by Equations 5.17 to 5.20, we can express the off-diagonals more compactly as

$$\chi^{(3)}_{ijkl} = \chi^{(3)}_{1122}\delta_{ij}\delta_{kl} + \chi^{(3)}_{1212}\delta_{ik}\delta_{jl} + \chi^{(3)}_{1221}\delta_{il}\delta_{jk}, \tag{5.21}$$

where Equation 5.21 holds when $i \neq j$. However, as we show below, Equation 5.21 also holds for $i = j$.

5.2.3 45° Rotation Symmetry

This final symmetry comes from the fact that the properties of an isotropic liquid do not depend on the orientation of the liquid. As such, we demand that the tensor is invariant by a rotation of 45° about the $\hat{3}$-axis. To do this rotation, we use a simplifying trick that is based on the definition of a tensor. Recall that a tensor cannot be any arbitrary $3 \times 3 \times 3 \times 3$ matrix. Rather, it needs to be expressible as a product of vectors so χ_{ijkl} can be expressed by the product of the vector components of four vectors, or

$$\chi_{ijkl} = q_i t_j v_k w_l. \tag{5.22}$$

If we rotate the coordinate system by 45° around the 3-axis, each vector **v** transforms as $v_1 \rightarrow \frac{1}{\sqrt{2}}(v_1 + v_2)$, so Equation 5.22 becomes

$$\begin{aligned}\chi_{1111} &= \frac{1}{\sqrt{2}}(q_1+q_2) \cdot \frac{1}{\sqrt{2}}(t_1+t_2) \cdot \frac{1}{\sqrt{2}}(v_1+v_2) \cdot \frac{1}{\sqrt{2}}(w_1+w_2) \\ &= \frac{1}{4}(q_1 t_1 v_1 w_1 + (q_1 t_1 v_1 w_2 + 3 \text{ permutations}) + \cdots + q_2 t_2 v_2 w_2) \\ &= \frac{1}{4}(\chi_{1111} + (\chi_{1112} + 3 \text{ permutations}) + \cdots + \chi_{2222}).\end{aligned} \tag{5.23}$$

But, we already know from 90° degree rotations that $\chi_{1111} = \chi_{2222}$ and that third-order susceptibilities elements with an odd number of indices vanish, so after some simple rearrangement of terms, Equation 5.23 becomes

$$\chi_{1111}^{(3)} = \chi_{1122}^{(3)} + \chi_{1221}^{(3)} + \chi_{1212}^{(3)}, \tag{5.24}$$

which is identical to Equation 5.21 with $i = j$. The third order susceptibility tensor thus has only three independent tensor components so Equation 5.21 is the most general form of the third-order susceptibility in an isotropic medium.

5.2.4 Intrinsic Permutation Symmetry

We can go further if there are more symmetries. Since the rest of this chapter deals with the intensity-dependent refractive index of a single beam, we will consider the third-order susceptibility $\chi_{ijkl}^{(3)}(\omega;\omega,\omega,-\omega)$. Full permutation symmetry allows us to interchange any two indices when also interchanging the corresponding frequencies. In this case, since the two interior frequencies are ω, interchanging the indices j and k leaves the dispersion fixed, so

$$\chi_{ijkl}(\omega;\omega,\omega,-\omega) = \chi_{ikjl}(\omega;\omega,\omega,-\omega). \tag{5.25}$$

and $\chi_{1212}^{(3)} = \chi_{1122}^{(3)}$. Therefore Equation 5.21 gives

$$\chi_{ijkl}^{(3)} = \chi_{1122}^{(3)}(\delta_{ij}\delta_{kl} + \delta_{ik}\delta_{jl}) + \chi_{1221}^{(3)}\delta_{il}\delta_{jk}. \tag{5.26}$$

This leaves two independent tensor components, from which all others can be calculated.

To determine how light propagates in a material, we need the polarization. For an isotropic material, the third order polarization with the appropriate pre-factor is given by

$$P_i^{(3)} = \frac{3}{4}\chi_{ijkl}^{(3)} E_j E_k E_l^*. \tag{5.27}$$

Substituting the susceptibility from Equation 5.26 into Equation 5.27 yields

$$P_i^{(3)} = \frac{3}{4}\left[\chi_{1122}^{(3)}\left(E_j(\mathbf{E}\cdot\mathbf{E}^*)\delta_{ij} + (\mathbf{E}\cdot\mathbf{E}^*)E_k\delta_{ik}\right) + \chi_{1221}^{(3)}(\mathbf{E}\cdot\mathbf{E})E_l^*\delta_{il}\right]. \tag{5.28}$$

Defining

$$A = \frac{3}{2}\chi_{1122}^{(3)} \quad \text{and} \quad B = \frac{3}{2}\chi_{1221}^{(3)}, \tag{5.29}$$

Equation 5.28 becomes

$$\mathbf{P}^{(3)} = A\,(\mathbf{E}\cdot\mathbf{E}^*)\,\mathbf{E} + \frac{1}{2}B\,(\mathbf{E}\cdot\mathbf{E})\,\mathbf{E}^*. \tag{5.30}$$

As will become apparent shortly, it is convenient to rewrite Equation 5.30 as

$$\mathbf{P}^{(3)} = \left(A - \frac{1}{2}\right)(\mathbf{E}\cdot\mathbf{E}^*)\,\mathbf{E} + \frac{1}{2}B\left((\mathbf{E}\cdot\mathbf{E})\,\mathbf{E}^* + (\mathbf{E}\cdot\mathbf{E}^*)\,\mathbf{E}\right), \tag{5.31}$$

where we have taken a bit from the first term and added it to the second one.

It is useful to write P_i in terms of the effective susceptibility

$$P_i = \chi_{ij}^{\text{eff}} E_j. \tag{5.32}$$

That we can do so is easy to see if we reexpress the second term on the right-hand side of Equation 5.31 in component form. The ith component is given by

$$\left[(\mathbf{E}\cdot\mathbf{E})\mathbf{E}^* + (\mathbf{E}\cdot\mathbf{E}^*)\mathbf{E}\right]_i = E_j E_j E_i^* + E_j E_j^* E_i = \left(E_i^* E_j + E_i E_j^*\right) E_j, \tag{5.33}$$

where summation convention is implied.

Using Equations 5.32 with the help of Equation 5.31 and the component form given by Equation 5.33, we finally get

$$\chi_{ij}^{\text{eff}} = \chi_{ij}^{(1)} + A'\,(\mathbf{E}\cdot\mathbf{E}^*)\,\delta_{ij} + \frac{1}{2}B'\left(E_i E_j^* + E_i^* E_j\right). \tag{5.34}$$

Here $A' = A - \frac{1}{2}B$ and $B' = B$. Note that the trace is the sum over the diagonals, and yields

$$Tr\left(\chi_{ij}^{\text{eff}} - \chi_{ij}^{(1)}\right) = 3\left(A' + B'\right)\mathbf{E} \cdot \mathbf{E}^*. \tag{5.35}$$

The next section describes how A' and B' depend on the mechanisms of $\chi^{(3)}$. A measurement of these coefficients gives the relative contributions from the mechanisms.

5.3 Measurements of the Intensity-Dependent Refractive Index

In this section we will discuss the use of linearly, circularly and/or elliptically polarized light to measure the intensity-dependent refractive index of a liquid to determine the tensor elements of $\chi^{(3)}$.

We begin by recasting the wave equation in terms of left- and right-handed circularly polarized light. To that end, we define the unit vectors

$$\hat{\sigma}_\pm = \frac{\hat{x} \pm i\hat{y}}{\sqrt{2}}, \tag{5.36}$$

which are complex. $\hat{\sigma}_-$ describes righthand circular polarization and $\hat{\sigma}_+$ describes left-hand circular polarization. Some properties of the unit vectors include:

$$\hat{\sigma}_\pm \cdot \hat{\sigma}_\pm = 0 \tag{5.37}$$

$$\hat{\sigma}_\pm^* = \hat{\sigma}_\mp \tag{5.38}$$

$$\hat{\sigma}_\pm \cdot \hat{\sigma}_\pm^* = \hat{\sigma}_\pm \cdot \hat{\sigma}_\mp = 1. \tag{5.39}$$

The fact that dot product of the unit vector with itself vanishes seems curious since unit vectors are usually orthonormal. However, the complex conjugate of the unit vector $\hat{\sigma}_\pm$ gives $\hat{\sigma}_\mp$. These properties imply that a projection onto a particular unit vector requires that one take it's complex conjugate followed by a dot product, or

$$E_\pm = \hat{\sigma}_\pm^* \cdot \mathbf{E}. \tag{5.40}$$

Since real unit vectors are the complex conjugates of themselves, the complex conjugation is not required. However, when working with complex unit vectors, it is important that the conjugation operation be performed before taking the dot product when calculating the projection.

In electrodynamics, classical mechanics, and statistical mechanics, all quantities are real. In quantum mechanics, wavefunctions are complex, although measured values are always real. The complex quantities introduced here are used as a computational tool while the actual measurements will always yield real values.

The electric field can be written in terms of the unit vectors as a linear combination of right and left circular polarization, or

$$\mathbf{E} = E_+ \hat{o}_+ + E_- \hat{o}_-. \tag{5.41}$$

The dot products therefore yield

$$\mathbf{E} \cdot \mathbf{E} = 2E_+ E_- \tag{5.42}$$
$$\mathbf{E}^* \cdot \mathbf{E} = |E_+|^2 + |E_-|^2, \tag{5.43}$$

where $|E_+|^2$ and $|E_-|^2$ are the intensities of each of the two polarizations. The dot product of the electric field vector with itself thus gives a cross term. The dot product of the electric field vector times its complex conjugate gives the total intensity. This is consistent with the definition given by Equation 5.40.

We can evaluate Equation 5.30 for fields expressed in terms of the circular polarization basis with the help of Equation 5.42 and 5.43, yielding

$$\mathbf{P}^{NL} = A(|E_+|^2 + |E_-|^2)\mathbf{E} + B(E_+ E_-)\mathbf{E}^*, \tag{5.44}$$

where we recall that A and B are constants that can be calculated from the $\chi^{(3)}$ tensor. The nonlinear polarization can also be expressed in terms of the circular polarization unit vectors, or

$$\mathbf{P}^{NL} = P_+^{NL} \hat{o}_+ + P_-^{NL} \hat{o}_-. \tag{5.45}$$

Setting Equation 5.45 equal to Equation 5.44 and projecting out the right and left-hand polarizations yields

$$P_+^{NL} = A|E_+|^2 E_+ + A|E_-|^2 E_+ + B E_+ E_- E_-^* \tag{5.46}$$

and

$$P_-^{NL} = A|E_+|^2 E_- + A|E_-|^2 E_- + B E_+ E_- E_+^*. \tag{5.47}$$

Since $E_\pm E_\pm^* = |E_\pm|^2$, we can simplify Equations 5.46 and 5.47 by combining these terms, yielding

$$\mathbf{P}^{NL} = A|E_+|^2 E_+ \hat{o}_+ + (A+B)|E_-|^2 E_+ \hat{o}_+ \\ + A|E_-|^2 E_- \hat{o}_- + (A+B)|E_+|^2 E_- \hat{o}_-. \tag{5.48}$$

Since $P_\pm^{NL} = \chi_\pm^{NL} E_\pm$, the nonlinear part of χ^{eff} is χ^{NL}, so

$$\chi_\pm^{NL} = A|E_\pm|^2 + (A+B)|E_\mp|^2. \tag{5.49}$$

Nonlinear Optics: A student's perspective

From the nonlinear susceptibility, we can write the refractive index as a function of the square of the electric field magnitude, as we did leading up to Equation 5.6 for two linearly polarized beams and Equation 2.109 for a single linearly polarized beam, yielding

$$n_\pm \approx n_0 + \frac{2\pi}{n_0}[A|E_\pm|^2 + (A+B)|E_\mp|^2], \qquad (5.50)$$

where n_0 is the isotropic zero-field refractive index.

Recall that for an isotropic material such as a liquid, Equation 5.26 holds for the single-beam intensity-dependent refractive index, and the two independent tensor components are χ_{1122} and χ_{1221}. Circularly polarized light probes both components by inducing birefringence in the material. As we will show below, linear combinations of left and right circularly-polarized light probe different combinations of the tensor components.

5.3.1 Plane Waves

We assume that the light wave can be approximated with a plane wave and that the effect of the intensity-dependent refractive index is to affect the polarization. The most general form of a plane wave written in terms of right and left circular polarization unit vectors is

$$\mathbf{E}(z) = A_+ \exp\left[\frac{in_+\omega z}{c}\right]\hat{\sigma}_+ + A_- \exp\left[\frac{in_-\omega z}{c}\right]\hat{\sigma}_-, \qquad (5.51)$$

where A_+ and A_- are the amplitudes of the left and right circularly-polarized components, which propagate at phase velocities $\frac{c}{n_+}$ and $\frac{c}{n_-}$.

Equation 5.51 can be expressed in the more convenient form

$$\mathbf{E}(z) = \left(A_+ e^{i\theta}\hat{\sigma}_+ + A_- e^{-i\theta}\hat{\sigma}_-\right)e^{ik_m z}, \qquad (5.52)$$

where

$$\theta = \frac{1}{2}\Delta n \frac{\omega}{c} z \qquad (5.53)$$

$$k_m = \left(n_- + \frac{1}{2}\Delta n\right)\frac{\omega}{c} \qquad (5.54)$$

$$\Delta n = n_+ - n_- = \frac{2\pi B}{n_0}\left(|E_-|^2 - |E_+|^2\right). \qquad (5.55)$$

We will use the parameters defined in Equations 5.53 through 5.55 to describe self-action effects for three special cases: linearly-, circularly-, and elliptically-polarized light.

Linear Polarization

Linear polarization consists of an equal mixture of left and righthanded circularly polarized light, or $E_- = E_+$, so the refractive index difference according to Equation 5.55 yields $\Delta n = 0$. Then, Equation 5.50 gives

$$\delta n_{\text{linear}} \equiv n_\pm - n_0 = \frac{2\pi}{n_0}\left[A|E_+|^2 + (A+B)|E_+|^2\right] = \frac{2\pi}{n_0}\left[\left(A+\frac{B}{2}\right)|E|^2\right]. \quad (5.56)$$

Circular Polarization

Let's consider left circular polarized light, where $E_- = 0$. The electric field is then given by

$$\mathbf{E}(z) = A_+ e^{\frac{in_+\omega z}{c}}\hat{\sigma}_+. \quad (5.57)$$

The same can be done for righthanded circularly polarized light. Using Equation 5.50, the refractive index change for left and righthanded circular polarization is

$$\delta n_{\text{circular}} = \frac{2\pi}{n_0}\left[A|E_\pm|^2\right] \quad (5.58)$$

$$= n_\pm - n_0. \quad (5.59)$$

A determination of the two independent susceptibility tensor components requires two independent measurements. For example, one can use linearly-polarized light to measure $n_\pm$, which through Equation 5.56 gives $(A+\frac{B}{2})$. Then, using circular polarization to determine $n_\pm$ gives A from Equation 5.58. Thus, the coefficients A and B are determined.

A particular mechanism is often characterized by a unique ratio of A and B. In this way, the mechanism is uniquely determined when only one mechanism is acting. Multiple mechanisms acting together will gave a tensor ratio not characteristic of a particular phenomena. However, other parameters can be varied, such as the laser pulse width and the temperature, each which affect various mechanisms differently. For example, when the temporal pulse width is made ultrashort, only the electronic response contributes so the ratio observed is the one indicative of it. As the pulse width is increased, A and B smoothly change, making it possible to determine the ratio between the two contributing contributions. When more mechanisms are present, the de-convolution process becomes more involved.

Elliptical Polarization

Finally, we consider the propagation of an elliptically polarized beam. The electric field vector can be written as Equation 5.52 but with $A_+ \neq A_-$, $A_+ \neq 0$,

Nonlinear Optics: A student's perspective

and $A_- \neq 0$. In the first term, we can write the unit vector $\hat{\sigma}_+$ in terms of x and y unit vectors as a function of phase angle,

$$\hat{\sigma}_+ e^{i\phi} = \frac{\hat{x} + i\hat{y}}{\sqrt{2}}(\cos\theta + i\sin\theta). \tag{5.60}$$

Expanding the $\hat{x}$ and $\hat{y}$ terms yields

$$\hat{\sigma}_+ e^{i\phi} = \frac{1}{\sqrt{2}}[\hat{x}\cos\theta - \hat{y}\sin\theta + i(\hat{x}\sin\theta + \hat{y}\cos\theta)]. \tag{5.61}$$

These terms appear to define the rotations

$$\hat{x}' = \hat{x}\cos(\theta) - \hat{y}\sin(\theta) \tag{5.62}$$

and

$$\hat{y}' = \hat{x}\sin(\theta) + \hat{y}\cos(\theta), \tag{5.63}$$

which can be expressed as the the action of the rotation matrix

$$\begin{bmatrix} \hat{x}' \\ \hat{y}' \end{bmatrix} = \begin{bmatrix} \cos\theta & -\sin\theta \\ \sin\theta & \cos\theta \end{bmatrix} \begin{bmatrix} \hat{x} \\ \hat{y} \end{bmatrix}. \tag{5.64}$$

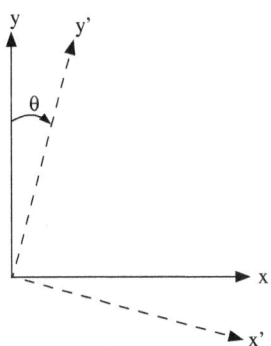

Figure 5.8: Rotation of axes $\hat{x}$ and $\hat{y}$ by an angle θ to give new axes $\hat{x}'$ and $\hat{y}'$.

$\hat{x}$ and $\hat{y}$ are the principle axes that define the ellipse of polarization. The same rotation can be applied to the unit vector $\hat{\sigma}_-$. The matrix transformation rotates the axes in a clockwise direction, as shown in Figure 5.8. The origin of the rotation lies in the phase shift between the fields along the two principle axes. The net effect according to Equation 5.53 is that the axes of the ellipse rotate as the light propagates through the material in proportion to Δn. Thus, a measure of the degree of rotation within a sample of fixed length determines Δn, which through Equation 5.55 can be used to determine B.

5.4 Intensity-Dependent Refractive Index Mechanisms

This section is concerned with determining the two-independent tensor components of $\chi^{(3)}$ in an isotropic liquid for various mechanisms, from which the A and B coefficients can be determined.

5.4.1 Electronic Response

The electronic response originates in the distortion of the electron cloud in response to light. Consider the classical anharmonic oscillator in three dimensions, which can be modelled by the potential

$$U(\mathbf{r}) = \frac{1}{2} m \omega_0^2 |\mathbf{r}|^2 - \frac{1}{4} b |\mathbf{r}|^4. \tag{5.65}$$

Being centrosymmetric, this potential gives $\chi^{(2)} = 0$. Using classical perturbation theory as presented in Section 2.1.3, the solution for $\chi^{(3)}$ in three dimensions is given by

$$\chi^{(3)}_{ijkl}(-\omega;\omega,\omega,-\omega) = \frac{Nbe^4 \left[\delta_{ij}\delta_{kl} + \delta_{ik}\delta_{jl} + \delta_{il}\delta_{jk}\right]}{3m^3 D^3(\omega) D(-\omega)}, \tag{5.66}$$

where $D(\omega) = \omega_0^2 - \omega^2 - 2i\omega\Gamma$.

Substituting Equation 5.66 into Equation 5.29 to solve for A and B yields

$$A = \frac{3}{2}\chi^{(3)}_{1122} = \frac{Nbe^4}{2m^3 D^3(\omega) D(-\omega)} \tag{5.67}$$

$$B = \frac{3}{2}\chi^{(3)}_{1221} = \frac{Nbe^4}{2m^3 D^3(\omega) D(-\omega)}. \tag{5.68}$$

Thus $A = B$ and the ratio of Equation 5.56 and Equation 5.59 yields

$$\frac{\delta n_{\text{linear}}}{\delta n_{\text{circular}}} = 1 + \frac{B}{2A} \tag{5.69}$$

$$= \frac{3}{2}. \tag{5.70}$$

Notice that this ratio is independent of ω because the denominator is the same for all of the tensor components.

If a combination of $\chi^{(3)}$ mechanisms is acting, there will be a convolution of the ratios, so the mechanisms may not be simple to isolate. In this case we can use measurements of varying temporal pulse widths that probe the time scales characteristic of each mechanism.

The electronic response modelled here is classical, next we move on to the quantum calculation.

Secular Divergence

Before moving on to the quantum calculation, we first need to deal with what are called secular divergences. These arise from perturbation theory when the sums of the frequencies vanish, leading to zeros in the denominator. As we see below, pairs of such terms diverge, but when they are summed, the divergence cancels, leading to a finite nonzero term.

Recall that $\chi^{(3)}$ is written as

$$\chi^{(3)}_{kjih}(-\omega_\sigma;\omega_r,\omega_q,-\omega_p) =$$
$$\frac{N}{\hbar^3}\mathcal{P}_F \sum_{lmn}\left[\frac{\mu^k_{0n}\mu^j_{nm}\mu^i_{ml}\mu^h_{l0}}{(\omega_{n0}-\omega_\sigma)(\omega_{m0}-\omega_q-\omega_p)(\omega_{l0}-\omega_p)}\right], \quad (5.71)$$

where $\mathcal{P}_F$ is the full permutation symmetry operator over the frequencies ω_σ, ω_q and ω_p. The outgoing frequency is a sum over the incoming frequencies

$$\omega_\sigma = \omega_r + \omega_q + \omega_p. \quad (5.72)$$

which for the intensity-dependent refractive index leads to $\omega_r \to \omega$, $\omega_q \to \omega$, $\omega_p \to -\omega$. The term in the sum includes the the ground state, for which $\omega_{00} = 0$. Unlike a resonance, which occurs when the frequencies of the light add up to a Bohr frequency, this one term diverges, independent of frequency. Thus, Equation 5.71 appears to be infinite at all frequencies.

To understand how this divergence goes away, we would write out all 24 terms that contribute to $\chi^{(3)}$, and would find two types of terms that when added would be of the form

$$\frac{1}{(X+Y)Y} + \frac{1}{(X+Y)X}. \quad (5.73)$$

Each term diverges separately when $X = -Y$. However the sum yields

$$\frac{1}{(X+Y)Y}\frac{X}{X} + \frac{1}{(X+Y)X}\frac{Y}{Y} = \frac{X+Y}{XY(X+Y)} \quad (5.74)$$
$$= \frac{1}{XY} = \frac{-1}{X^2}, \quad (5.75)$$

which is finite and negative.

One- and Two-Photon States

When adding all 24 terms and combining diverging terms as described above, we get

$$\chi^{(3)}_{kjih}(-\omega_\sigma;\omega_r,\omega_q,-\omega_p)$$

$$= \frac{N}{\hbar^3}\mathcal{P}_F\left[\sum_{lmn}{}' \frac{\mu^k_{0n}\mu^j_{nm}\mu^i_{ml}\mu^h_{l0}}{(\omega_{n0}-\omega_\sigma)(\omega_{m0}-\omega_q-\omega_p)(\omega_{l0}-\omega_p)} \right.$$

$$\left.- \sum_{ln}{}' \frac{\mu^k_{0n}\mu^j_{n0}\mu^i_{0l}\mu^h_{l0}}{(\omega_{n0}-\omega_\sigma)(\omega_{l0}-\omega_q)(\omega_{l0}-\omega_p)}\right]. \quad (5.76)$$

The summation $\sum'$ indicates that the ground state is excluded.

If the molecule is centrosymmetric, the wavefunctions will be parity eigenstates. For illustration, let's consider the 1111 tensor element. The ground state is typically symmetric, so is of even parity. The transition moment from the ground state to state l, $-ex_{l0}$, is calculated according to

$$x_{l0} = \int dx\, \phi_l^*(x) x \phi_0(x), \quad (5.77)$$

where $\phi_1(x)$ is an energy eigenfunction. Equation 5.77 is therefore nonzero only when the parity of state l is odd. This can be understood by recalling that x is of odd parity, so the two wavefunctions must be of opposite parity to get a nonzero result. The 4 transition moments in the first term of Equation 5.76 are thus nonzero when state m is of even parity and state is n odd. Using the same argument, the second term requires that state l and n be of odd parity.

The left-hand side of Figure 5.9 is an energy-level diagram that represents the nonzero transitions that together yield a nonzero first term. The righthand diagram shows the nonzero contributions to the second term. The parities required to get nonzero contributions to the third-order susceptibility are labelled. States that are of opposite parity to the ground state, such as states n and l, are called one-photon states because it takes one photon to excite the system between states 0 to n or n to m by absorbing a photon or from states m to l or l to 0 by emitting a photon. The state m is called a two-photon state because a one-photon transition is disallowed between states 0 and m, which are of the same parity. However the state can be excited through an intermediate one-photon state as shown.

As we have seen above, both terms in Equation 5.76 contribute to $\chi^{(3)}_{1111}$ so both one- and two-photon states are represented. For the $\chi^{(3)}_{1221}$ tensor element, the first term in Equation 5.76 does not vanish but the second term

Nonlinear Optics: A student's perspective

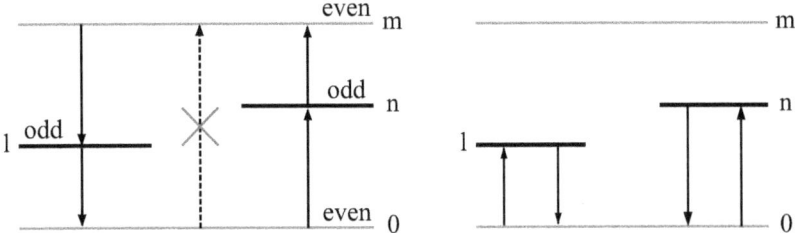

Figure 5.9: An energy level diagram for transitions that contribute to the third-order susceptibility tensor. State n is called a one-photon because it takes one excitation to reach it. State m is called a two-photon state because a single excitation from state 0 to m is disallowed; but a transtion with two excitations are allowed. The left and right diagram corresponds to the first term and the righthand one the second term in Equation 5.76.

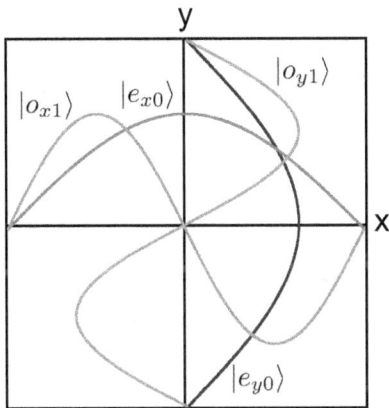

Figure 5.10: The ground state and first excited state wave functions in the x and y directions are depicted, where the ground state is even and has no nodes and the first excited state is odd with one node. The wavefunctions are labelled $|e_{x0}\rangle$, $|o_{x1}\rangle$, $|e_{y0}\rangle$ and $|o_{y1}\rangle$.

does. This can be illustrated for wavefunctions inside a two-dimensional box as shown in Figure 5.10. The ground state will be an even-parity state in both the x and y directions and the first excited state will have one even-parity state along the short length of the box, which has a higher state energy, and an odd-parity state along the long longer edge of the box. Breaking the wavefunctions into products along x and y, the even-parity ground state can be written as

$$|\psi_0\rangle = |e_{x0}\rangle|e_{y0}\rangle, \qquad (5.78)$$

and the odd-parity first excited states will take the form

$$|\psi_1\rangle = |o_{x1}\rangle|e_{y0}\rangle, \tag{5.79}$$

where e and o refer to the even and odd parity state vectors, respectively, and the numerals are the quantum numbers along the axes specified. Note that in this case, we assumed that the longer edge is along x. If the edges are not too disparate, the next higher-energy excited state will be

$$|\psi_2\rangle = |e_{x0}\rangle|o_{y1}\rangle. \tag{5.80}$$

To illustrate how the transition moments are calculated, we determine $\langle\psi_0|x|\psi_1\rangle$ as follows:

$$\langle\psi_0|x|\psi_1\rangle = \langle e_{x0}|\langle o_{y0}|x|o_{x1}\rangle|e_{y0}\rangle = \langle e_{x0}|x|o_{x1}\rangle\langle e_{y0}||e_{y0}\rangle = \langle e_{x0}|x|o_{x1}\rangle \neq 0, \tag{5.81}$$

where we have used the fact that the x operator acts only on state vectors along x and that the state vectors are orthonormal. We can apply the same steps to calculate the transition moments of y, which yields

$$\langle\psi_2|y|\psi_1\rangle = \langle o_{x0}|\langle e_{y1}|y|e_{x1}\rangle|e_{y0}\rangle = \langle o_{x0}||e_{x1}\rangle\langle e_{y1}|y|e_{y0}\rangle = 0. \tag{5.82}$$

Using the kinds of relationships developed in Equations 5.81 and 5.82, we can show show that $\chi^{(3)}_{1221}$ cannot have one-photon character as given by the second term in Equation 5.76, as follows. The first two transition moments (read from right to left) are of the form $\langle\psi_0|y|\psi_n\rangle\langle\psi_n|x|\psi_0\rangle$. The transition from state 0 to n along x requires state n to be of odd parity along x while the wavefunction along y is unchanged so remains in the ground state, similar to what happens in Equation 5.81. This transition is non-zero to any odd-parity excited state along x. The next transition from state n to 0 is along y, so the final state is the ground state for both the x and y directions. Since the y transition is between states of the same parity, it vanishes. Also, since the transition along x is between two different states, orthogonality demands that it too vanish. Thus, the one-photon term must always vanish.

To summarize, $\chi^{(3)}_{1221}$ has only two-photon character while $\chi^{(3)}_{1122}$ has both one- and two-photon character. The ratio of the $\chi^{(3)}$ components, $\frac{\chi^{(3)}_{1221}}{\chi^{(3)}_{1122}}$ is frequency-dependent so we cannot uniquely determine A and B, and so we cannot learn about the mechanisms using any combination of polarized measurements.

In the off-resonance case, $\omega \to 0$ and all frequencies vanish. Then, the two-photon terms are non-negative and the one-photon terms are non-positive. Kleinman symmetry, which allows any two indices to be exchanged without affecting the third-order susceptibility, leads to $A = B$. Then we get

$$\frac{\delta n_{\text{linear}}}{\delta n_{\text{circular}}} = 1 + \frac{B}{2A} = \frac{3}{2}. \tag{5.83}$$

Thus, polarization-dependent measurements can be used to determine if the mechanism is electronic.

Problem 5.4-1a: Calculate the x and y transition moments between all possible states given by Equations 5.78 through 5.80. This will give you a three-by-three matrix for x and y.

(c): Use the results that you got in Part (a) to show that the two-photon term, i.e. the first term in Equation 5.76, is not required to vanish.

(c): Show that $A = B$ when Kleinman symmetry holds.

5.5 Molecular Reorientation

We begin by considering the process of molecular reorientation, which is slower only than the electronic mechanism. We will assume the material to be composed of a collection of freely rotating anisotropic molecules. Common examples of such materials are liquids and gases. Figure 5.11a diagrams a "cigar"-shaped molecule where the electric field is applied along the z-axis. Here, α is the polarizability tensor, **E** is the electric field, **P** is the polarization, and θ is the angle between the long axis of the molecule and the direction of the electric field. Note that the induced dipole moment, **p**, is not necessarily along **E** or the principle axes of α.

An anisotropic molecule aligns with an applied electric field, increasing the refractive index along the direction of the field and decreasing it perpendicular to the electric field. Molecular reorientation is induced by a torque due to the electric field, τ, which is given by

$$\boldsymbol{\tau} = \mathbf{p} \times \mathbf{E} \tag{5.84}$$

and the internal energy per molecule is

$$U = -\mathbf{p} \cdot \mathbf{E}. \tag{5.85}$$

Equation 5.85 shows that the energy is lowest when the molecule's long axis is aligned with the electric field. For a molecule that lacks a permanent

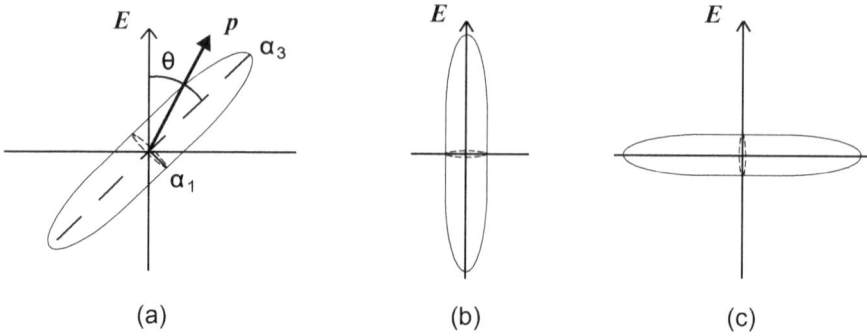

Figure 5.11: (a) An anisotropic molecule in an electric field with anisotropic polarizability. Associated with the semi-major and semi-minor axes are the polarizabilities, α_3 and α_1, respectively. (b) The minimum energy configuration and (c) the maximum energy configuration

dipole moment, there is no preferred polar orientation; that is, up and down orientation is equivalent – called axial order. Figures 5.11b and 5.11c shows these lowest and highest energy configurations.

The polarization of an anisotropic molecule is given by

$$P_i = \alpha_{ij} E_j. \tag{5.86}$$

When the polarizability is anisotropic, the internal energy is found by integrating the right hand side of Equation 5.85, yielding

$$U = -\int_0^{\mathbf{E}} \mathbf{p} \cdot d\mathbf{E}. \tag{5.87}$$

When a molecule is uniaxial as shown in Figure 5.11, the refractive index ellipsoid is described by two refractive indices. The refractive index component along the semi-minor axis is related to α_1 and refractive index component along the semi-major axis is related to α_3. Equation 5.87 for a uniaxial molecule can be evaluated with the help of Equation 5.86, giving

$$U = -\int_0^{\mathbf{E}} \alpha_3 E_3' \, dE_3' - \int_0^{\mathbf{E}} \alpha_1 E_1' \, dE_1'. \tag{5.88}$$

E_3' and E_1' are the electric field components along the direction of the semi-major and semi-minor axis of the molecule, where $E_3' = E\cos\theta$ and $E_1' = E\sin\theta$. Upon integration, Equation 5.88 becomes

$$U = -\frac{1}{2}E^2 \left(\alpha_3 \cos^2\theta + \alpha_1 \sin^2\theta\right) = -\frac{1}{2}E^2 \left[\alpha_1 + (\alpha_3 - \alpha_1)\cos^2\theta\right]. \tag{5.89}$$

Figure 5.12 shows $U(\theta)$ given by Equation 5.89.

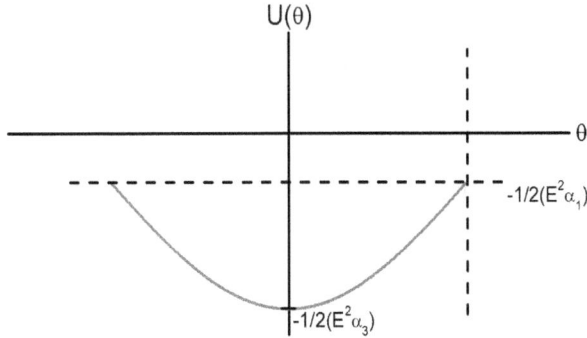

Figure 5.12: The energy per anisotropic molecule in an electric field, **E**.

The electric field of a plane wave is given by

$$E = E_0 \cos(\omega t), \qquad (5.90)$$

where E_0 is the amplitude and ω is the angular frequency. Then the internal energy is proportional to $\cos^2(\omega t)$. By using a trigonometric identity, we can rewrite this as

$$U = \frac{E_0}{4}[1 + \cos(2\omega t)]\left[\alpha_1 + (\alpha_3 - \alpha_1)\cos^2\theta\right]. \qquad (5.91)$$

Since the reorientational response time is on the order of 10^{-12} seconds, an electric field of frequency greater than the reciprocal of the response time will not lead to significant reorientation. An optical cycle is on the order of 10^{-15} seconds – much shorter than reorientational time scales – so the time average over an optical cycle of the energy U will yield the observed effective potential

$$\overline{U} = -\frac{1}{4}E_0^2\left[\alpha_1 + (\alpha_3 - \alpha_1)\cos^2\theta\right]. \qquad (5.92)$$

As usual, the linear refractive index is given by $n_{ii}^2 = 1 + 4\pi\chi_{ii}^{(1)}$, so the reorientational contribution to the intensity-dependent refractive index is given by

$$n_{ii}^2 = 1 + 4\pi N \langle \overleftrightarrow{\alpha} \rangle_{ii}, \qquad (5.93)$$

where the brackets signify an orientational ensemble average of the linear molecular susceptibility – as we calculate below, and summation convention is not implied here.

The linear molecular susceptibility in the lab frame α_{zz} can be found in terms of the polarizability tensor in the frame of reference fixed to the principle axes of the molecule by projecting components from one frame to the other. These projection factors are incorporated in the rotation matrix a, yielding

$$\alpha_{zz} = \sum_{IJ} \alpha_{IJ} a_{Iz}(\Omega) a_{Jz}(\Omega), \tag{5.94}$$

where Ω represents the triplet of Euler angles θ, ϕ, and ψ. The upper case indices represent the molecular frame coordinates and the lower case ones represent the lab frame coordinates. Equation 5.94 is general, in that α_{IJ} can be of any form, not just a uniaxial molecule as assumed in the derivation leading to Equation 5.92.

We define $P(\Omega)$ as the probability density of finding a molecule at the angle Ω from Boltzmann statistics. Therefore,

$$P(\Omega) = \frac{e^{-\beta \overline{U}}}{\int d\Omega e^{-\beta \overline{U}}}, \tag{5.95}$$

where $\beta = kT$ with k denoting the Boltzmann constant, and T denoting the temperature. Since no intermolecular forces are being considered in the energy, the orientational ensemble average of the linear molecular susceptibility is

$$\langle \overleftrightarrow{\alpha} \rangle_{ii} = \frac{1}{Z} \int d\Omega \alpha_{ii}(\Omega) e^{-\beta \overline{U}}, \tag{5.96}$$

where $d\Omega = d(\cos\theta) d\phi d\psi$. Z is the partition function defined as

$$Z = \int d\Omega e^{-\beta \overline{U}}. \tag{5.97}$$

5.5.1 Zero Electric Field

We will start by examining the susceptibility when no electric field is present. α_{xx} and α_{zz} for a uniaxial molecule are then given by Equation 5.94, yielding

$$\alpha_{xx} = \alpha_1 + (\alpha_3 - \alpha_1)\sin^2\theta \cos^2\phi, \tag{5.98}$$
$$\alpha_{zz} = \alpha_1 + (\alpha_3 - \alpha_1)\cos^2\theta. \tag{5.99}$$

From Equation 5.92, when $\mathbf{E} = 0$, then $\overline{U} = 0$ and $P(\Omega) = 1$. Substituting Equations 5.98 and 5.99 into Equation 5.96, we get

$$\langle \overleftrightarrow{\alpha} \rangle_{xx} = \frac{\int_0^{2\pi} \int_{-1}^{1} \int_0^{2\pi} d\psi\, d(\cos\theta)\, d\phi\, [\alpha_1 + (\alpha_3 - \alpha_1)\sin^2\theta \cos^2\phi]}{\int_0^{2\pi} \int_{-1}^{1} \int_0^{2\pi} d\psi\, d(\cos\theta)\, d\phi},$$
(5.100)

$$\langle \overleftrightarrow{\alpha} \rangle_{zz} = \frac{\int_0^{2\pi} \int_{-1}^{1} \int_0^{2\pi} d\psi\, d(\cos\theta)\, d\phi\, [\alpha_1 + (\alpha_3 - \alpha_1)\cos^2\theta]}{\int_0^{2\pi} \int_{-1}^{1} \int_0^{2\pi} d\psi\, d(\cos\theta)\, d\phi}.$$
(5.101)

After integration, the two non-zero linear molecular susceptibilities are,

$$\boxed{\langle \overleftrightarrow{\alpha} \rangle_{xx} = \frac{1}{3}\alpha_3 + \frac{2}{3}\alpha_1}$$
(5.102)

$$\boxed{\langle \overleftrightarrow{\alpha} \rangle_{zz} = \frac{1}{3}\alpha_3 + \frac{2}{3}\alpha_1}$$
(5.103)

Equations 5.102 and 5.103 are equivalent to one another as would be expected in an isotropic media consisting of anisotropic molecules which are randomly oriented. Therefore, the refractive index given by Equation 5.93 can be written as

$$n_0^2 \equiv n_{ii}^2 = 1 + \frac{4}{3}\pi N (\alpha_3 + 2\alpha_1).$$
(5.104)

5.5.2 Non-Zero Electric field

Next we consider the case when an electric field is present. When $\mathbf{E} \neq 0$, the time-averaged energy per molecule due to an external field is greater than zero. This means that the probability of finding a molecule at an angle Ω is no longer uniform. We define

$$J = \frac{1}{2}\beta \overline{E^2}(\alpha_3 - \alpha_1),$$
(5.105)

which is independent of the Euler angles and will thus remain after an integration is performed to calculate the ensemble average.

The orientational ensemble average of the polarizability will be used to determine the reorientational response. By, using Equations 5.99, 5.96, and

5.105, we get

$$\langle \overleftrightarrow{\alpha} \rangle_{zz} = \frac{\int_{-1}^{1} d(\cos\theta)\left[\alpha_1 + (\alpha_3 - \alpha_1)\cos^2\theta\right] e^{J\cos^2\theta}}{\int_{-1}^{1} d(\cos\theta) e^{J\cos^2\theta}}. \quad (5.106)$$

When the magnitude of the field is small enough so that the alignment energy is much smaller than thermal energies, we can expand the exponential term,

$$e^{J\cos^2\theta} \approx 1 + J\cos^2\theta + \cdots \quad (5.107)$$

and

$$\langle \overleftrightarrow{\alpha} \rangle_{zz} = \frac{1}{3}\alpha_3 + \frac{2}{3}\alpha_1 + \frac{2}{45}\beta\overline{E^2}(\alpha_3 - \alpha_1)^2. \quad (5.108)$$

Substituting Equation 5.108 into Equation 5.93, the refractive index in the z direction is

$$n_{zz}^2 = n_0^2 + \frac{8\pi N}{45}\beta\overline{E^2}(\alpha_3 - \alpha_1)^2. \quad (5.109)$$

We define $\Delta n = n_{zz} - n_0$. But $n_{zz}^2 - n_0^2 = (n_{zz} - n_0)(n_{zz} + n_0) \equiv (\Delta n)(2\overline{n})$. For a very small reorientationally-induced refractive index change, $\Delta n = \delta n_{zz}$, where $\overline{n} \approx n_0$, so we have

$$2n_0\Delta n = \frac{8\pi N}{45}\beta\overline{E^2}(\alpha_3 - \alpha_1)^2. \quad (5.110)$$

Therefore,

$$\delta n_{zz} = \frac{4\pi N}{45 n_0}\beta\overline{E^2}(\alpha_3 - \alpha_1)^2. \quad (5.111)$$

Defining $\delta n_{zz} = n_2\overline{E^2}$, and including the Lorentz local field correction $L(\omega)$, we can find the fourth rank tensor n_2^{zzzz} to be of the form

$$n_2^{zzzz} = L^4(\omega)\frac{4\pi N}{45 n_0}\beta(\alpha_3 - \alpha_1)^2. \quad (5.112)$$

We can calculate n_2^{xxzz} using the relationship

$$n_2^{xxzz} = -\frac{1}{2}n_2^{zzzz}. \quad (5.113)$$

Thus, the refractive index increases along the field and decreases perpendicular to it, as expected.

5.5.3 General Case

In the most general case, there are three independent linear polarizability tensor elements, which in the principle axis frame of the molecule can be represented by the diagonal elements of $\overleftrightarrow{\alpha}$ as

$$\overleftrightarrow{\alpha} = \begin{pmatrix} a & 0 & 0 \\ 0 & b & 0 \\ 0 & 0 & c \end{pmatrix}. \tag{5.114}$$

The orientational ensemble average of $\overleftrightarrow{\alpha}_{ij}$ is the sum of the non-zero field result and the zero field result, yielding

$$\langle \alpha_{ij} \rangle = Q\delta_{ij} + \gamma_{ij}\left(E^2\right), \tag{5.115}$$

where δ_{ij} is the Kronecker delta function, the first term on the righthand side of Equation 5.115 is from the generalized zero field solution and the second term is reorientational third order susceptibility contribution. Here, we state without proof that

$$Q = \frac{1}{3}(a + b + c). \tag{5.116}$$

The second term on the righthand side of Equation 5.115 is calculated analogously to the derivation leading to Equation 5.112, yielding

$$\gamma_{ij} = C\left(3\delta_{ik}\delta_{jl} - \delta_{ij}\delta_{kl}\right) E_k^{\text{loc}} E_l^{\text{loc}}. \tag{5.117}$$

Here, the superscript "loc" refers to the local electric field. The constant C is then of the form

$$C = \frac{\beta}{90}\left[(a-b)^2 + (b-c)^2 + (a-c)^2\right]. \tag{5.118}$$

We can represent the electric field as

$$E_k(t) = \frac{1}{2} E_k e^{-i\omega t} + \text{c.c.} \tag{5.119}$$

The polarization will be given by

$$P_i^{(3)} = N \sum_j \gamma_{ij} E_j. \tag{5.120}$$

Using Equations 5.117 - 5.119, the polarization can be found from Equation 5.120. In vector form, the polarization derived from Equation 5.120 is given by

$$\mathbf{P}^{(3)} = A\left(\mathbf{E}\cdot\mathbf{E}^*\right)\mathbf{E} + \frac{1}{2} B\left(\mathbf{E}\cdot\mathbf{E}\right)\mathbf{E}^*, \tag{5.121}$$

where $A = NC/4$ and $B = 3NC/2$. Equation 5.121 is derived in the same way as was Equation 5.30.

Recalling that Equation 5.83 gives

$$\frac{\delta n_{\text{linear}}}{\delta n_{\text{circular}}} = 1 + \frac{B}{2A}, \qquad (5.122)$$

then the fraction we get here is $\frac{\delta n_{\text{linear}}}{\delta n_{\text{circular}}} = 4$. This is larger than the 3/2 value of the electronic response, and therefore the two mechanisms are easily separable.

Figure 5.13 shows a schematic diagram of the fraction $\frac{\delta n_{\text{linear}}}{\delta n_{\text{circular}}}$ as a function of laser pulse width. When the pulse width is shorter that the reorientational response time, only the electronic response contributes and the ratio is 3/2. For longer pulses, the reorientational response kicks in. If it is larger than the electronic response, the measured ratio will give 4. If the two are comparable, a value between these two will be observed.

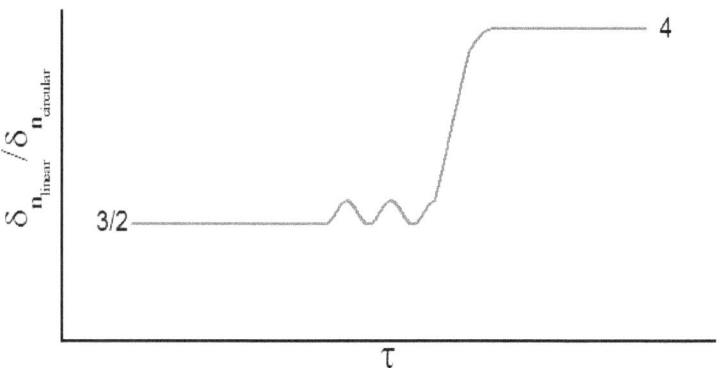

Figure 5.13: $\frac{\delta n_{\text{linear}}}{\delta n_{\text{circular}}}$ as a function of the laser pulse width τ.

Chapter 6

Applications

This chapter applies the fundamentals of what we have learned to interesting phenomena such as optical phase conjugation and time reversal, self focusing, negative temperature, single-molecule nonlinear optics, optical multistability and epsilon-near-zero materials.

6.1 Optical Phase Conjugation and Time Reversal

The idea of travelling back in time is a romantic one. A more practical way to reverse time is to limit ourselves to the propagation of light. It is well known that time-reversed waves are also solutions to Maxwell's equations. For example, a focused beam prorogating to the right is the time reversed version of a beam that is spreading to the left. The interesting question pertains to how a wave can be converted to its time-reversed one.

A time-reversed beam is generated from a weak beam by reflecting it from what's called a phase conjugate mirror. Such a mirror is not passive like a silver film on a piece of glass, but rather needs to be pumped by two powerful beams. Through a third-order susceptibility, a phase conjugate wave is generated, which is the time-reversed version of the incident wave.

To appreciate the process, we start by considering reflections from a surface, then describe the peculiar time-reversed reflection from a phase-conjugate mirror. Formally, we then show how a phase-conjugate wave is equivalent to a time-reversed one. Then we dive into the coupled equations that govern this process.

Let us consider a plane wave incident on a mirror at normal incidence.

The incident plane wave field is given by

$$\mathbf{E}_{in} = \hat{e}\frac{A(\mathbf{r})}{2}e^{i(kz-\omega t)} + \text{c.c.} \tag{6.1}$$

The polarization $\hat{e}$ can be circular, linear, or elliptical. Maxwell's equations demand that the tangential component of the electric field be continuous at an interface between two materials. The electric field on the vacuum side of the mirror is the sum of the incident wave and the reflected wave, ($\mathbf{E}_{in} + \mathbf{E}_r$), and inside the mirror the electric field vanishes. Thus, the continuity condition for an interface at $z = 0$ yields

$$(\mathbf{E}_r + \mathbf{E}_{in})|_{z=0} = 0. \tag{6.2}$$

Equation 6.2 implies that the reflected wave's polarization is opposite to the incident wave's polarization, yielding

$$\mathbf{E}_r = -\hat{e}\frac{A(\mathbf{r})}{2}e^{i(-kz-\omega t)} + \text{c.c.} \tag{6.3}$$

The negative wave vector comes about because because the reflected wave propagates opposite to the incident one.

Next we consider the time-reversed version of the incident wave given by Equation 6.1, which is obtained simply by reversing the sign of the time, yielding

$$\mathbf{E}_c \equiv \mathbf{E}_{in}(-t) = \left[\hat{e}\frac{A(\mathbf{r})}{2}e^{ikz}\right]e^{i\omega t} + \text{c.c.}, \tag{6.4}$$

where we have placed in brackets the time-independent part. This is the final result. However, it is more convenient to re-express Equation 6.4 by interchanging the first and second terms in Equation 6.4 (i.e. writing the + c.c. part first), yielding

$$\mathbf{E}_c = \left[\hat{e}\frac{A(\mathbf{r})}{2}e^{ikz}\right]^* e^{-i\omega t} + \text{c.c.} = \hat{e}^*\frac{A^*(\mathbf{r})}{2}e^{-ikz}e^{-i\omega t} + \text{c.c.} \tag{6.5}$$

The exponent in Equation 6.5 is the same as for the reflected wave from a mirror given by Equation 6.3 with the exception that the amplitude is the complex conjugate. As such, the time-versed wave is called the phase conjugate wave and a material that converts the a wave to its phase conjugate wave is called a phase conjugate mirror. Section 6.2 describes how a phase conjugate mirror is made from a third-order nonlinear-optical material.

Before proceeding, we discuss some of the properties of phase conjugate waves and how they compare to reflection from a mirror. Upon reflection, a

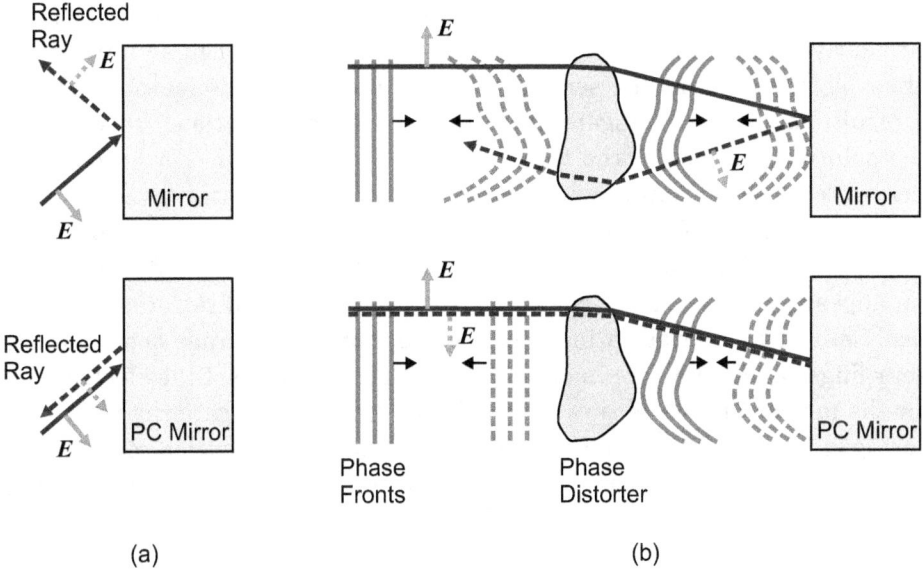

Figure 6.1: In a phase conjugate mirror (a) the beam retraces its incident path in reverse; and, (b) the phase fronts do not invert as they do for a regular mirror. Solid lines are incident phase fronts and rays and dashed lines the reflected phase fronts and rays.

regular mirror inverts the normal component of the wave vector and maintains the tangential component. In contrast, a phase conjugate beam's wave vector changes sign, so the ray will retrace it's path as shown in Figure 6.1a. In a mirror, the polarization of a linearly-polarized beam flips its tangential component and maintains its normal component, but the phase conjugate beam retains the same polarization. As such, the phase conjugate wave retraces its path with the same polarization.

For a circularly polarized beam, complex conjugation turns a righthand circularly-polarized beam into a lefthanded one. In a regular mirror, the phase shifts by π but the sense of circular polarization is maintained. As a result, the phase conjugate beam retraces its propagation. To visualize the polarization, picture the tip of the electric field vector. For an observer that watches a righthand circularly polarized wave recede (wave vector pointing away from the observer), the tip rotates clockwise. The phase conjugate beam, on the other hand, travels toward the observer. Being lefthanded, the tip appears again to rotate clockwise. To see that the tip retraces the identical helix, grab a large spring, like the one found in a garage door, and run your finger along the wire in one direction. That's the tip of the electric field for the indicant ray. The tip's motion for the phase conjugate ray traces the spring in reverse. Its the same spring, illustrating how the time-reversed ray traces the same path and the polarization traces the same helix. This will be generally true for any arbitrary polarization.

Figure 6.1b shows the phase fronts, a representative ray, and the electric field vector of an incident plane wave as solid lines. After the wave passes through a phase distorter (picture a blob of glass) the phase fronts bend. The phase fronts, electric field and ray after reflection are shown as dashed lines. In the case of a regular mirror, the phase fronts get inverted upon reflection, so when the wave passes through the distorter on the return trip, the phase front become more deformed and the ray gets deflected in a complex way.

In a phase conjugate mirror, the phase fronts do not flip, the ray retraces its path as does the electric field. As such, the distortion of the phase fronts are returned to parallel planes and the ray retraces the original path all the way back to the source. In this way, a phase conjugate mirror will repair distortions provided that there are no losses in the system because absorbed light will not be regenerated.

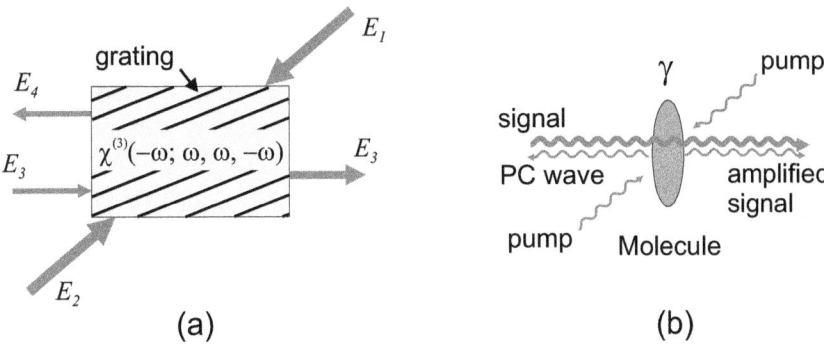

Figure 6.2: (a) A phase conjugate mirror reflects a conjugate wave E_4 and amplifies the signal E_3. (b) The microscopic picture of phase conjugation.

6.2 Phase Conjugate Mirror

We begin this section by showing that degenerate four-wave mixing, mediated by the third-order susceptibility, leads to a phase conjugate wave. Then the coupled equations are solved to yield the intensities of the waves.

6.2.1 A Physical Picture

Now we are ready to design the phase conjugate mirror using degenerate four wave mixing, a process that is mediated by $\chi^{(3)}(-\omega;\omega,\omega,-\omega)$. We begin with two pump beams (E_1, E_2), a probe beam E_3 and the reflected beam E_4 in the configuration shown in Figure 6.2a. The source of phase conjugation can be understood by recognizing that the probe beam and one of the pump beams interfere and form a refractive index grating from which the other pump scatters. Figure 6.2a shows the grating formed by E_2 and E_3, from which E_1 scatters to form beam E_4. We could have just as easily chosen the interference pattern to be due to beam E_3 and E_4, from which pump beam E_2 would scatter to form E_4.

We assume that the fields are approximately plane waves of the form

$$E_i(\mathbf{r},t) = \frac{A_i(\mathbf{r})}{2}e^{i(\mathbf{k}_i\cdot\mathbf{r}-\omega t)} + \text{c.c.}. \tag{6.6}$$

Recall that the subscripts label the beams. To simplify the derivation, we assume the fields to be scalars, so $\chi^{(3)}$ will also be a scalar. This simplification ignores the tensor nature of the nonlinear response, but will not change the character of the results provided that $\chi^{(3)} \neq 0$ for the components of interest.

Next we assume that the pump beams are much more intense than the probe and reflected beams, so that pump beam depletion is negligible. The largest contribution to the nonlinear polarization in the direction of E_4 is

$$E_4 \propto (P^{(3)}) = \frac{3}{2}\chi^{(3)} E_1 E_2 E_3^*, \qquad (6.7)$$

where we have used that fact that the scattered field is proportional to the polarization.

Substituting Equation 6.6 into Equation 6.7 yields

$$E_4 \propto \frac{3}{2}\chi^{(3)} A_1 A_2 A_3^* e^{i(\mathbf{k}_1+\mathbf{k}_2-\mathbf{k}_3)\cdot\mathbf{r}}. \qquad (6.8)$$

Because the two pump beams are counter propagating, $\mathbf{k}_1 = -\mathbf{k}_2$. Equation 6.8 then becomes

$$E_4 \propto \frac{3}{2}\chi^{(3)} A_1 A_2 A_3^* e^{-i\mathbf{k}_3\cdot\mathbf{r}}, \qquad (6.9)$$

which shows that $E_4 \propto E_3^*$ and E_4 propagates in the $-\mathbf{k}_3$ direction, meeting it the criteria for a phase conjugate wave. As we will see in the derivation that follows, E_3 will be amplified.

Figure 6.2b shows the process at the microscopic level. Let's first assume that only the two pump photons are incident. The total energy between the two photons is $2\hbar\omega$. Through a $\chi^{(3)}$ process, these two photons will interact, emitting two counter-prorogating photons which have no net momentum and energy $2\hbar\omega$. The pair can be emitted in any direction provided that the third-order susceptibility tensor is non-zero for those polarizations. Consequently, the intensity will be low for any given emissions direction. However, when the signal beam is turned on, it induces stimulated emission in the direction of $\mathbf{k}_3$, thus amplifying the signal. To conserve momentum, the other photon must be emitted in the direction $-\mathbf{k}_3$ and by necessity will be the phase conjugate wave since it meets all the criteria.

6.2.2 Solving for the Intensities

To solve the problem in detail we consider the polarizations to first order in E_3 and E_4 as they are much smaller than E_1 and E_2. The third order polarizations are

$$P_1 = \frac{3}{2}\chi^{(3)}\left[E_1^2 E_1^* + 2E_1 E_2 E_2^*\right], \tag{6.10}$$

$$P_2 = \frac{3}{2}\chi^{(3)}\left[E_2^2 E_2^* + 2E_2 E_1 E_1^*\right], \tag{6.11}$$

$$P_3 = \frac{3}{2}\chi^{(3)}\left[2E_3 E_1 E_1^* + 2E_3 E_2 E_2^* + 2E_1 E_2 E_4^*\right], \tag{6.12}$$

and

$$P_4 = \frac{3}{2}\chi^{(3)}\left[2E_4 E_1 E_1^* + 2E_4 E_2 E_2^* + 2E_1 E_2 E_3^*\right]. \tag{6.13}$$

To solve for E_4, we substitute the polarizations into the nonlinear wave equation and solve it for the backward scattered wave to third-order in the field at frequency ω using the slowly varying envelope approximation. The result for A_1 is obtained using Equation 6.10

$$\left(-k_1^2 + 2ik_1\frac{d}{dz'} + \frac{\varepsilon\omega^2}{c^2}\right)A_1 = \frac{-4\pi}{c^2}\frac{1}{2}\chi^{(3)}\left(|A_1|^2 + 2|A_2|^2\right)A_1, \tag{6.14}$$

where z' is along $\mathbf{k}_1$. Remembering that $k_1 = \frac{\varepsilon\omega^2}{c^2}$, and that A_1 and A_2 are counter propagating, we find the differential equations for A_1 and A_2

$$\frac{dA_1}{dz'} = \frac{i\pi\omega}{nc}\chi^{(3)}\left(|A_1|^2 + 2|A_2|^2\right)A_1 = i\kappa_1 A_1, \tag{6.15}$$

$$\frac{dA_2}{dz'} = \frac{-i\pi\omega}{nc}\chi^{(3)}\left(|A_2|^2 + 2|A_1|^2\right)A_2 = -i\kappa_2 A_2, \tag{6.16}$$

where κ_1 and κ_2 are defined as

$$\kappa_1 = \frac{\pi\omega}{nc}\chi^{(3)}\left(|A_1|^2 + 2|A_2|^2\right), \tag{6.17}$$

$$\kappa_2 = \frac{\pi\omega}{nc}\chi^{(3)}\left(|A_2|^2 + 2|A_1|^2\right). \tag{6.18}$$

In the non-depletion approximation, the square of the amplitudes $|A_1|^2$ and $|A_2|^2$ are constant making κ_1 and κ_2 constant, giving the solutions for A_1 and A_2

$$A_1(z') = A_1(0)e^{i\kappa_1 z'}, \tag{6.19}$$

$$A_2(z') = A_2(0)e^{i\kappa_2 z'}. \tag{6.20}$$

For real $\chi^{(3)}$, $|E_1|^2$ and $|E_2|^2$ are constants and $E_1 E_2$ takes the form

$$E_1 E_2 = A_1(0) A_2(0) e^{i(\kappa_1 - \kappa_2) z'}. \tag{6.21}$$

Substituting Equation 6.21 into the nonlinear wave equation with the polarizations given by Equations 6.13 and 6.13 yields

$$\frac{dA_3}{dz'} = \frac{2i\pi\omega}{nc} \chi^{(3)} \left[\left(|A_1|^2 + |A_2|^2 \right) A_3 + A_1 A_2 A_4^* \right] \tag{6.22}$$

$$\frac{dA_4}{dz'} = \frac{-2i\pi\omega}{nc} \chi^{(3)} \left[\left(|A_1|^2 + |A_2|^2 \right) A_4 + A_1 A_2 A_3^* \right]. \tag{6.23}$$

When the pump intensities are equal, or $|A_1| = |A_2|$, then $\kappa_1 = \kappa_2$ and Equations 6.19 and 6.20 imply that $A_1 A_2$ is constant. Then Equations 6.22 and 6.23 take the form

$$\frac{dA_3}{dz} = i\kappa_3 A_3 + i\kappa A_4^*, \tag{6.24}$$

and

$$\frac{dA_4}{dz} = -i\kappa_3 A_4 - i\kappa A_3^*. \tag{6.25}$$

Equations 6.24 and 6.25 are coupled differential equations which can be decoupled in many ways. Here we suggest one of them.

First, add Equations 6.24 and 6.25. This will give a differential equation for the sum of the two field amplitudes in terms of the difference between the two of them. Next take the difference, which will yield a differential equation that relates the difference between the two fields to the sum of the two fields. Finally, differentiate one of the equations with respect to z and use the other equation to get a second-order differential equation in terms of just the sum or the difference between the fields. Note that you will need to separately solve for the real and imaginary parts of these equations. Finally, you can use these solutions to get the fields A_3 and A_4. The results of this procedure yields

$$A_3'^*(L) = \frac{A_3'^*(0)}{\cos(|\kappa|L)}, \tag{6.26}$$

and

$$A_4'^*(0) = \frac{i\kappa}{|\kappa|} \tan(|\kappa|L) A_3'^*(0), \tag{6.27}$$

where L is the length of material traversed by the incident signal field.

The microscopic view of phase conjugation as shown in Figure 6.2b can aid in interpreting Equations 6.26 and 6.27. First we note that only some of the incident signal photons will give a conjugate photon as diagrammed, and the rest will pass through without nonlinear interaction. As such, the phase conjugate wave will be of lower intensity than the incident wave. If the material is lossless, the signal wave intensity can be no lower than the incident signal intensity since photons are added to it when phase conjugate photons are created.

Recall that κ is obtained by comparing Equations 6.22 and 6.24, so is given by

$$\kappa = \frac{2\pi\omega}{nc}\chi^{(3)}A_1^2. \qquad (6.28)$$

Since nonlinear processes must be small if Equation 2.85 is to be a good approximation, $\chi^{(3)}A_1^2 \ll \chi^{(1)}$. In the discussions that follow, we thus assume that $|\kappa|L \lesssim 1$. Above this level, higher-order processes will kick in.

Equation 6.26 shows the expected gain, where in the limit of $|\kappa|L \to 0$ (i.e. no pump intensity), the signal intensity remains unchanged, and according to Equation 6.27, no conjugate beam is produced. As the pump intensity is increased, the intensity of the signal is amplified and the phase conjugate wave turns on.

The microscopic pictures shows that the phase conjugate field amplitude cannot exceed the signal field amplitude, so Equations 6.26 and 6.27 yield $\tan(|\kappa|L) < 1$.

We conclude this section by pointing out a peculiarity of phase conjugation. Recall that a regular mirror meets our expectations when the part of the wave that hits the surface first gets reflected first, causing the phase fronts to be inverted. In a phase conjugate mirror, though, the phase fronts are not reversed so it may appear that the wave reflects before it reaches the surface.

The fact that Equation 6.27 depends on the thickness of the material, L, is a clue at the resolution of this paradox. The phase conjugate beam is produced from the radiation produced by the whole material. Since even a short pulse is made of a superposition of monochromatic waves, the waves are interacting within the material before the pulse reaches it, creating the reflected pulse at the time the incident pulse arrives. A pictorial description of this phenomenon for fresnel reflections is presented in Section 2.2. Even in that case, the full volume of the material is responding to the light.

> **Problem 6.2.2-1:** Solve the coupled differential equations given by Equations 6.24 and 6.25 to get Equations 6.26 and 6.27.

6.3 Self Focusing

A high-intensity beam of light will alter the refractive index profile that follows the intensity profile according to Equation 5.9. For a beam with a Gaussian intensity profile, positive n_2 will increase the refractive index at the beam's center more than at the periphery. The increased optical path through the center will result in self focusing. Diffraction, on the other hand acts to spread the beam in inverse proportion to the beam width. When the the effect of self-focusing balances the effects of diffraction, an equilibrium beam waist results.

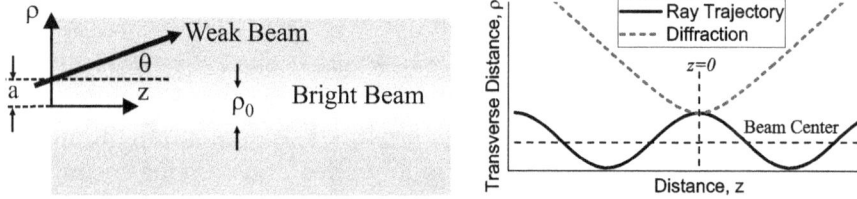

Figure 6.3: (left) A weak probe beam is launched into the region where a strong pump beam has self-focused into a Gaussian beam of width ρ_0. The probe beam is launched at an angle θ to the pump beam axis and enters at a distance a from the axis. (right) The trajectory of the ray when $\theta = 0$ and $a \neq 0$ (solid curve); and, and the trajectory of the beam waist from diffraction(dashed curve).

How self focusing operates can be appreciated with a hand-waving calculation of this balance between the action of self focusing and diffraction. We consider self focusing of a powerful beam with a Gaussian intensity profile (call it the pump beam) by first treating the propagation of a probe ray in the induced refractive index profile established by the pump, then associating the extrema of the probe ray trajectory with the powerful beam's waist. Subsequently, we will balance the curvature of the probe ray at the extrema with the curvature associated with the diffraction of the beam waist to get the equilibrium beam diameter.

Consider a ray launched into a material with a Gaussian refractive index profile of width ρ_0 as shown in Figure 6.3(left). Problem 6.3-1 is concerned

Nonlinear Optics: A student's perspective

with solving for the ray's trajectory. Here we quote the result for the special case when the ray is launched into the material parallel to the z-axis and a distance a away from it, where $a \ll \rho_0$. The trajectory is given by

$$\rho(z) = a \cos\left(\frac{z}{z_0}\right), \tag{6.29}$$

where

$$z_0 = \sqrt{\frac{n_0}{2n_2 I_0}} \rho_0, \tag{6.30}$$

and where n_0 is the linear refractive index and I_0 is the peak intensity given by Equation 6.37. Thus, the period of oscillation decreases with increased intensity, as one would expect; and, the turning points are at $\rho = a$.

Next we determine the effect of diffraction in the absence of self focusing. An analysis of Gaussian wave propagation shows that the beam waist $w(z)$ as a function of propagation distance z from the position of the minimum beam waist is related to the minimum beam waist radius w_0 via

$$w(z) = w_0 \sqrt{1 + \left(\frac{\lambda z}{\pi w_0^2}\right)^2}. \tag{6.31}$$

Figure 6.3(right) shows a plot of Equation 6.29 (solid curve) – the trajectory of the ray due to the Gaussian gradient index profile, and Equation 6.31 (dashed curve) – the Gaussian beam waist as a function of distance from the position of the minimum beam waist.

Now we assume that there is just one beam, and it is acted upon by self focusing and diffraction. The probe ray is now one ray within this single beam. If the beam is self trapped so that the diameter does not change, this ray will remain parallel to the axis. We set the minimum beam width to be ρ_0 and trace a ray that is originally parallel to the beam axis at coordinate $a = \rho_0$. The beam is said to be self-trapped when diffraction is balanced by self focusing. Note that we are after an order-of-magnitude estimate of the conditions for self trapping, so we are not concerned with the fact that we are considering a ray at the beam width even though Equation 6.29 holds only when the ray is well within the beam width.

At the beam waist, the ray trajectory and the beam waist change due to diffraction have opposite curvatures. Self trapping occurs when the curvatures are of equal magnitude and of opposite sign. The curvatures are proportional to the second derivative of Equations 6.29 and 6.31, which are given by

$$\left.\frac{d^2\rho(z)}{dz^2}\right|_{z=0} = -\frac{a}{z_0^2} \tag{6.32}$$

and

$$\left.\frac{d^2w(z)}{dz^2}\right|_{z=0} = \frac{2}{\rho_0^3}\left(\frac{\lambda}{\pi}\right)^2, \tag{6.33}$$

respectively. Setting the sum of Equations 6.32 and 6.33 to zero and using Equation 6.31 for z_0 yields

$$\rho_0 = \lambda\sqrt{\frac{n_0}{\pi^2 n_2 I_0}}. \tag{6.34}$$

Equation 6.34 shows that the equilibrium area within the beam waist depends inversely on the intensity at the peak and n_2, which is what we expect since more powerful beams should produce a smaller beam diameter by more strongly overcoming the effects of diffraction. Since $\rho^2 I_0$ is proportional to the power, Equation 6.34 gives a condition for the required power for self trapping. The power is calculated by integrating the intensity given by Equation 6.37 over the plane perpendicular to the propagation direction, or

$$P = \int_0^\infty I(\rho) 2\pi\rho\, d\rho = \pi I_0 \rho_0^2. \tag{6.35}$$

Using Equation 6.35, Equation 6.34 can be re-expressed as

$$P = \frac{n_0 \lambda^2}{\pi n_2}. \tag{6.36}$$

Equation 6.36 is the condition for the power required for self trapping. Given this amount of power, a wide beam will self trap at a lower peak intensity than a narrow beam. If the power is less that this critical amount, the beam will spread and if greater than this amount, will self focus to a point, overwhelming the effects of diffraction. However, higher-order nonlinearities kick in before the light can focus to a point.

When the power is a multiple of the critical power, the beam can break up into multiple self-trapped beams, each carrying the critical power. Imperfections in the intensity profile of a beam can seed the process so that each little peak acts like an independent self-focusing beam.

Problem 6.3-1: A high-power laser beam as shown in Figure 6.3(left) is launched into a material with an intensity dependent refractive index n_2. It self focuses until it reaches an equilibrium transverse beam profile of the form

$$I(\rho) = I_0 \exp\left[-\left(\frac{\rho}{\rho_0}\right)^2\right], \qquad (6.37)$$

where ρ is the coordinate perpendicular to the beam propagation direction and ρ_0 is the beam waist. If a weak laser beam is launched into this region at $\rho = a$ and at an angle θ to the z-axis, calculate the path of the weak beam $\rho(z)$ in the parabolic approximation $\rho(z) \ll \rho_0$. Recall that the universal law of refractive bending is given by

$$\frac{d^2\rho}{dz^2} = \frac{1}{n(\rho)}\frac{dn(\rho)}{d\rho}. \qquad (6.38)$$

You may assume that the probe beam is a classical ray.

6.4 Nonlinear Optics at Negative Temperature

The temperature scale bottoms out at zero degrees Kelvin, called absolute zero. At absolute zero, all the atoms, molecules etc are in their grounds states, so have no more energy to emit. One method for determining the temperature is to measure the populations of states and compare them with predictions from the Boltzmann distribution, as follows.

For simplicity, we consider the two lowest energy states of a system with energies E_0 and E_1. The fraction of the molecules in there ground state is P_0 and in the first excited state is P_1, where $E_1 > E_0$. We thus call the P_is the populations. Then, the Boltzmann distribution predicts that the population ratio of the two states is

$$\frac{P_1}{P_0} = \frac{e^{-E_1/kT}}{e^{-E_0/kT}} = e^{-E_{10}/kT}, \qquad (6.39)$$

where $E_{10} = E_1 - E_0$ and k is Boltzmann's constant. Solving for the temperature T, we get

$$T = \frac{E_{10}}{k \ln\left(\frac{P_0}{P_1}\right)}. \qquad (6.40)$$

Thus, an observation of the relative populations gives the temperature.

When the material is in its ground state, all the molecules will be in their ground states so $P_0/P_1 \to \infty$. Then, Equation 6.40 gives $T = 0$, as is expected. When the system is heated to infinite temperature, all states are equally populated, so $P_0/P_1 = 1$ and the temperature determined from the populations approaches infinity. Thus, it would seem that the allowable range of temperature is $0 < T < \infty$.

What if we use light to add energy to the system? If the system starts in its ground state, all the "molecules" in the system are in their ground states. An absorbed photon will excite a molecule. As more photons are absorbed, the excited state population grows provided that the molecules are excited on time scales far shorter than their radiative lifetimes. However, as we saw in Section 4.2.4, a molecule in its excited state can be de-excited by stimulated emission. When the two states are equally populated, the probability of an excitation equals the probability of stimulated emission. No matter how intensely the material is pumped, the excited state population can not exceed the ground state one.

There are tricks to fully populating the excited state. For example, if the two states are due to magnetic splitting of a spin up and spin down state, one can quickly flip the field, thus inverting the population. Then, a system originally with a population in its ground state will have all its population in an excited state. Using Equation 6.40, we see that a slightly positive temperature becomes a slightly negative one. In the limit when the upper-state population is a tad greater than the lower state population, $T \to -\infty$. On the other hand, when the excited state is a tad less populated, $T \to +\infty$. So, there is a discontinuity of the temperature when the two states are equally populated.

Here we investigate the consequences of negative temperature on the nonlinear-optical properties. The first-order susceptibility depends on population according to Equation 4.248, which in one dimension and for a two-level system is of the form

$$\chi_{xx}^{(1)} = \frac{N}{\hbar} \sum_{n=0}^{1} \sum_{m=0}^{1} \rho_{mm}^{(0)} \left[\frac{\mu_{mn}^i \mu_{nm}^j}{\omega_{nm} - \omega_p - i\gamma_{nm}} + \frac{\mu_{nm}^i \mu_{mn}^j}{\omega_{nm} + \omega_p + i\gamma_{nm}} \right]. \quad (6.41)$$

At zero temperature, $\rho_{00} = 1$ and $\rho_{11} = 0$, so we get the usual expression given by Equation 4.182 for this special case, or

$$\chi_{xx}^{(1)}\bigg|_{T=0^+} = \frac{N}{\hbar} \sum_{n=0}^{1} \left[\frac{\mu_{0n}^i \mu_{n0}^j}{\omega_{n0} - \omega_p - i\gamma_{n0}} + \frac{\mu_{n0}^i \mu_{0n}^j}{\omega_{n0} + \omega_p + i\gamma_{n0}} \right], \quad (6.42)$$

where $T = 0^+$ is the positive side of zero temperature.

Nonlinear Optics: A student's perspective

When $\rho_{00} = 0$ and $\rho_{11} = 1$, only the excited state is fully populated and Equation 6.41 becomes

$$\chi_{xx}^{(1)}\Big|_{T=0^-} = \frac{N}{\hbar}\sum_{n=0}^{1}\left[\frac{\mu_{n0}^i \mu_{0n}^j}{\omega_{0n}-\omega_p-i\gamma_{0n}} + \frac{\mu_{0n}^i \mu_{n0}^j}{\omega_{0n}+\omega_p+i\gamma_{0n}}\right], \quad (6.43)$$

where $T = 0^-$ is the negative side of zero temperature. Note that because $\omega_{n0} = -\omega_{0n}$ and $\gamma_{n0} = \gamma_{0n}$, the first term of Equation 6.42 is the negative of the second term of Equation 6.43 and the first term of Equation 6.43 is the negative of the second term of Equation 6.42. As such,

$$\chi_{xx}^{(1)}\Big|_{T=0^-} = -\chi_{xx}^{(1)}\Big|_{T=0^+}. \quad (6.44)$$

A moment's reflection confirms that Equation 6.44 makes sense. Recall that the imaginary part of the linear susceptibility is always positive for positive temperature, so light is always absorbed by the material as it gets excited. However, when the temperature is negative, there is more excited state population than ground state population, so rather than being absorbed, the incident photon stimulates an emission, thus adding photons to the beam. As a result, the incident beam is amplified.

It is easy to show that at infinite temperature, when the two populations are the same, stimulated emission balances absorption, and the linear susceptibility vanishes. The material then becomes transparent. In the problems below, you will calculate the nonlinear susceptibilities for negative and positive temperatures.

The dependence of the first-order susceptibility on temperature is given by Equation 6.41, which can be recast in the more compact form

$$\chi_{xx}^{(1)}(T) = (\rho_{00}-\rho_{11})\chi_{xx}^{(1)}\Big|_{T=0^+} = \chi_{xx}^{(1)}\Big|_{T=0^+}\tanh\left(\frac{E_{10}}{kT}\right). \quad (6.45)$$

An interesting paper on the debate about negative temperature can be found in the American Journal of Physics.[20] Spoiler alert: A careful analysis shows that negative temperature is not prohibited as some papers have claimed.

Problem 6.4-1: Verify Equation 6.45.

Problem 6.4-2: Determine the hyperpolarizability and the second hyperpolarizability for a two-level system at $T = 0^+$, $T \to \infty$, $T \to -\infty$ and $T = 0^-$ and argue why the behavior is reasonable or unreasonable.

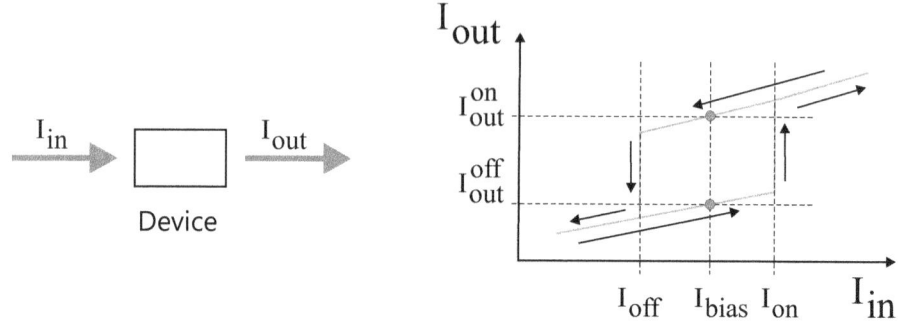

Figure 6.4: The output versus input plot of an optically bistable device is double valued.

6.5 Optical Bistability and Multistability

A bistable device can emit two distinct output intensities for a single input intensity. Then, the output versus input function is double valued for some range of input intensities. Figure 6.4 shows an example of such a response. With increased input intensity below I_{off}, the output rises, as shown by the bottom arrow. At I_{on}, the output jumps to the upper branch. Above I_{on}, the function is single valued. As the intensity is decreased, the output follows the upper path below I_{on}. When the input intensity falls below I_{off}, the output intensity drops abruptly to the lower branch.

This behavior makes an optical memory, where an output intensity of I_{out}^{off} is the zero state and I_{out}^{on} is the one state. This device is powered with a bias intensity I_{bias} that falls halfway between I_{off} and I_{on}. The bias beam enters the device with the input intensity. If the device is in its zero state, it turns on with an input pulse whose intensity exceeds $I_{on} - I_{bias}$. When in the on state, it is turned off by providing a transient drop of the bias intensity below I_{off}.

A Fabry-Perot interferometer with an intensity-dependent refractive index material between the parallel reflectors is bistable. Bistability is a special case of multistability, where many output states are possible for one input intensity. Multistable devices operate on feedback, where the output light is feed back into the material and interacts through the optical nonlinearity with the incident light.

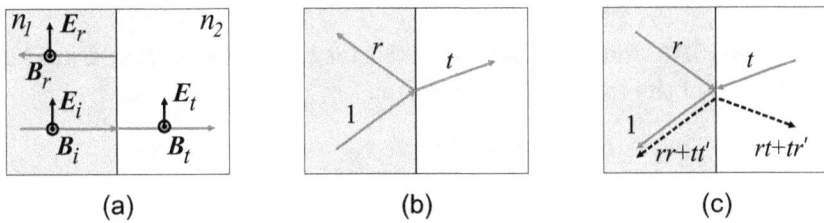

Figure 6.5: (a) Reflected and transmitted plane waves at an interface from an incident plane wave at normal incidence. (b) Ray at at arbitrary incident angle with reflected (r) and transmitted (t) rays. (c) Time-reversed rays (solid), where the dashed ray on the right must vanish and the dashed ray on the left must add to unity by reciprocity.

6.5.1 Reflections from an interface

As a prequel for understanding bistability, we make a small diversion to treat the linear Fabry-Perot interferometer. We begin by calculating the reflectivity and transmittance from an interface between two materials. Figure 6.5a shows an interface between two materials with refractive indices n_1 and n_2 with a ray incident at normal incidence. All the electric field vectors are arbitrarily chosen to point up while the magnetic fields point out of the page. There is no need to pick the correct relative directions because the calculations will provide them.

Since the tangential component of the electric field is continuous across an interface, and at normal incidence all electric fields are tangential, we get

$$E_i + E_r = E_t, \tag{6.46}$$

where we have equated the total field on the left to the total field on the right. The magnetic field is related to the electric field through Maxwell's Equations, so applying Equation 1.20 to a plane wave yields

$$\boldsymbol{B}_j = i n_j \hat{\boldsymbol{k}} \times \boldsymbol{E} \tag{6.47}$$

for the jth medium. Continuity of the tangential component of the magnetic field from Equation 6.48 yields

$$n_1 E_i - n_1 E_r = n_2 E_t, \tag{6.48}$$

where the negative sign comes from the cross product because the reflected wave travels to the left.

Equations 6.46 and 6.48 can be solved for the reflected and transmitted fields in terms of the incident one, yielding

$$E_r = \frac{n_1 - n_2}{n_1 + n_2} E_i \equiv rE_i \quad \text{and} \quad E_t = \frac{2n_1}{n_1 + n_2} E_i \equiv tE_i, \quad (6.49)$$

In a lossless medium, the refractive indices are real, so the reflected electric field's phase shifts by π when $n_2 > n_1$ but the transmitted field's phase remains unchanged. Also, r and t are real.

> **Problem 6.5.1-1:** Calculate the intensities of the three fields with the help of Equations 6.49 and use them to show that the flow of energy through the interface is conserved.

6.5.2 Reciprocity

Consider a wave of unit electric field incident on the interface at arbitrary angle as shown in Figure 6.5b. A transmitted and reflected ray results as shown. Figure 6.5c shows the time-reversed rays (solid lines), which must be the correct solutions to Maxwell's Equations – a property called reciprocity. Thus, launching fields with amplitudes r and t at the interface results in only one wave of unit field.

Alternatively, we can assume ignorance of the fact that the solid outgoing ray of unit amplitude in Figure 6.5c is the solution and ask what the reflected and transmitted rays are for the two incident rays. If the transmittance of the ray traveling from the left is t' and the reflectance is r', then the sum of the contributions of the two incident rays gives the dashed rays. Since there is no outgoing wave in the righthand medium, $rt + tr' = 0$. Since the outgoing wave in the left-hand medium must be of unit amplitude, $r^2 + tt' = 1$. Putting these together yields

$$r' = -r \quad \text{and} \quad tt' = 1 - r^2. \quad (6.50)$$

Equations 6.50 are useful when evaluating the reflections from the interfaces of a slab of material.

6.5.3 The Fabry-Perot Interferometer

The two surfaces of a single slab of material act as reflectors, from which a Fabry-Perot interferometer can be made. The simplicity of using a single slab

Nonlinear Optics: A student's perspective

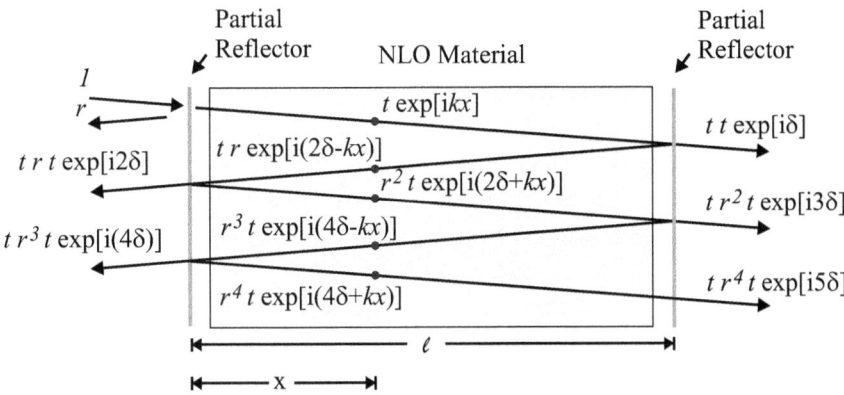

Figure 6.6: A Fabry-Perot interferometer with a nonlinear material between partial reflectors. Shown is the incident ray of unit field strength as well as a few of the reflected and transmitted rays. Each reflector has a transmittance of t and a reflectance of r. The points show the field at position x on each ray.

is outweighed by the fact that high reflectivity requires that the material be of high refractive index, so only a small fraction of the incident electric enters the cavity. A better design uses individual partial reflectors that straddle the material.

Figure 6.6 shows a Fabry-Perot interferometer. We chose to use partial reflectors and a nonlinear material within rather than using surface reflections from the nonlinear material. As such, each reflector has a reflectance r and a transmittance t for both left- and right-going beams. If we had instead used the surface reflections from a slab, then r and t would each be different for a beam entering or exiting the material. Then the reciprocity relations given by Equation 6.50 would come in handy.

We only consider a lossless material, lossless reflectors, and rays at normal incidence, but the rays are angled in the figure so that they can be more easily seen. A portion of a ray exits upon each reflection, so the left- and right-going beams are a superposition of all the rays, whose amplitudes are determined from the number of reflections and phases determined from the number of traversals of the cavity.

We start by summing over all the rays to calculate the transmitted electric field, which yields

$$E_t = E_{in} t^2 \exp[i\delta]\left[1 + (r\exp[i\delta])^2 + (r\exp[i\delta])^4 + \ldots\right] \quad (6.51)$$

$$= \frac{t^2 \exp[i\delta]}{1 - r^2 \exp[2i\delta]} E_{in}, \quad (6.52)$$

where $\delta = 2\pi i n \ell/\lambda$ is the accumulated phase traversing the length ℓ once. The infinite series is summed using $\sum_{n=0}^{\infty} \epsilon^2 = 1/(1-\epsilon^2)$. The factor $t^2 \exp[i\delta]$ accounts for the fact that all rays that exit on the right are transmitted through front and back reflectors once, thus the factor of t^2 and the single traversal of the cavity gives the phase $\exp[i\delta]$. The second term, $(r\exp[i\delta])^2$, corresponds to the ray that has reflected once from the back and front reflector before exiting, the next term for two round trips, etc. Each term also accumulates an additional phase due to the round-trip traversal of the cavity.

Recalling that the intensity is proportional to the square of the magnitude of the electric field and the refractive index of the material, Equation 6.52 yields the transmitted intensity

$$\boxed{I_t = \frac{I_{in}}{1 + \left(\frac{2r}{1-r^2}\right)^2 \sin^2 \delta}}, \quad (6.53)$$

where we have used the fact that $1 - r^2 = t^2$.

The transmittance (I_t/I_{in}), given by Equation 6.53, is plotted as a function of the optical path length, $\delta = 2\pi \ell n/\lambda$ in Figure 6.7. The period of this function depends on the spacing between the mirrors, the refractive index of the medium, and the wavelength of the light. As the reflectance of the mirrors increases, the fringes become sharper and the contrast increases.

Problem 6.5.1-2: Determine the reflected intensity using the same approach that leads to Equation 6.53 and show that the power flowing into the cavity is balanced by the power flowing out into the transmitted and reflected beams.

6.5.4 Multistability of a Fabry-Perot Interferometer with End Reflectors

Multistability arises when the optical path length inside the material depends on the intensity of light inside, thus providing nonlinear feedback. The most common demonstrations of optical multistability are for systems where the refractive index depends on the intensity. The results will be the same if the length of the material depends on the intensity.[21] Here, we focus on the intensity-dependent refractive index.

The intensity inside the interferometer varies with position because of the interference between the left- and right-going waves. The electric field at

Nonlinear Optics: A student's perspective 297

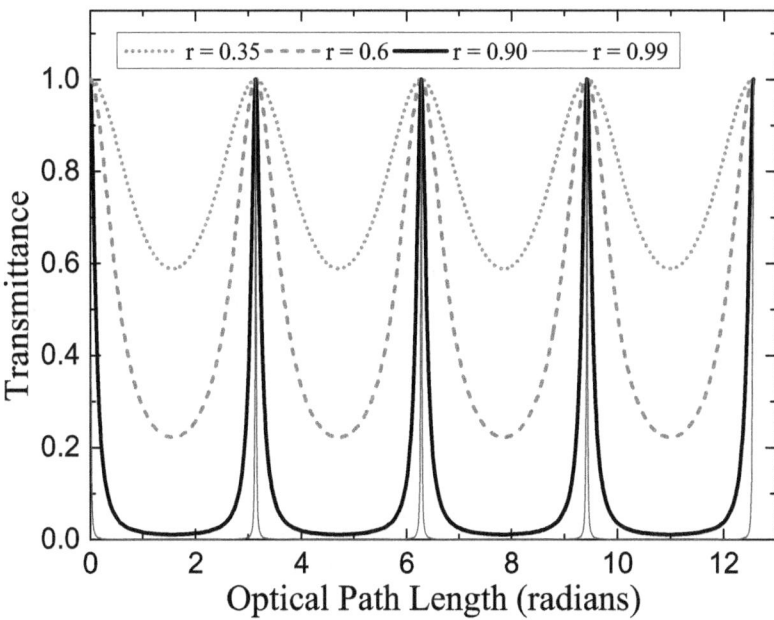

Figure 6.7: Transmittance of a Fabry Perot interferometer as a function of optical path length between the reflectors for four reflectance values.

some point x inside the material is obtained by summing over all the rays, which yields

$$\begin{aligned} E_c &= E_{in} t \left(\exp[ikx] [1 + (r \exp[i\delta])^2 + (r \exp[i\delta])^4 + \ldots] \right., \\ &+ \left. r \exp[2i\delta] \exp[-ikx] [1 + (r \exp[i\delta])^2 + (r \exp[i\delta])^4 + \ldots] \right), \end{aligned}$$
(6.54)

where x is the distance from the first reflector and $k = 2\pi n/\lambda$.

Equation 6.54 associates a ray with each term, as follows. The pre-factor t appears once since all of the light inside the interferometer is transmitted through the first reflector only once (to get inside from the left). The first line in Equation 6.54 represents the right-going rays and the second line the left-going ones. The factor $\exp[ikx]$ is the right-going wave (the factor $-i\omega t$ in the exponent is omitted for the sake of clarity and will be averaged over when calculating the intensity). The first term in the sum (the "1") represents the transmitted ray. The next term in the sum represents the ray that has reflected from the right reflector, then the left one; hence, the term goes as the square of the reflectance. The round trip phase upon this double bounce is 2δ where $\delta = k\ell = 2\pi n\ell/\lambda$.

The left-propagating rays (second line in Equation 6.54) suffer an extra reflection, hence the extra r pre-factor. The first term in the brackets (the "1" term) corresponds to the ray that has reflected from the right reflector once, the next term corresponds to two turns, etc. The pre-factor $\exp[2i\delta]\exp[-ikx]$ represents the accumulated phase of a ray at position x after one bounce inside the interferometer, which has travelled a distance ℓ to the right and a distance $\ell-x$ to the left - leading to a total phase of $k(2\ell-x)$.

Summing the infinite series and combining terms, we get the electric field between the reflectors:

$$\frac{E_c}{E_{in}} = \frac{t}{\exp[-i\delta] - r^2 \exp[i\delta]} \times [\exp[i(kx-\delta)] + r\exp[-i(kx-\delta)]]. \tag{6.55}$$

Taking the square of the modulus of Equation 6.55 and multiplying by the refractive index, we get the ratio of the intensity inside to the incident intensity:

$$\boxed{\frac{I_c}{I_{in}} = \frac{n|E_c|^2}{|E_{in}|^2} = \frac{n}{1-r^2}\left[\frac{(1+r)^2 - 4r\sin^2(kx-\delta)}{1+\left(\frac{2r}{1-r^2}\right)^2 \sin^2\delta}\right].} \tag{6.56}$$

When the reflectivity approaches unity, the intensity inside the cavity diverges, which can be understood from the fact that it takes an infinite number of bounces for the light to leave the cavity. As a consequence, it also takes an infinite time for the light to approach infinite intensity. Keep in mind that the solutions presented here are the steady-state values.

A Fabry-Perot cavity is therefore a light intensifier, which makes it useful as part of a device whose response depends on the intensity. Many laser designs use a Fabry-Perot cavity filled with a gain medium. Anyone who has accidentally placed a finger inside the cavity of a modestly-power laser can attest to the fact that the power inside is much larger than the emitted beam's power.

It is also convenient to relate the intensity inside the cavity to the transmitted intensity. Using Equation 6.53 to eliminate I_{in}, Equation 6.56 yields

$$\boxed{I_c = \frac{nI_t}{1-r^2}\left[(1+r)^2 - 4r\sin^2\left(\frac{2\pi n}{\lambda}(x-\ell)\right)\right],} \tag{6.57}$$

where we have explicitly displayed $k = 2\pi n/\lambda$ and $\delta = 2\pi n\ell/\lambda$ for purposes of evaluating the integrals that follow. Note that Equation 6.57 also diverges for $r = 1$.

The first term in brackets is a constant while the second term describes the variation in intensity due to interference between the counter-propagating waves in the interferometer. The mean intensity inside the interferometer is given by:

$$\begin{aligned}\bar{I}_c &= \frac{1}{\ell}\int_0^\ell I_c dx = \frac{nI_t}{(1-r^2)\ell}\left[(1+r^2)x - \frac{\lambda r}{2\pi n}\sin\left(\frac{4\pi n}{\lambda}(x-\ell)\right)\right]\Big|_0^\ell \\ &= \frac{nI_t}{1-r^2}\left[(1+r^2) + \frac{r}{2\pi n}\left(\frac{\lambda}{\ell}\right)\sin\left(\frac{4\pi n \ell}{\lambda}\right)\right],\end{aligned} \quad (6.58)$$

where we have used the trigonometric identity $\sin^2\theta = (1-\cos 2\theta)/2$. Clearly, by n we mea

For a long device, where $\ell \gg \lambda$, the second term in brackets in the second line of Equation 6.58 will be much smaller than the first term, which is on the order of unity, so the mean intensity is then

$$\boxed{\bar{I}_c = n\left(\frac{1+r^2}{1-r^2}\right)I_t.} \quad (6.59)$$

When $r = 0$, Equation 6.59 gives a mean intensity inside the cavity that equals the transmitted intensity, aside for the factor of n which accounts for the different phase velocity inside the cavity than outside. When $r = 1$, the intensity inside diverges. Placing a material in a cavity can thus enhance the nonlinear-optical response through the intensification of the intensity.

Consider a material whose refractive index depends on the intensity, which is placed in a spatially-modulated intensity of the form

$$I(x) = \bar{I} + A\sin(kx). \quad (6.60)$$

For a refractive index change that is linearly proportional to the intensity,

$$n(I) = n + n_2 I(x), \quad (6.61)$$

the average refractive index change over the length of the material is

$$\delta\bar{n} = n(I) - n_0 = \frac{1}{\ell}\int_0^\ell dx\, n_2\left(\bar{I} + A\sin(kx)\right). \quad (6.62)$$

Evaluating the integral in Equation 6.62 yields

$$\delta\bar{n} = n_2\left(\bar{I} + \frac{2A}{k\ell}\sin^2\left(\frac{k\ell}{2}\right)\right). \quad (6.63)$$

When $k\ell \gg 1$, Equation 6.63 for the intensity in the material takes the simple form
$$\delta \bar{n} = n_2 \bar{I}_c. \tag{6.64}$$
For a thin cavity where $k\ell \ll 1$, Equation 6.63 becomes
$$\delta \bar{n} = n_2 \left(\bar{I} + \frac{k\ell A}{2} \right) \to n_2 \bar{I}, \tag{6.65}$$
the same result as for the thick limit given by Equation 6.64. Finally, when $k\ell = (2m+1)\pi$, where m is an integer, Equation 6.63 becomes
$$\delta \bar{n} = n_2 \left(\bar{I} + \frac{2A}{k\ell} \right). \tag{6.66}$$
The calculations that follow assume the long cavity limit since that is the regime that yields the largest effect.

Our goal is to get an expression that relates the transmitted intensity – the output of the device – to the incident intensity. However, since the output intensity is a muti-valued function of the input intensity, our strategy is to express the input intensity in terms of the output intensity, then take the inverse. To make it possible to get the inverse, it is most convenient to express the intensity inside the cavity in terms of the transmitted intensity. This may sound confusing, but it will make more sense as we work through the details.

Substituting Equation 6.59 into Equation 6.64 gives the average refractive index change in the cavity, which when when substituted into Equation 6.53 yields
$$\boxed{I_t = \frac{I_{in}}{1 + \left(\frac{2r}{1-r^2}\right)^2 \sin^2\left(\frac{2\pi \ell n_0}{\lambda}\left[1 + n_2 \left(\frac{1+r^2}{1-r^2}\right)I_t\right]\right)}}, \tag{6.67}$$
where n_0 is the refractive index of the material in the cavity when $I = 0$ and n_2 is the intensity-dependent refractive index. We can easily express Equation 6.67 in terms of I_{in} as a function of I_t by multiplying both sides by the messy denominator, and the result is a single-valued function. However, it is not possible to solve for I_t as a function of I_{in}.

There are two ways to attack Equation 6.67. The simplest is to numerically evaluate I_{in} as a function of I_t, then plot the values of I_t as a function of I_{in}. The second method is graphical. We start with the former. The inset in Figure 6.8 shows the plot of I_{in} as a function of I_t and the main figure shows a plot of the inverse. Clearly, the transmittance is a multi-valued function of the incident intensity.

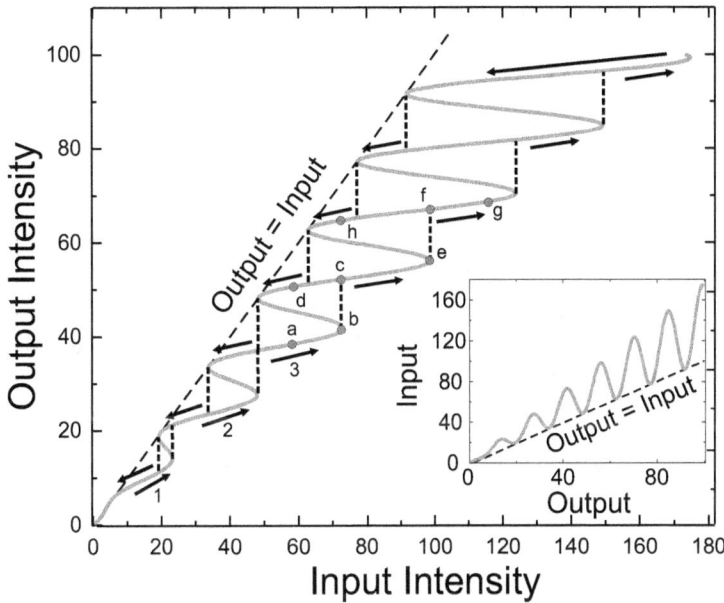

Figure 6.8: The transmitted intensity as a function of incident intensity for a multistable device. The inset shows the incident intensity as a function of the transmitted intensity.

Now, we consider in some detail how the transmitted intensity depends on the incident intensity. If we start at zero intensity and turn up the incident intensity, the transmitted intensity will follow the arrow labelled 1 in Figure 6.67, until the output intensity makes a vertical jump along the dotted line. Then the output intensity follows the curve above arrow 2, making another vertical jump, etc. As long as the incident power is being *increased*, the output will go through a series of steps.

Next consider the output at point a. If the incident intensity is increased to just beyond point b, the output will jump to point c. If now the intensity is *decreased*, the output will follow the curve from points c to d. If the intensity is continually decreased from this point, the output intensity will make a series of steps that follow the upper arrows. Clearly, this type of response leads to a series of loops, each schematically taking the form shown in Figure 6.4.

While the very first loop (near arrow 1) resembles Figure 6.4, the higher loops are connected. For example, above point b, there are two additional branches along points c and h, so in this region, we observe tri-stability. As input intensity is increased to ever higher values, it is possible to have more and more stable output states for a single incident intensity. The number of stable states in the multi-stable regions can be arbitrarily large.

Even when a system appears to be bistable, it may be multi-stable. For example, if the system is in a state given by point f, a change of the input intensity that takes the system between points h and g keeps it on the same branch. If the intensity falls far enough below h the system jumps down to the d-c-e branch. Furthermore, observing a single loop is not a guarantee that other loops are not lurking above at higher intensities for that same input intensity range. For example, the loop defined by the branches a-b and c-d could be traced out in an experiment without seeing the higher-intensity stable output at point h.

When multistability is due to the intensity-dependent refractive index, it is not possible to access a state inside of the loop because the only controllable parameter is the incident intensity. Thus, it is not be possible to get the system to a state corresponding to the curve between points a and d or points c and h. However, if the effect is due to an intensity-dependent length change mechanism, it may be possible to mechanically shock the system to bring it to one of the interior curves. Photomechanical effects are beyond the scope of this book, but deserve further study.

Problem 6.5.1-3: Repeat the multi-stability calculation for a slab of a lossless material of refractive index n surrounded by vacuum where the reflections come from the fresnel factors given by Equation 6.49. You may assume that the fresnel factors are independent of intensity, which of course they are not. Plot your results in the form of Figure 6.8. Repeat the calculation using Python and include the effect of intensity-dependent Fresnel reflection/transmission. Plot the results on the same graph as before to assess the approximation that the nonlinear Fresnel contributions are negligible.

6.5.5 Graphical Solution to Transcendental Equations

Solving an equation "by hand" with paper and pencil often leads to a deeper understanding than blindly plugging equations into a computer for numerical evaluation as we have done in generating Figure 6.8. Here we demonstrate how to solve Equation 6.67 with a graphical method. Equation 6.67 is of the form:

$$\frac{I_t}{I_{in}} = f(I_t), \qquad (6.68)$$

where $f(I_t)$ is a periodic function of I_t. The idea behind the graphical solution is to plot the lefthand and righthand sides of Equation 6.68 as a function of I_t. The intersections between these two curves yields the solution.

Figure 6.9a shows a schematic representation of the periodic function $f(I_t)$ and I_t/I_{in} for several values of the incident intensity I_{in}. At low intensity, the slope of the line given by I_t/I_{in} versus I_t is large and the oscillating curve intersects this line at one point, labelled a. In the vicinity of point a, the solution is single-valued. For a higher intensity, the slope of the line decreases and there are two intersections at points g and h. This critical line corresponds to the incident intensity that has two solutions and is the start of the bistable regime as shown in Figure 6.9b. At higher intensities as represented by the smaller sloped line bf, there are three intersections. Points b and f are the bistable solutions and the open point corresponds to the unstable solution, which is inside the hysteresis loop as shown in the Figure 6.9b.

As the intensity is increased further, another critical line is reached with intersection points c and e, above which the function appears single valued. In this regime, the first hysteresis loop closes, but another loop opens (not shown) due to intersections at higher I_t that is outside the range of the curves

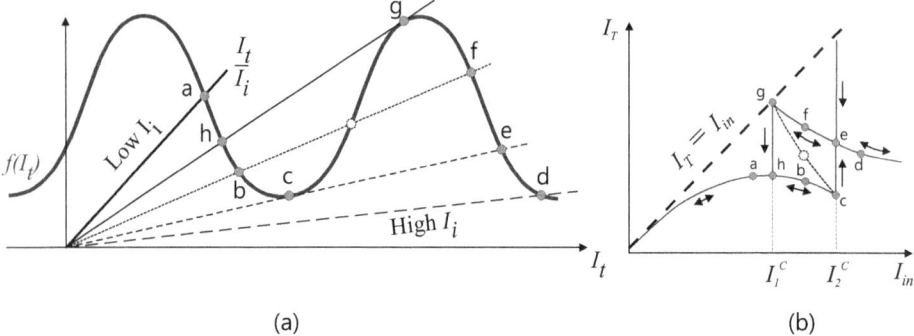

Figure 6.9: (a) $f(I_t)$ and I_t/I_{in} as a function of I_{in} for five different intensities. (b) The corresponding intersection points on the bistable curve. The dashed line inside the loop shows the disallowed region.

drawn in the figure. The number of intersections increases at higher intensities and the system becomes multistable with ever more tangent loops since the smaller-sloped lines will intersect more periods of the oscillating function. The graphical method can be used to obtain a sketch of the hysteresis loop, as shown in Figure 6.9b. Try convincing yourselves that this really works using paper, pencil and ruler.

> **Problem 6.5.1-4:** Implement the graphical technique using Python and use it to trace out several loops, plotting the results in a graph similar to Figure 6.9b.

6.6 Evaluating Materials for Applications and Devices

6.6.1 Figure of Merit

A device usually has certain required characteristics that are dictated by applications. While one might think that the magnitude of the nonlinear-optical response is the most important property for making a device, it is not the only factor. For example, consider an intensity modulator, which can be used to pass or block a laser beam in response to a voltage. Such devices require an electro-optic material, which modulates the phase of a beam of light. When built into a Fabry-Perot cavity, a π phase change of the light in the cavity will result in a switching from the fully transparent state to the blocked state.

The phase shift of the light beam due to the electro-optic effect is given by

$$\delta\phi = \frac{2\pi}{\lambda} n_1 E d, \qquad (6.69)$$

where λ is the wavelength of the light, n_1 is proportional to the second-order susceptibility $\chi^{(2)}(-\omega;\omega,0)$ as given by Equation 2.102, E is the applied electric field and d the distance between the reflectors.

A phase shift $\Delta\phi = \pi$ to switch the light on and off for fixed n_1 can be accomplished with various combinations of the applied electric field and device length. However, it is not so simple because other constraints limit the magnitude of the applied electric field and the length of of the device. These constraints can be fundamental, where basic physics might provide the roadblock, or technological, where new tools need to be developed to make further improvements possible.

For example, large enough electric fields to induce a π phase shift might cause catastrophic currents to course through the sample like little lightening bolts, damaging the optical quality and increasing the conductivity – traits that render the device inoperable. Impurities provide a conduction path that leads to damage, making it a technological constraint based on the efficacy of the best purification processes available. A pristine material that is free from impurities may have a larger threshold for breakdown, but is limited by intrinsic properties of the material that cannot be further optimized by processing.

Making the device longer allows for lower electric fields to make it work, but as length is increased, optical loss increases. Optical loss also has intrinsic and extrinsic contributions. The scattering loss from impurities and inhomogeneity from clumping of molecules can be minimized through purification and inhomogeneity can be reduced through processing such as annealing. When such extrinsic properties are fully removed, the material itself will still absorb light. This intrinsic loss cannot be further decreased short of changing the material used to make the device.

For device requirements to be met often requires a holistic approach that takes advantage of all the relevant materials properties. A single quantity that accounts for all the factors for a specific device is called the figure of merit (FOM), which when optimized gives the optimum combination. In our example of a modulator, one could imagine a figure of merit FOM_{EO} of the form

$$FOM_{EO} = \frac{n_1 E_D}{\alpha_L}, \qquad (6.70)$$

where n_1 is the nonlinearity, and E_D the electric field at the damage threshold. A large number of materials can then be made and tested for the three critical properties and selected based on the figure of merit.

Let's assume for the sake of argument that all of the extrinsic effects are removed by purification, processing, etc. Then, all of the parameters in Equation 6.70 are the intrinsic ones. Next we seek to understand how the intrinsic parameters are related to the fundamental ones. All theories at some point are reduced to fundamental assumptions. Recall that a material is described as a collection of charges in which a polarization P is induced in response to an electric field E according to Equation 1.24. The electric displacement D is obtained by taking the sum of the applied electric field and the contribution from the polarization using Equation 1.24. The electric displacement is then used to get Maxwell's wave equation, which is obtained from Equations 1.20 and 1.21.

The fundamental quantities that define the material are thus the susceptibilities $\chi^{(n)}$. At the quantum level, the induced dipole moment comes from the distortions of the electron clouds of the molecules from which the material made. The molecular response, in turn, is expressed as a function of the dipole moment matrix and the energy spectrum, as given by the sum-over-states expressions as in Equation 4.187. Mossman et al used this approach to define a figure of merit if the electro-optic effect for a dye-doped polymer in terms of the quantum parameters. Using the theory of fundamental limits of the susceptibilities, the authors were able to cast the figure of merit in terms of quantum parameters to determine the quantum ceiling of the performance of an electro-optic device.[22]

To summarize, the linear and nonlinear susceptibilities are the fundamental macroscopic properties that describe how a material interacts with light, and the quantum properties from which they derive are the dipole matrix elements and the energy eigenvalues . One might argue that certain mechanisms are classical in nature, so the quantum origins are not relevant. That may be true in some regimes, but if we are interested in the largest possible nonlinear-optical response, it must originate in the quantum realm since constructive interference between wave functions is needed to reach the maximum and the classical response by definition lacks such interference.[19]

6.6.2 Epsilon Near Zero

An example of a non-fundamental composite property is the intensity-dependent refractive index. Boyd and coworkers have shown indium tin oxide

(ITO) to have a large intensity-dependent refractive index due to the dielectric constant being near zero (ENZ).[23] The underlying principles can be understood by considering the dielectric constant, ϵ, which depends on the electric field E according.

$$\epsilon = \epsilon_0 + \epsilon_1 E + \epsilon_2 E^2 + \ldots \qquad (6.71)$$

Ignoring the linear electro-optic effect, which vanishes in a centrosymmetric material, the lowest-order nonlinearity is due to ϵ_2. Then, the refractive index is calculated from Equation 6.71, leading to

$$n = \sqrt{\epsilon} = \sqrt{\epsilon_0}\left(1 + \frac{\epsilon_2 E^2}{\epsilon_0}\right)^{1/2} \equiv n_0\left(1 + \frac{\epsilon_2 E^2}{n_0^2}\right)^{1/2}, \qquad (6.72)$$

where n_0 is the zero-intensity refractive index. Since the change in the refractive index is typically small compared with n_0, Equation 6.72 can be approximated to order E^2, yielding

$$n \approx n_0 + \frac{\epsilon_2 E^2}{2n_0}. \qquad (6.73)$$

Note that Equation 6.73 requires that $\frac{\epsilon_2 E^2}{2n_0} \ll n_0$. Otherwise, higher-order terms come into play.

Equation 6.73 seems to imply an infinite refractive index change when $n_0 \to 0$. This limiting case violates the assumption that the second term in Equation 6.73 is small compared with the first. As long as the small refractive index change condition holds, a considerable enhancement is possible provided that n_0 can be varied while keeping ϵ_2 fixed. For example, if the refractive index can be lowered from 1.5 to 0.15, a tenfold enhancement is found. The lower the refractive index, the greater the enhancement.

The series expansion leading to Equation 6.73 needs to be re-evaluated when the refractive index change is large compared to the refractive index. In this case, where the refractive index vanishes so that $\frac{\epsilon_2 E^2}{n_0^2} \gg 1$, Equation 6.72 is approximated by

$$n \approx \sqrt{\epsilon_2 E^2}. \qquad (6.74)$$

Thus, the refractive index change for small refractive index is independent of the refractive index and increases linearly with the electric field.

When the real part of the relative permittivity vanishes in indium tin oxide (ITO), $\epsilon = 0.4i$, so the magnitude of the refractive index remains nonzero. The degree of enhancement is thus not arbitrarily large. To understand how one might take best advantage of the enhancements is to study how the material nonlinearity and the refractive index depend on the underlying quantum parameters (transition moments and energies), since all optical properties are given by Equations of the form of Equation 4.187. Even when the fundamental nonlinearity γ is bounded, the refractive index and the intensity-dependent refractive index are not. It is thus possible for ratios of quantities to give infinite results when the denominator vanishes.

This is an important point that is missed in the quest for making materials with a large nonlinear-optical susceptibility. The nonlinear susceptibility alone may not be the quantity of interest. Rather, composite properties such as the figure of merit for an application may be a more appropriate parameter for optimization. In fact, a figure of merit may be large when the nonlinearity is small. Thus, materials design would be more effective when the figure of merit is the quantity targeted. This ground-up approach for electro-optic modulators shows that low optical loss may be more important than high nonlinearity.[24]

6.7 Single Molecule Nonlinear Optics

The Feynman diagrams of Section 4.6 describe photons interacting with a quantum system, and are usually applied to molecules. Here we will stress nano-scale quantum systems, which occupy that fuzzy region where quantum effects transition to classical behavior. The simplest all-optical process is second harmonic generation with two incident photons, each at frequency ω, and one emitted photon at frequency 2ω, as shown in Figure 4.9 with $\omega_1 = \omega_2 = \omega$. This section investigates the experimental detection of such a process.

Rather than considering a specific molecule, we will use the sum rules to determine the largest possible off-resonant hyperpolarizability, as we did for $\chi^{(1)}$ using Equation 4.277. Since resonance enhancement can considerably increase the hyperpolarizability, we are presenting a somewhat conservative estimate of a ceiling. This will develop intuition for the intensity required to detect signal.

The calculation of the upper limit of the hyperpolarizability β_{max} is somewhat tedious,[25] but in the same spirit as the approach used to getting the limits of the linear susceptibility given by Equation 4.277. We thus state the result without derivation,[26]

$$\beta_{max} = \sqrt[4]{3}\left(\frac{e\hbar}{\sqrt{m}}\right)^3 \frac{N_{el}^{3/2}}{E_{10}^{7/2}}, \tag{6.75}$$

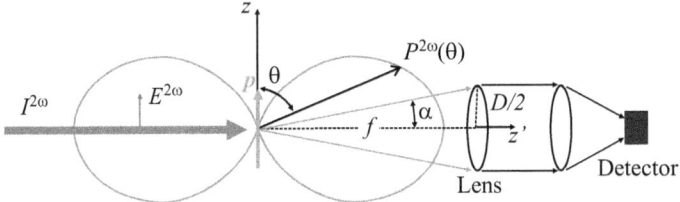

Figure 6.10: The second harmonic power from a single molecule is collected with a lens and focused into a detector.

where N_{el} is the number of participating electrons and $E_{10} = E_1 - E_0$ is the energy difference between the first excited state and the ground state, m the electron mass and $-e$ the electron charge. β_{max} is the largest tensor component, which also must be a diagonal one.

Optimizing the nonlinear-optical response is an exercise in getting all the electrons to respond fully to an oscillating electric field. Another somewhat involved calculation shows that a one-dimensional system gives the largest nonlinearity because the quantum states transverse to the polarization of the light are not excited so those electrons are wasted.[19] This is not surprising since the hyperpolarizability is defined in the electric dipole approximation, which is by definition a vector.

Equation 6.75 can be converted to the more experimentally-friendly form

$$\beta_{max}[cm^5/statcoul] = 1186 \times 10^{-30} \frac{N_{el}^{3/2}}{E_{10}^{7/2}[eV]}, \quad (6.76)$$

where the energy is in electron volts – units familiar to a physicist, or

$$\beta_{max}[cm^5/statcoul] = 559 \times 10^{-30} \cdot N_{el}^{3/2} \cdot \lambda^{7/2}[\mu m], \quad (6.77)$$

where λ is the wavelength of maximum absorption in micrometers – units that are familiar to spectroscopists.

We are now ready to calculate the detected second harmonic power from a single molecule. The angular dependence of the radiated power, $P^{2\omega}$, from an electric dipole oscillating at frequency 2ω, $\boldsymbol{p}^{2\omega}$, is given by[10]

$$\frac{dP^{2\omega}}{d\Omega} = \frac{1}{8\pi c^3}(2\omega)^4 |\boldsymbol{p}^{2\omega}|^2 \sin^2\theta, \quad (6.78)$$

where θ is the polar angle. Figure 6.10 shows a polar plot of the radiated power and the rays that are captured by a lens at $\theta = \pi/2$.

The induced dipole moment at frequency 2ω is the source of the second harmonic light, which is a response to a beam with an electric field that oscillates at frequency ω through Equation 1.32. The relevant term here is

$$p^{2\omega} = \frac{1}{2}\beta(E^\omega)^2, \tag{6.79}$$

where the factor of 1/2 comes from the degeneracy perfector given by Equation 2.75. Since the second-harmonic signal is the largest when the incident electric field E^ω is polarized along the long axis of the molecule, which is also the direction of the induced dipole moment as shown in Figure 6.10, we express all vectors and tensors as scalars and all tensors are understood to be the magnitude of that largest diagonal component.

The intensity I^ω is related to the magnitude of the electric field according to Equation 2.112, so Equation 6.79 can be written in terms of the intensity, yielding

$$p^{2\omega} = \frac{4\pi}{c}\beta I^\omega, \tag{6.80}$$

where in vacuum, $n_0 = 1$. Equation 6.80 is more convenient than Equation 6.79 because the intensity is an easy-to-measure property of a light beam, and the electric field is not.

We use Equation 6.78 to determine the power collected by the detector at $\theta = \pi/2$, where the radiated power is highest. A lens of diameter D is placed a focal-length distance f from the source. All the light in the solid angle Ω caught by the lens reaches the detector. We can calculate the solid angle subtended by the lens more easily using a new spherical coordinate system with symmetry axis z' shown in Figure 6.10. The solid angle is then given by

$$\Omega = \int_{-\alpha}^{+\alpha} d(\sin\beta) \int_0^{2\pi} d\gamma = 4\pi \sin\alpha = \frac{4\pi}{\sqrt{1+\left(\frac{2f}{D}\right)^2}}, \tag{6.81}$$

where the last step expresses $\sin\alpha$ from the triangle defined in Figure 6.10. Integrating Equation 6.78 over the solid angle and using Equation 6.81, we finally get the power at the detector

$$P^{2\omega} = \frac{(2\omega)^4}{2c^3} \frac{|p^{2\omega}|^2}{\sqrt{1+\left(\frac{2f}{D}\right)^2}}, \tag{6.82}$$

where we assume that the power is approximately uniform over the lens.

Nonlinear Optics: A student's perspective

Substituting Equation 6.80 into Equation 6.82, we get the power in the detector as a function of the incident intensity of the fundamental

$$P^{2\omega} = \frac{128\pi^2}{c}\left(\frac{\omega}{c}\right)^4 \frac{\beta^2 (I^\omega)^2}{\sqrt{1+\left(\frac{2f}{D}\right)^2}}. \qquad (6.83)$$

Equation 6.83 is the final result. When signal is small, photon counting techniques must be used. In such cases, Equation 6.83 is more useful if expressed in terms of the photon emission rate. Also, β will be expressed in terms of the limits of the hyperpolarizability.

Since we are interested in a ballpark estimate, we make the following assumptions, which should be attainable. The numerical aperture, defined by $N.A = D/2f$, determines the collection efficiency of the lens. Since we need only ballpark numbers, we assume that the numerical aperture is $N.A. = \sqrt{3}$, since that simplifies the final expression. The number of signal photons captured in a time interval Δt is given by

$$N_c = \frac{P^{2\omega}}{2\hbar\omega}\Delta t, \qquad (6.84)$$

where dividing by 2ω converts the power to the second harmonic photon emission rate. Substituting Equation 6.83 into Equation 6.84, using $N.A. = \sqrt{3}$ and Equation 6.77 for the hyperpolarizability, we get

$$N_c = 3.12 \times 10^{-55} \cdot \frac{32\pi^2}{\hbar\omega c}\left(\frac{\omega}{c}\right)^4 \cdot N_{el}^3 \cdot \lambda^7[\mu m] \cdot (I^\omega)^2 \Delta t. \qquad (6.85)$$

For a fundamental at $\lambda = 1\mu m$, and recalling that $\omega = ck = 2\pi c/\lambda$, Equation 6.85 becomes

$$N_c = 2.6 \cdot N_{el}^3 \cdot \left(I^\omega[GW/cm^2]\right)^2 \Delta t, \qquad (6.86)$$

where the intensity was converted into the experimental-friendly units of Gigawatts per square centimeter.

If multiple molecules or nanorod structures are laid side-by-side in an area that is smaller than the wavelength of the light, the total radiated electric field will add coherently. The total dipole moment of the n nanorods is then $np^{2\omega}$ so the radiated power given by Equation 6.82 will go as the square of the number of nanowires. Thus, for n nanowires, the number of photons produced will be given by

$$\boxed{N_c = 2.6 n^2 \cdot N_{el}^3 \cdot \left(I^\omega[GW/cm^2]\right)^2 \Delta t,} \qquad (6.87)$$

We first evaluate Equation 6.87 for a single molecule that has what is considered an exceptionally large hyperpolarizability. Assuming that 30 electrons are excited by a light beam of $100 GW/cm^2$ peak intensity and $30ps$ duration, we get $N_c = 0.02$, so it would take 50 pulses before one second harmonic photon would be detected. However, this result assumes that this molecule uses it's electrons with ultimate efficiency. The hyperpolarizability of the best molecules fall a factor of thirty below the fundamental limits, and since the second harmonic intensity is proportional to the square of the second hyperpolarizability, the second harmonic efficiency falls by a factor of 900, leading to $N_c = 2.3 \times 10^{-5}$. A laser that emits 50 pulses per second would require a runtime of 20 minutes before observing a second harmonic photon. Given the noise levels in a typical detector, it would be challenging – but possible – to observe this photon.

Next we consider nanorods. A cylindrical nanorod will not generate second harmonic light because it is centrosymmetric. The symmetry can be broken by adding a notch, bump, or small branch near one end. For illustration, consider a gold nanorod with an optical damage threshold of about $100 GW/cm^2$.[27] Somewhat higher thresholds have been observed by Johnson and coworkers, who measured second harmonic from sections of a nanorod using a $160 GW/cm^2$ intensity fundamental beam, making a second-harmonic image.[28] Such imaging enhances contrast that may not be visible when observing the image from scattered or transmitted light, as is usually the case in an optical microscope. Campagnola and coworkers used second harmonic imaging to detect structural proteins in biological tissue.[29]

The imaging studies of Johnson were applied to 70nm diameter wires made of zinc oxide. These are too large to observe quantum effects and the nonlinear-optical response is best described using classical models of the material as a dielectric. To illustrate that the classical response is suboptimal, we can estimate the expected signal from the number of electrons in the $(100 nm)^3$ probed volume if all electrons are used to maximum efficiency. As a ballpark estimate, assuming one electron per cubic angstrom, $N_{el} = 10^9$. For $I = 160 GW/cm^2$, the intensity they used, $N_c = 5 \times 10^{18}$ second harmonic photons generated per pulse. The laser used emits 1,000 pulses per second, so the 5×10^{21} second harmonic photons generated per second greatly exceeds the incident photon flux and the emitted power of $1,600 W$ greatly exceeds the incident power.

This calculation is clearly nonsensical because the second harmonic beam exceeds the fundamental power by at least six orders of magnitude. The paradox is resolved by realizing that the upper bound of the nonlinearity is so large that perturbation theory breaks down and the theory of the nonlinear susceptibility becomes invalid. Given the small signal they observed, the actual nonlinearity is many-orders of magnitude smaller than the fundamental limit, as we would expect of a nonlinear response that originates from a classical process.

Quantum behavior appears to transition to a classical response in silver spheres in the diameter range between $2nm$ to $20nm$.[30] Consider, then a $3nm \times 3nm \times 15nm$ conducting rod, which falls into this transition regime. It has about 100,000 available electrons. If excited by a laser pulse of $30ps$ duration and $100GW/cm^2$ intensity, the rod will produce about one second harmonic photon per pulse. A measurement of 100 nanorods would yield about 10,000 photons, and if the laser runs at 50Hz, that corresponds to 5,000,000 photons per second, easily detected with a photomultiplier tube. Thus, there is plenty of leeway to see smaller nanorods with fewer electrons or at lower laser power, making it possible to study the quantum to classical transition by determining the size-dependence of the nonlinear response.

Such back-of-the-envelop calculations might lead to an optimistic estimate of thresholds for measuring the nonlinear response due to the fact that the quantum efficiency of a detector is less than unity, fewer photons are collected because of geometrical constraints, the nonlinearity is well below the limit and the damage threshold might be lower than expected. Furthermore, background second harmonic form the surface of the substrate might be larger than the signal.

The background contributions can be subtracted by measuring the second harmonic signal as a function of the polarization of the light beam. The maximum signal will result when the light is polarized along the long axis of the rod and minimum when perpendicular to it. If the transverse component is negligible, the difference between the second harmonic efficiencies between the two polarizations yields the nanorod nonlinearity.

The most catastrophic damage event is melting, which can be induced when the increase in temperature that results from absorbed light elevates the temperature above the melting transition of the material. If the nanorod is isolated from the environment, heat can escape only through radiative cooling. If the light is absorbed at a faster rate than the material can cool, melting will result. The long-term buildup of heat can be minimized by placing the nanorod on a heat-conducting surface such as sapphire.

Problem 6.7-1: Using a particle-in-a-box model of the wave functions in a nanorod (make the rod a rectangular box in two dimensions for simplicity), argue why the off-resonance hyperpolarizability is larger for light polarized along the long dimension.

Problem 6.7-2: What is the temperature increase of a $3nm$-diameter and $15nm$-long silver nanorod when it absorbs a photon of wavelength $\lambda = 1\mu m$? Use this to determine the required number of absorbed photons to melt the nanorod if it is initially at room temperature. Can an $80fs$ pulse of $1mW$ peak power melt such a nanorod? Silver has a specific heat of $c = 2.3 \times 10^6 erg/gm \cdot K$, a density of $\rho = 10.5 g/cm^3$, and a melting temperature of $T = 962°C$.

List of Tables

4.1 The local field factors in one, two, and three dimensions; and, the relationship between the field inside and outside an infinite sheet, cylinder and sphere. 236

5.1 The magnitude of $\chi^{(3)}$ and the response times of several mechanisms. 254

List of Figures

1.1 Kerr observed the change in transmittance through a sample between crossed polarizers due to an applied voltage. Inset at the bottom left shows the orientations of the polarizers and the applied electric field due to the static voltage. 2
1.2 (left) A system is excited by a photon if its energy matches the difference in energies between two states. (right) Two-photon absorption results when two photons, each of energy $(E_2 - E_1)/2$, are sequentially absorbed. 3
1.3 The experiment used by Franken and coworkers to demonstrate second harmonic generation. The inset shows an artistic rendition of the photograph that recorded the two beams. 4
1.4 In the Optical Kerr Effect a strong pump laser is polarized 45° to the weak probe beam, causing a rotation of its polarization. . 5
1.5 Polarizer set perpendicular to the original wave propagation direction (the first defines the polarization of the incident wave.) . 6
1.6 Plot of the electric field across the interface of a polarized material and a diagram of the field, **E**, bound charges, and polarization, **P**. 11
1.7 Light inducing a polarization inside a material. The light source is a monochromatic plane wave. 13
1.8 A small volume element of material in an electric field 14
1.9 (a) Macroscopic view of second harmonic generation; and, (b) the quantum view. 21
1.10 (a) Macroscopic view of difference frequency generation; and, (b) the quantum view. 22
1.11 (a) Quantum view of reverse sum frequency generation; and, (b) difference frequency generation as a stimulated process of reverse sum frequency generation. 22
1.12 The intensity as a function of position on N cells and the discretized variables. 25

1.13 Plots made in Python. 30
1.14 Ratio of transmitted to incident intensity as a function of time over a 2.5s data run (points). The smooth curve is a fit to the data. 31

2.1 Charge On A Spring 36
2.2 Harmonic Oscillator Free Body Diagram 36
2.3 Susceptibility Divergence 38
2.4 Harmonic Oscillator Potential 38
2.5 Real and Imaginary Susceptibility 42
2.6 Plot of the real and imaginary parts of $\chi^{(1)}$. ω_0 is the resonant frequency of the system. 51
2.7 Resonances of $\chi^{(2)}$ in the ω_1–ω_2 plane 52
2.8 Illustration of the microscopic origin of the refractive index. The six solid vertical lines represent phase fronts of the external electric field in and outside the material and the shorter dashed lines represent the electric field generated by the induced dipoles fields (dotted curves) inside the sample. The superposition of the induced electric dipole fields and plane waves inside the sample leads to a plane wave with a decreased phase velocity, which defines the refractive index. 53
2.9 Optical Kerr Effect 70
2.10 The contour integral is composed of three parts: (1) The semicircular part of radius $R \to \infty$; (2) the real axis, excluding the singularity at $\omega = \omega'$; and, (3) the semicircular part of radius $\delta \to 0$, which is integrated using Cauchy's Integral Formula. . . . 73
2.11 The principal value of an integral is calculated by adding the areas of the trapezoids (top) which is equivalent to adding the areas of the rectangles (bottom) when the interval $\Delta\omega$ is sufficiently small. 76
2.12 (a) A molecule made of 5 atoms in the form of a pentagon is not centrosymmetric because no center of inversion point exists. (b) A hexagon, such as the benzene molecule, which is made of 6 carbon atoms and six hydrogen atoms , is centrosymmetric because ever atomic coordinate represented by a vector (shown as a solid arrow) is also the coordinate of an atom when the inversion operation is applied (dashed arrow). 82
2.13 (a) Two crystals in series through a $\chi^{(2)}$ process mimic $\chi^{(3)}$. (b) Two molecules in series through a β process mimic γ. 85

2.14 (a) Two side-by side molecules, each have polarizability α; but, only the left one has a permanent dipole moment μ. (b) Two molecules each with polarizability α in an electric field. The induced dipole moments and dipole fields are not shown. 86

2.15 A monopole, dipole and quadrupole in one dimension. A current loop is also a magnetic dipole, where its direction is given by the righthand rule. 93

2.16 (a) A sheet of dipoles act as two sheets of opposite surface charge density, which for a small-enough gap, give a uniform field between them. (b) When such sheets are stacked – as shown here in side view, the charge densities cancel for all internal sheets, so the net field is due to the two outer ones. 94

2.17 A vertical slice of material made of quadrupolar layers where the quadrupole moment increases linearly with each layer. ... 95

2.18 (a) Magnetization from an ensemble of magnetic dipoles at finite temperature and (b) induced dipole moment of a wire loop due to magnetic induction and subsequent rotation from magnetic torque. 98

2.19 (a) An electric quadrupole modeled as charges connected with two unequal springs. The moments change under the application of an electric field with a field gradient of $\delta E/a$. 101

2.20 (a) A helical molecule modeled as a resistive coil with conducting ends. (b) The equivalent circuit of the molecule, where the coil is modeled as an inductor and resister in series, and the spherical conductors as a capacitor. 105

3.1 Slowly varying envelope approximation: $A(z)$ varies negligibly over a wavelength. 115

3.2 Through the second-order susceptibility, two incident light waves of frequencies ω_1 and ω_2 interact to generate light at frequency $\omega_3 = \omega_1 + \omega_2$. Most of the incident light passes through the sample without being converted. 117

3.3 The intensity of the sum frequency light as a function of sample length ℓ under the phase-matching condition when $L_c \to \infty$ (thick solid curve), and when $L_c < \ell$ (thin solid curve and dashed curve). The period and amplitude increases with the coherence length, as seen in the sequence of three curves. 122

3.4 A layered material in which the second-order susceptibility depends on position as shown in the plot below. If the layers are much thinner than all wavelengths involved in the process, the smooth curve (dashed) is well approximated by the discrete function (solid). 124

3.5 Two sheets of a material that together yield a second-order susceptibility that depends on position as shown in the bottom plot. 125

3.6 A layered material in which the second-order susceptibility depends on position as shown in the plot below under optimum phase matching 127

3.7 In the normal dispersion part of the spectrum, the refractive index increases with frequency, so if all three waves are in the normal dispersion regime (dashed vertical lines), $\Delta k < 0$. If the ω_3 wave is in the anomalous regime, where the refractive index of the sum-frequency wave is lower than for the incident waves (solid vertical lines), $\Delta k > 0$ 129

3.8 Sum frequency generation in the small depletion regime is represented by the nonlinear interaction of two incident waves of frequencies ω_1 and ω_2 within the material, to generate a photon of frequency ω_3. Downstream, this generated photon and a photon of frequency ω_2 interact and create a photon of frequency ω_1. The inset shows how the energy is conserved in the process of the destruction of a photon of frequency ω_3 and creation of two photons of frequencies ω_1 and ω_2. 132

3.9 The Manley-Rowe equation expresses the fact that the absolute change in the number of photons at each frequency is the same. For example, the destruction of one photon at ω_1 and one photon at ω_2 is accompanied by the creation of one photon at frequency ω_3. 137

3.10 Comparison of the intensities of the three beams of frequency ω_1, ω_2, and ω_3 in sum frequency generation. I_1 and I_2 have their minima when I_3 peaks. 138

3.11 The intensities at frequencies ω_1 and ω_3 plotted as a function of propagation distance. 140

3.12 In a second-order nonlinear-optical material, the pump beam at frequency ω_1 and the amplified beam at frequency ω_2 results in the generation of a difference frequency beam at frequency $\omega_3 = \omega_1 - \omega_2$. 142

3.13 The intensities at frequencies ω_2 and ω_3 as a function of depth, z, into the material. 145

3.14 (left) In second harmonic generation, two incident photons of frequency ω interact with a nonlinear medium, generating an output photon of frequency 2ω. (right) Down conversion – where the second harmonic light interacts with the fundamental – is difference frequency generation, as described in the previous section. 146

3.15 Phase-mismatching reduces the efficiency of generating second harmonic wave significantly. 147

3.16 (left) The fundamental wave is depleted as it is converted to the second harmonic wave. At $z = 0$ the slope of the amplitude of the fundamental wave vanishes and the second harmonic amplitude's slope is nonzero. (right) When both fundamental and second harmonic intensities are incident on the material, they oscillate in the absence of phase-matching. 148

3.17 Rotating a quartz sample changes the path length ℓ that the light travels in the material. Consequently, the intensity of the second harmonic varies as the orientation of the quartz sample changes. 149

3.18 The intensity of the second harmonic wave as a function of the angle of refraction, ϕ, of the fundamental wave. Here we neglect the birefringence of the sample, the tensor nature of $\chi^{(2)}$ and the Fresnel reflections at the sample's surfaces. 150

3.19 The observed angular dependence of the second harmonic intensity in quartz. 151

3.20 (a) Uniaxial birefringence is characterized by an ellipsoid having the ordinary refractive index n_o for any electric field polarization in the xy plane and the extraordinary refractive index n_e along z. (b) A wave incident in the yz plane and a cross-section of the ellipse in the plane perpendicular to $\mathbf{k}$. (c) The cross-section of the ellipsoid shown with $\mathbf{k}$ pointing out of the page for $\theta = 0$, $0 < \theta < 90°$ and $\theta = 90°$. 152

3.21 The dispersion of the refractive index for several angles θ for (a) a positive uniaxial crystal and (b) a negative uniaxial crystal. Phase matching demands that the fundamental and second harmonic are of opposite types. If the fundamental is an extraordinary ray, the second harmonic is an ordinary ray, or vice versa, as shown at the indicated points. 153

3.22 The transverse intensity profile of the two lowest modes of an optical fiber. 156

3.23 Figure for practical problem. 157

3.24 Figure for practical problem. 158

4.1 (a) An excited molecule (b) emits a photon in a process called spontaneous emission. (c) An excited molecule in a sea of photons (d) emits a photon of the same frequency by stimulated emission. 179

4.2 Particle trajectories in an inelastic (left) and elastic (right) collision. 198

4.3 Space-time diagram for process shown in Figure 4.2. 198

4.4 A Feynman diagram that represents the interaction between two electrons (solid lines) that is mediated by the exchange of a photon (wiggly line). 199

4.5 A molecule (solid line) in it's ground state absorbs a photon and is excited to state n. 199

4.6 The two Feynman diagrams for the linear susceptibility. 200

4.7 A vertex where a photon is absorbed (left) or emitted (right) by a molecule in its ground state to create a virtual superposition of states (represented by a vertical line). These diagrams must be part of a larger one since they do not individually conserve energy. 202

4.8 Applying the propagator rule to second-order in the field between the two crosses. 204

4.9 All possible Feynman diagrams for sum frequency generation. . 205

4.10 (left) A more detailed drawing of one Feynman diagram for sum frequency generation and (right) an outgoing photon in the propagator. 206

4.11 Real excitations occur when the photon energy matches the energy difference between two states of the material system. If the photon energy is less than this energy difference, the system cannot be excited and the state of the material system can be thought of as being in a superposition of states. Two-photon absorption can be mediated by an excitation through a virtual state. 207

4.12 (a) Second hyperpolarizability and (b) the cascading second hyperpolarizability that results from two interacting molecules through the first hyperpolarizability. 209

4.13 The output spectrum observed when an intense beam at frequency ω_1 is launched into a sample with $\chi^{(2)} \neq 0$. 209

4.14 A Feynman diagram for a process that includes a static electric field. 210

4.15 A plot of the imaginary part of the polarizability α as a function of energy peaks for state n of width Γ_{n0}. 211

4.16 The light transmitted by a monomolecular gas is measured with a spectrometer. 211

4.17 a) The absorption spectrum of three different molecules with different velocities (red), and the Boltzmann distribution of peak positions (dashed curve). b) The spectrum shifts when nearby molecules interact with each other through their coulomb fields. c) The sum of the individual contributions in an ensemble (red peaks) weighted by the Boltzmann factor leads to the aggregate inhomogeneously-broadened peak (dashed curve). 212

4.18 (a) A material made of molecules (b) is modeled as a uniform dielectric of permittivity ϵ. A uniform electric field is applied, which induces charges on the surfaces. The dashed circle represents a spherical volume that contains one molecule and the dashed rectangle is the side view of a Gaussian box. 230

4.19 An electric field $\mathbf{F}$ applied to a dielectric sphere induces charges on its surface, leading to a dipole field $\mathbf{E}_P$. The total field $\mathbf{E}$ is a sum of the applied field and the dipole field. 231

4.20 If the cavity radius is much smaller than the plane wave's wavelength, the electric field will be spatially uniform inside the cavity that contains a molecule. The Lorentz-Lorenz local field model for static fields then applies. 233

4.21 In a dye-doped polymer or molecular crystal, the volume swept out by a molecule V_2 is less than the volume occupied per molecule $V_1 = 1/N'$. 240

4.22 The position of the surface plasmon peak changes with the shape of the metal particle. 244

5.1 A pump beam of light influences the propagation of a probe beam. 247

5.2 Electronic mechanism of $\chi^{(3)}$. A molecule with a nuclear framework represented by the thick arrow, and electrons represented by the surrounding dots. The electron cloud is deformed by the electric field, changing the optical refractive index of the molecule. Note that the degree of deformation is greatly exaggerated. 249

5.3 Reorientational mechanism of $\chi^{(3)}$. The nuclei of each molecule are shown as arrows. 250

5.4 (a) An optical field is polarized along the $\hat{z}$ axis and induces a dipole moment in a molecule along its long axis $\hat{z}'$. (b) The electric field inside and outside of a dielectric slab is polarized perpendicular to its surface as shown. 251

5.5 Electrostriction mechanism of $\chi^{(3)}$ in solution. The effective width of the light beam is shown by the dashed lines and the dots represent molecules in solution, which are attracted to regions of higher intensity. 251

5.6 Saturated absorbtion mechanism of $\chi^{(3)}$, where the molecule in its excited state has a different refractive index than in its ground state. 252

5.7 Some light from a laser beam is absorbed by a material, depositing heat in its path, and thus creates a thermal gradient. The temperature gradient acts like a prism, bending the light. . . 253

5.8 Axes Transformation 263

5.9 Transitions for One- and Two-Photon States 267

5.10 Wavefunctions in x and y 267

5.11 (a) An anisotropic molecule in an electric field with anisotropic polarizability. Associated with the semi-major and semi-minor axes are the polarizabilities, α_3 and α_1, respectively. (b) The minimum energy configuration and (c) the maximum energy configuration 270

5.12 The energy per anisotropic molecule in an electric field, **E**. 271

5.13 $\frac{\delta n_{\text{linear}}}{\delta n_{\text{circular}}}$ as a function of the laser pulse width τ. 276

6.1 Phase Conjugate Mirror 279

6.2 Phase Conjugate Mirror 281

6.3 (left) A weak probe beam is launched into the region where a strong pump beam has self-focused into a Gaussian beam of width ρ_0. The probe beam is launched at an angle θ to the pump beam axis and enters at a distance a from the axis. (right) The trajectory of the ray when $\theta = 0$ and $a \neq 0$ (solid curve); and, and the trajectory of the beam waist from diffraction(dashed curve). 286

6.4 The output versus input plot of an optically bistable device is double valued. 292

6.5 (a) Reflected and transmitted plane waves at an interface from an incident plane wave at normal incidence. (b) Ray at at arbitrary incident angle with reflected (r) and transmitted (t) rays. (c) Time-reversed rays (solid), where the dashed ray on the right must vanish and the dashed ray on the left must add to unity by reciprocity. 293

6.6 A Fabry-Perot interferometer with a nonlinear material between partial reflectors. Shown is the incident ray of unit field strength as well as a few of the reflected and transmitted rays. Each reflector has a transmittance of t and a reflectance of r. The points show the field at position x on each ray. 295

6.7 Transmittance of a Fabry Perot interferometer as a function of optical path length between the reflectors for four reflectance values. 297

6.8 The transmitted intensity as a function of incident intensity for a multistable device. The inset shows the incident intensity as a function of the transmitted intensity. 301

6.9 (a) $f(I_t)$ and I_t/I_{in} as a function of I_{in} for five different intensities. (b) The corresponding intersection points on the bistable curve. The dashed line inside the loop shows the disallowed region. 304

6.10 The second harmonic power from a single molecule is collected with a lens and focused into a detector. 309

Bibliography

[1] J. Kerr, "A new relation between electricity and light: dielectrified media birefringent," Phil. Mag. S. **50**, 337–348 (1875).

[2] J. Kerr, "Electro-optic observations on variaou liquids," Phil. Mag. **8**, 85–102,229–245 (1875).

[3] J. Kerr, "Electro-optic observations on various liquids," J. Phys. Theor. Appl. **8**, 414–418 (1879).

[4] T. H. Maiman, "Stimulated optical radiation in ruby," nature **187**, 493–494 (1960).

[5] M. Goeppert-Mayer, "Über Elementarakte mit zwei Quantensprüngen," Annalen der Physik **401**, 273–294 (1931).

[6] P. A. Franken, A. E. Hill, C. W. Peters, and G. Weinreich, "Generation of Optical Harmonics," Phys. Rev. Lett. **7**, 118–119 (1961).

[7] G. Mayer and F. Gires, "Action of an intense light beam on the refractive index of liquids," Comptes Rendus, Acad. Sci. Paris **258**, 2039 (1964).

[8] R. W. Boyd, *Nonlinear Optics* (Academic Press, 2009), 3rd edn.

[9] P. A. M. Dirac, "Quantised Singularities in the Electromagnetic Field," Proc. Roy. Soc. A **133**, 60 (1931).

[10] J. D. Jackson, *Classical Electrodynamics* (Wiley, New York, 1996), 3rd edn.

[11] R. A. Norwood and G. Khanarian, "Quasi-Phase-Matched Frequency Doubling Over 5mm in Periodically Poled Polymer Waveguide," Electron. Lett. **26**, 2105–06 (1990).

[12] P. D. Maker, R. W. Terhune, M. Nisenhoff, and C. M. Savage, "Effects of Dispersion and Focusing on the Production of Optical Harmonics," Phys. Rev. Lett. **8**, 21–22 (1962).

[13] M. Reichert, P. Zhao, J. M. Reed, T. R. Ensley, D. J. Hagan, and E. W. Van Stryland, "Beam deflection measurement of bound-electronic and rotational nonlinear refraction in molecular gases," Opt. Express **23**, 22 224–22 237 (2015).

[14] J. Zyss and J. Oudar, "Relations between microscopic and macroscopic lowest-order optical nonlinearities of molecular crystals with one-or two-dimensional units," Physical Review A **26**, 2028 (1982).

[15] M. G. Kuzyk, K. D. Singer, H. E. Zahn, and L. A. King, "Second Order Nonlinear Optical Tensor Properties of Poled Films Under Stress," J. Opt. Soc. Am. B **6**, 742 (1989).

[16] M. G. Kuzyk and C. W. Dirk, *Characterization techniques and tabulations for organic nonlinear optical materials* (Marcel Dekker, 1998).

[17] K. Clays and A. Persoons, "Hyper-Rayleigh Scattering in Solution," Phys. Rev. Lett. **66**, 2980–2983 (1991).

[18] M. Joffre, D. Yaron, J. Silbey, and J. Zyss, "Second Order Optical Nonlinearity in Octupolar Aromatic Systems," J. Chem. Phys. **97**, 5607–5615 (1992).

[19] M. G. Kuzyk, "A path to Ultralarge Nonlinear-Optical Susceptibilities," J. Opt. Soc. Am. B **33**, E150 (2016).

[20] D. Frenkel and P. B. Warren, "Gibbs, Boltzmann, and negative temperatures," Am. J. Phys. **83**, 163–170 (2015).

[21] M. G. Kuzyk, *Polymer Fiber Optics: materials, physics, and applications*, Vol. 117 of Optical science and engineering (CRC Press, Boca Raton, 2006).

[22] S. Mossman, R. Lytel, and M. G. Kuzyk, "Fundamental limits on the electro-optic device figure of merit," J. Opt. Soc. Am. B **33**, E109 (2016).

[23] M. Zahirul Alam, I. De Leon, and R. W. Boyd, "Large optical nonlinearity of indium tin oxide in its epsilon-near-zero region," Science **352**, 795 (2016).

[24] S. Mossman, R. Lytel, and M. G. Kuzyk, "Dalgarno–Lewis perturbation theory for nonlinear optics," J. Opt. Soc. Am. B **33**, E31–E39 (2016).

[25] M. G. Kuzyk, "Physical Limits on Electronic Nonlinear Molecular Susceptibilities," Phys. Rev. Lett. **85**, 1218 (2000).

[26] M. G. Kuzyk, J. Perez-Moreno, and S. Shafei, "Sum Rules and Scaling in Nonlinear Optics," Phys. Rep **529**, 297–398 (2013).

[27] A. M. Summers, A. S. Ramm, G. Paneru, M. E. Kling, B. N. Flander, and Trallero-Herrero, "Optical damage threshold of Au nanowires in strong femtosecond laser fields," Optics Express **22**, 4235 (2014).

[28] J. Johnson, H. Yan, R. Schaller, P. Petersen, P. Yang, and R. Saykally, "Near-field imaging of nonlinear optical mixing in single zinc oxide nanowires," Nano Lett. **2**, 279–283 (2002).

[29] P. Campagnola, A. Millard, M. Terasaki, P. Hoppe, C. Malone, and W. Mohler, "Three-dimensional high-resolution second-harmonic generation imaging of endogenous structural proteins in biological tissues," Biophysical Journal **82**, 493–508 (2002).

[30] J. A. Scholl, A. L. Koh, and J. A. Dionne, "Quantum plasmon resonances of individual metallic nanoparticles," Nature **483**, 421 (2012).

Index

3-level terms, 225

absolute zero, 289
absorption coefficient, 3
accumulated phase, 296, 298
acoustical wave, 14, 252
alignment energy, 274
anharmonic oscillator, 264
anharmonic potential, 9
anisotropic material, 233, 239
anisotropic medium, 5
anisotropic molecule, 269–271
anisotropic molecules, 269, 273
annihilation operator, 175, 176, 182–184, 197
anomalous dispersion, 129, 130, 151
Anomalous Dispersion Phase Matching, 151
asymmetric potential, 225, 226
axial orientation, 270

background second harmonic, 313
basis vectors, 162, 163, 165, 225
beam waist, 286–289
beam width, 286, 287
bias intensity, 292
biological tissue, 312
birefringence, 2, 5
birefringence phase matching, 151
bistability, 292
bistable device, 292
Blue Ray players, 156

Bohr frequencies, 80, 134, 193, 194
Bohr frequency, 191, 194, 203, 217, 265
Boltzmann distribution, 211, 212, 289
Boltzmann statistics, 272
bound charge, 9–12, 94, 112

cascading, vii, ix, 86, 87, 208, 209, 239
catastrophic damage, 313
cavity, 295, 296, 298
centrosymmetric material, 59, 71, 82, 225, 307
centrosymmetric potential, 224, 225
centrosymmetric systems, 254
circular polarization, 259–262, 280
circular polarized light, 262
circularly polarized beam, 280
circularly polarized wave, 280
classical limit, 20
classical spring model, 241
Clausius-Mossotti relation, 241
closure, 164, 165, 169, 170, 215, 220, 229
coherent state, 183–185
Collisional losses, 194
column vector, 165
complex Bohr frequency, 194
conducting rod, 313
conductivity, 305
constitutive relationship, 255

continuum approximation, 230
Coulomb gauge, 171, 178
Coulomb's Law, 7
counter-prorogating photons, 282
coupled differential equations, 284, 286
creation operator, 175, 184
crossed polarizers, 2, 6
cylindrical cavity, 236

damage threshold, 306, 312, 313
degeneracy perfector, 310
degenerate four-wave mixing, 281
density fluctuations, 239
density matrix, 220
density matrix time evolution, 220
dichroic mirror, 5
dielectric constant, 118, 133, 230–232, 234, 237, 241–243, 245, 252, 307
dielectric function, 55, 57, 233, 237
dielectric sphere, 231, 232, 237
difference frequency generation, 21, 22, 67, 141–144, 146
difference frequency photon, 23
diffraction, 286–288
dilute gas approximation, 37
dimensional analysis, 13
dipole approximation, 9, 12, 13, 181, 186
dipole matrix elements, 306
dipole operator, 19, 195, 219
Dirac notation, vii, 161, 162, 164, 165, 169, 170, 213
dispersion, 48–50, 74, 77, 119, 121, 129, 130, 135, 150, 151, 153–156, 197, 239, 257
dressed dipole moment, 234
dressed molecular susceptibility, 234
dressed polarization, 234

dye-doped polymer, 37, 59, 130, 240, 241, 306

effective potential, 101, 271
effective refractive index, 248
effective susceptibility, 258
elastic collision, 198
electret, 19, 233
electric dipole approximation, 9, 309
electric dipole field, 9, 53
electric displacement, 10–12, 54, 57, 92, 93, 96, 112, 116, 118, 230, 231, 306
electric field intensifiers, 235
electric moment, 9, 104, 105
electric polarization, 12
electric quadrupole, 9, 93, 95, 101, 104, 105
electro-optic device, 306
electro-optic effect, 57, 305–307
electro-optic material, 304
electro-optic modulator, 308
electrodynamics, 1
electron cloud, 3, 97, 249, 252, 264, 306
electron cloud shape, 249
electron in a box, 240
electron volts, 309
electronic mechanisms, 247
electronic response, 249, 262, 264, 276
Electrostriction, 251, 252
elliptical polarization, 261
elliptically polarized beam, 262
elliptically-polarized light, 261
energy conservation, 63, 65, 77, 132, 136, 141, 148, 179, 199, 202, 203, 206, 207
energy density, 79, 172
energy dissipation, 194

energy eigenstate, 3, 163, 165, 170, 188, 194, 207, 214, 219
energy eigenstate vector, 170
energy eigenvalues, 165, 179, 306
energy-level diagram, 266
ensemble average, 213, 272, 273, 275
epsilon-near-zero materials, 277
equilibrium beam diameter, 286
equilibrium beam waist, 286
Euler angle, 272, 273
even-parity ground state, 267
even-parity state, 267
even-parity state vectors, 268
excited state population, 290, 291
expectation value, 3, 166, 168, 182–185, 188, 192, 194, 195, 214–217, 219, 222
extraordinary ray, 152–155

Fabry-Perot cavity, 298, 305
Fabry-Perot interferometer, 292–295
Faraday's Law, 7, 99
feedback, 292, 296
Feynman diagram, 86, 198–203, 205–207, 210, 239, 308
Feynman-like diagrams, 161, 198
field amplitude per photon, 174
field gradient, 10, 12, 101–103, 252
figure of merit, viii, 254, 305, 306, 308
free charges, 10–12, 94, 111, 171
free currents, 10, 55
free electrons, 10
fresnel factors, 303
fresnel reflection, 285
full permutation symmetry, 79, 80, 134, 142, 265
fundamental limits, 229, 306, 312
fundamental limits of susceptibilities, 228, 229

gain medium, 298
Gaussian intensity profile, 286
Gaussian units, 7, 8, 10, 11, 19, 54, 88, 92, 96, 106, 107, 120
gold, 312
graphical method for hysteresis, 304
green laser pointer, 156
Green's function, 221, 222
ground state population, 291
group velocity of fiber mode, 155

Hamiltonian, 19, 165, 168–171, 173, 176–179, 181, 182, 185, 187–190, 193, 194, 210, 214–216, 220, 223, 224, 226, 229
harmonic conversion efficiency, 156
harmonic oscillator, 171, 175–177, 183, 185, 188
Helmholtz equation, 140
Hermitian Adjoint, 167, 168
hermiticity, 193, 194, 210, 217, 219
Hilbert space, 162–164, 175, 213, 215, 216
homogeneous equation, 221
homogeneous solution, 221
hydrogen atom, 20, 82, 97, 164
hyper Rayleigh scattering, 239
hyperpolarizability, off-resonant, 308
hysteresis, 303, 304
hysteresis loop, 303, 304

idler, 138
idler beam, 140
impurities, 305
indium tin oxide, 306, 307
inelastic collision, 198
infinite energy of plane waves, 172
infinite potential well, 18
infinite refractive index, 307
infinite temperature, 216, 290, 291

infinite universe, 172
inhomogeneous equation, 221
inhomogeneous solution, 221
inner product, 162, 163, 166, 167, 170, 183
intensity dependent absorption, 3
intensity dependent refractive index, 2, 289
intensity modulator, 304
intensity-dependent Fresnel reflection, 303
intensity-dependent Fresnel transmission, 303
intensity-dependent length change, 302
intensity-dependent refractive index, 59, 248, 249, 252, 254, 257, 259, 261, 265, 271, 292, 296, 302, 306–308
interference, 121, 281, 296, 299, 306
interferometer, 296–299
intermolecular forces, 272
intrinsic loss, 305
intrinsic permutation operator, 196
intrinsic permutation symmetry, 68, 79, 254, 257
intrinsic polarizability, 229
inversion operator, 254
inversion symmetry, 84, 254, 255
isotropic material, 130, 258, 261
isotropic media, 273
iterative method, 145

Jaynes-Cummings model, 187

Kerr, 1, 2
Kerr coefficient, 2
Kerr Effect, 1, 2
Kerr susceptibility, 249
ket-bra operator form, 165

lab frame, 250, 272
Langmuir-Blodgett films, 123
laser, 2, 4, 5, 24, 86, 154, 156, 157, 208, 210, 211, 249, 253, 262, 276, 289, 298, 304, 312, 313
layered structure, 123
left-handed circularly polarized light, 259
left-handed coordinate system, 108
light intensifier, 298
limits of the hyperpolarizability, 311
linear absorption coefficient, 3
linear harmonic oscillator, 35, 40, 334
linear polarization, 262
linear susceptibility, 9, 37, 72, 74, 99, 159, 193, 195, 200, 203, 224, 227, 229, 234, 238, 242, 244, 291, 308
linearly-polarized beam, 5, 280
liquid, 1, 59, 158, 232, 239, 247, 252–256, 259, 261, 264, 269
local electric field, 87, 90, 91, 230–232, 234, 235, 237, 243, 244, 275
local field, vii, 89, 232–239, 244, 274
local field enhancement, 243
local field factor, 233
Lorentz-Lorenz local field, vii, 232, 233
lossless material, 295, 303
lowering operator, 170, 171, 175–177

magnet, 233
magnetic dipole, vii, 9, 93, 97–100, 104, 106
magnetic moment, 9, 104
magnetic splitting, 290
Maria Goeppert-Meyer, 3
Maxwell's Equations, 10, 11, 111, 293, 294

mean intensity, 299
medical imaging, 156
melting, 313, 314
metal nanoparticles, 235
metallic sphere, 241, 244
microscopic mechanism, 33
minimum beam waist, 287
mode index, 155, 156
mode index phase matching, 155
molecular beam epitaxy, 123
molecular crystal, 59, 233, 240
molecular frame coordinates, 272
molecular monolayer, 123
molecular reorientation, 247, 269
molecular size, 20, 123
momentum basis, 166
monochromatic plane wave, 6, 7, 13
Monte Carlo, 229
multistability, viii, 292, 296, 302

nanorod, 311–314
nanowire, 311
negative temperature, viii, 277, 290, 291
non-depletion approximation, 123, 283
non-radiative losses, 194
nonlinear crystals, 156
nonlinear feedback, 296
nonlinear harmonic oscillator, 35, 197
nonlinear harmonic oscillator, 37
nonlinear phase shift, 248
nonlinear polarization, 9, 10, 55, 63, 81, 82, 112, 116, 118, 132, 135, 146, 260, 282
nonlinear refractive index, 247
nonlinear spring, vi, 35, 46, 49, 71
nonlinearity, upper bound, 312
normal dispersion, 129, 130, 151, 154

normalization, 184, 186, 187, 193, 195, 196, 206, 225
normalization factor, 194
numerical aperture, 311

octupolar molecules, 239
octupole, 12, 13, 93
odd-parity excited state, 268
odd-parity state, 267
odd-parity state vectors, 268
off-resonance hyperpolarizability, 314
one-dimensional harmonic oscillator, 35, 40
one-dimensional local field, 237
one-photon character, 268
one-photon states, 224, 266
Onsager local field, 232
optical bistability, 292
optical computing, 248
optical cycle, 271
optical damage threshold, 306
Optical Kerr Effect, 2, 4, 5, 59, 70, 247
optical loss, 244, 305, 308
optical memory, 292
optical microscope, 312
optical multi-stability, 277
optical path length, 296, 297
optical phase conjugation, 277
optical switching, 248
ordinary ray, 152–155
orientational average, 235
orientational ensemble average, 272

P. A. Franken, 4
parity, 223–227, 266–268
parity eigenstate, 266
parity invariance, 224
parity operator, 223

particle-in-a-box, 314
peak intensity, 287, 288, 312
periodic boundary condition, 172
permittivity, viii, 11, 55, 92, 230, 307
perturbation, 198
perturbation theory, 3, 43, 178, 197, 201, 204, 220, 249, 264, 265, 312
phase conjugate beam, 280, 285
phase conjugate mirror, 277–281, 285
phase conjugate photon, 285
phase conjugate wave, 277, 278, 280–282, 285
phase conjugation, viii, 281, 285
phase distorter, 280
phase fronts, 53, 279, 280, 285
phase matching, vii, 121, 124, 127, 129–131, 138, 142, 150, 151, 153–156
phase velocity, 53, 54, 299
photomechanical effect, 302
photomultiplier tube, 313
photon counting, 311
photon emission rate, 311
photon flux, 312
photothermal heating, 253
plasma frequency, 242, 245
polar orientation, 270
polarization field, 21
polarizer, 2, 5–7, 159
poled dye-doped polymer, 59
position basis, 166
position matrix element, 226, 229, 240
position operator, 268
positive temperature, 290, 291
Poynting's theorem, 120, 141, 173
principle axes, 5, 232, 263, 269, 272
prism, 1, 4, 253
probability decay, 194

probe beam, 5, 6, 247, 248, 281, 286, 289
projection factor, 272
projection operator, 163, 164
projector, 217
Propagator Rule, 203
propagator rules, 204
propagators, 201–204, 206
proteins, 312
pulse width, 262, 264, 276
pump beam, 5, 6, 142, 144, 145, 247, 248, 281, 282, 286
pump beam depletion, 282
Python, vi, 23, 28, 30, 49, 77, 145, 148, 197, 229, 303, 304

quadratic electrooptic effect, 2, 5
quadrupole moment, 12
quantization of photons, 161
quantum efficiency, 313
quantum limit, 21
quantum to classical transition, 313
quartz crystal, 4
quasi phase matching, vii, 127, 130, 131, 155

Rabi oscillations, vii, 187
radiative cooling, 313
raising operator, 175, 177, 185
ray trajectory, 154, 286, 287
real excitation, 207, 208
reciprocity, 293, 294
reflectance, 294–297
reflectivity, 293, 295, 298
reflector, 292
reflectors, viii, 292, 294–298, 305
refractive index, 1–3, 5, 6, 52–54, 58, 59, 70, 117, 119, 121, 123, 126, 129, 130, 150–156, 158, 233, 241, 247–250, 252, 253,

262, 269–271, 273, 274, 286, 287, 295, 296, 298–300, 302, 303, 307, 308
refractive index change, 307
refractive index ellipsoid, 151, 270
refractive index grating, 281
refractive index profile, 286
relative permittivity, 307
reorientational response, 271, 273, 276
reorientational response time, 271, 276
reorientational third order susceptibility, 275
response function, vi, 10, 15, 16, 35, 60–62, 64, 67, 68, 71, 72, 112, 133
response time, 253, 254, 271
right-handed circularly polarized light, 259
rotation matrix, 250, 263, 272
rotation symmetry, 256
row vector, 165
ruby, 4, 5

sapphire, 313
saturated absorbtion, 252, 253
second harmonic crystals, 156
second harmonic efficiency, 156, 312
second harmonic generation, 4, 9, 21, 50, 66, 69, 146, 308
second harmonic imaging, 312
second harmonic photon, 86, 147, 311–313
second harmonic tensor, 80
second hyperpolarizability, 16, 86, 91, 197, 209, 239, 291, 312
second quantization, 170, 175, 177, 201

second-order molecular susceptibility, 16
second-order susceptibility, vii, 9, 39, 46, 50, 65, 69, 79, 80, 82, 86, 117, 120, 121, 123–128, 130, 131, 137, 150, 154, 155, 196, 197, 203, 226, 227, 305
second-order susceptibility profile, 124
second-quantized electric field, 184
secular divergence, 265
self focusing, viii, 277, 286, 287
self trapping, 287, 288
SI units, 7, 8, 11, 20, 55, 92, 107
silver, 235, 277, 313, 314
simultaneously measurable observables, 183
single-molecule nonlinear optics, viii, 277
slowly varying envelope approximation, 119, 283
small-diameter optical fiber, 155
solid angle, 310
space-time diagram, 198
spatial overlap between modes, 156
spatially-modulated intensity, 299
spectrometer, 2, 211, 213
speed of light, 11, 97
sphere, 231, 232, 235–237, 239, 241–244, 313
spin states, 165
spring force, 37, 49, 101
standing wave, 172, 185
statcoulombs, 8
state vector, 166
stationary state vector, 194
statistical mechanics, 18, 98, 259
step function, 71, 222
stimulated emission, 22, 161, 171, 179, 182, 187, 282, 290, 291

stress-optical coefficient, 252
sum frequency generation, 22
sum rules, 228, 229, 308
sum-over-states, 306
summation convention, 8, 93, 113, 234, 258, 272
sunlight, 1
superposition, 11, 52–54, 101, 135, 183, 186, 207, 208, 213–215, 218, 222, 225, 227, 231, 285, 295
superposition of states, 3, 22, 187, 202, 207, 216, 218
superposition of the homogeneous solutions, 222
surface charge, 12, 94, 96, 252
surface gradient, 12
surface plasmon, vii, 235, 241–245
surface plasmon resonance, vii, 235, 242–245
surface plasmons, vii, 235, 241

temperature gradient, 253
temperature scale, 289
thermal effects, 253
thermal fluctuations, 220
thermal gradient, 253
thermal mechanism, 253
third-order molecular susceptibility, 16
third-order susceptibility, 39, 59, 227, 235, 240, 247, 254–257, 266, 267, 269, 277, 281, 282
three-level model, 227
threshold for breakdown, 305
time evolution operator, 166, 169, 170, 187
time harmonic function, 232
time reversal, 277
time-dependent perturbation, 189, 201, 249
time-revered rays, 294
time-reversed reflection, 277
time-reversed waves, 277
transcendental equations, 303
transition amplitudes, 171
transition energy, 3, 194, 224, 253
transition moment, 224, 225, 240, 266, 268, 269, 308
transition probability, 180–182
transmittance, 2, 6, 150, 293–296, 300
transverse wave, 152
tri-stability, 302
turning points, 287
two-dimensional box, 267
two-dimensional local field, 237
two-level model, 187, 188, 225, 227
two-photon absorption, 3, 5, 59, 207, 224, 337
two-photon absorption, 3
two-photon character, 268
two-photon states, 224, 266
two-state quantum system, 207

unbound states, 164
uncertainty principle, 199, 201
uniaxial crystal, 153, 154
uniaxial molecule, 270, 272
units of the dipole moment, 20
unstable solution, 303

vacuum hyperpolarizability, 233
vector potential, 96, 171–173, 175, 176, 181
virtual state, 3, 207

wave equation, vii, 5, 43, 56, 111, 112, 114, 115, 131–134, 136, 159, 171, 248, 259, 283, 284, 306

wave propagation, 1, 6, 33, 54, 287
wavelength of maximum absorption, 309

xenon flashlamp, 5

Young's modulus, 252

zinc oxide, 312